Nanostructure Science and Technology

Series Editor

David J. Lockwood, FRSC
National Research Council of Canada
Ottawa, ON, Canada

Nanostructure science and technology now forms a common thread that runs through all physical and materials sciences and is emerging in industrial applications as nanotechnology. The breadth of the subject material is demonstrated by the fact that it covers and intertwines many of the traditional areas of physics, chemistry, biology, and medicine. Within each main topic in this field there can be many subfields. For example, the electrical properties of nanostructured materials is a topic that can cover electron transport in semiconductor quantum dots, self-assembled molecular nanostructures, carbon nanotubes, chemically tailored hybrid magnetic-semiconductor nanostructures, colloidal quantum dots, nanostructured superconductors, nanocrystalline electronic junctions, etc. Obviously, no one book can cope with such a diversity of subject matter. The nanostructured material system is, however, of increasing significance in our technology-dominated economy and this suggests the need for a series of books to cover recent developments.

The scope of the series is designed to cover as much of the subject matter as possible – from physics and chemistry to biology and medicine, and from basic science to applications. At present, the most significant subject areas are concentrated in basic science and mainly within physics and chemistry, but as time goes by more importance will inevitably be given to subjects in applied science and will also include biology and medicine. The series will naturally accommodate this flow of developments in the sciences and technology of nanostructures and maintain its topicality by virtue of its broad emphasis. It is important that emerging areas in the biological and medical sciences, for example, not be ignored as, despite their diversity, developments in this field are often interlinked. The series will maintain the required cohesiveness from a judicious mix of edited volumes and monographs that while covering subfields in depth will also contain more general and interdisciplinary texts.

Thus the series is planned to cover in a coherent fashion the developments in basic research from the distinct viewpoints of physics, chemistry, biology, and materials science and also the engineering technologies emerging from this research. Each volume will also reflect this flow from science to technology. As time goes by, the earlier series volumes will then serve as reference texts to subsequent volumes.

Jean-Pierre Leburton
Editor

Solid State Nanopores

From Fabrication to Biosensing

Springer

Editor
Jean-Pierre Leburton
Department of Electrical and Computer
Engineering and Department of Physics
University of Illinois Urbana-Champaign
Urbana, IL, USA

ISSN 1571-5744 ISSN 2197-7976 (electronic)
Nanostructure Science and Technology
ISBN 978-3-031-42335-2 ISBN 978-3-031-42336-9 (eBook)
https://doi.org/10.1007/978-3-031-42336-9

This Springer imprint is published by the registered company Springer Nature Switzerland AG
The registered company address is: Gewerbestrasse 11, 6330 Cham, Switzerland

Paper in this product is recyclable.

Preface

The ability of engineering nanopores with desired sizes and shapes in ultra-thin solid-state membranes is nowadays a new field of research, with among other goals mimicking biological functions such as ion channels in living cells or nanofluidic channels in blood vessels. The resourcefulness of solid-state materials for their electronic and mechanical properties as well as their robustness to chemical and thermal environment provides a wide spectrum of ways to increase bio-detection sensitivity. In this perspective, the emergence of two-dimensional materials such as graphene and transition metal dichalcogenides with semiconducting properties adds to their capabilities to sense, read and manipulate biomolecules electronically. Indeed, the versatility of semiconductor materials in terms of conduction polarity by selective n- or p-doping, as well as their ability to tailor space charge at the nanopore liquid–solid interface, provides the necessary ingredients to tailor the electrostatic landscape close to the membrane while enabling control of the biological object motion to enhance the nanopore detection sensitivity.

Whereas approaches based on solid-state nanotechnology offers tremendous advantages in terms of cost, speed and reliability over traditional bio-chemical techniques of detection, the miniaturization capability of nanoscale materials provides the natural electronic and mechanical ingredients for new compact devices with a broad range of diagnostic or therapeutic applications in personal medicines. However, realization and investigation of such bio-nano-electronic systems is not straightforward, neither from an experimental nor a theoretical viewpoint as one deals with a non-homogenous multi-phase material system consisting of bio-molecule–electronic interacting at nanoscale solid–liquid interfaces.

This book offers a collection of works addressing the latest developments in the science and technology of solid-sate nanopores, ranging from their fabrication and characterization to nanofluidics and detection, and ultimately to modeling techniques to understand the complexity of the solid-state/biological system interaction through a liquid medium. Hence, K. Briggs et al. at the University of Ottawa, Canada, introduce recent developments in the fabrication of solid state nanopores for single molecule detection, while H. Li et al., at the University of Arkansas, address the detection of DNA-binding sites of regulation proteins with ion beam-sculpted silicon nitride nanopores. M. Niu et al. at Zhejiang University, China, discuss emerging abnormal phenomena in confined nanofluidics through nanopore in graphene, and K. Chen and U. Keyser at the University of Cambridge, UK, address new techniques to design DNA to enhance their sensing by nanopore pipettes. R. J. Yu et al. at Nanjing University, China, investigate the use of confined solid-state nanopores for protein analysis. D. V. Melnikov and M. Gracheva, at Clarkson University, NY, provide self-consistent Brownian dynamics simulations for the study of ionic current blockade in solid state nanopores. Finally, M. Xiong et al. at University of Illinois at Urbana-Champaign describe a comprehensive molecular dynamics-device modeling approach that anticipates powerful electronic bio-sensing with 2D semiconductor nanopores.

I would like to thank all the authors for their contribution to the scope and quality of book, more specifically to R. Chakraborty and M. Xiong, who helped me with different aspects of its assembly. It is our hope that the review articles presented here will provide the necessary background and basic knowledge to scientists and engineers in the broad fields of living and physical sciences with interest in the interaction between bio-molecules and solid-state nanopores.

Urbana, IL, USA Jean-Pierre Leburton

Contents

Chapter 1
Fabricating Solid-State Nanopores for Single-Molecule Sensing

Kyle Briggs, Matthew Waugh, and Vincent Tabard-Cossa

Abstract Solid-state nanopores are a technology poised to disrupt multiple critical fields, including but not limited to nucleic acid sequencing, proteomics, in vitro diagnostics, drug discovery, and next-generation information storage. While several proof-of-concept studies have been demonstrated the feasibility of many of these promising applications, work involving solid-state nanopores has been limited to research labs with specific nanofabrication expertise due to challenges in fabrication methods that have made them difficult to use outside of highly specialized academic settings. Over the past two decades, numerous methods of nanopore fabrication have arisen to tackle the problem of producing individual nanopores in the 1–30 nm range with angstrom precision in thin solid-state membranes. In this Chapter, we review the main methods of solid-state nanopore fabrication that have been developed so far to drive research in the field. We review nanopore formation by drilling with a transmission electron microscope, by ion beam milling, by pipette pulling, and provide a special emphasis on nanopore fabrication by controlled breakdown, which shows promise to expand the user base and accelerate progress in the field. We compare and contrast each method in terms of the accessible range of pore sizes, the precision of size control, the shape of the resulting nanopores, and present an outlook for each method in terms of its potential to address the overarching scalability challenges on the way toward solid-state nanopore technologies.

Keywords Nanopores · Solid-state · Membrane · Nanofabrication · Nanopipettes · Controlled breakdown · Single molecule · Sensors

1.1 Introduction

Following the realization that biological channel proteins could be modified and used as sensitive single-molecule detectors [1, 2], the last two decades have seen a push toward fabricating abiotic analogs called solid-state nanopores, for which several

K. Briggs · M. Waugh · V. Tabard-Cossa (✉)
Department of Physics, University of Ottawa, Ottawa, ON, Canada
e-mail: tcossa@uOttawa.ca

J.-P. Leburton (ed.), *Solid State Nanopores*, Nanostructure Science and Technology,
https://doi.org/10.1007/978-3-031-42336-9_1

fabrication techniques have been developed [3–5]. In this chapter, the commonly used techniques will be reviewed. The nanopore field moves quickly, and the fabrication of solid-state nanopores remains an active sub-area of research, with novel fabrication methods and modifications to protocols frequently reported in the literature. The work discussed here represents a subset of all the studies published on this topic, focused on those which we believe to be the most interesting and impactful, with a special emphasis on nanopore fabrication by controlled breakdown.

The first two techniques to be discussed involve the direct drilling of materials with tightly focused beams of ions or electrons. Prior to their development, it was known that beams of high energy particles could induce controllable material changes, including the drilling of holes, in dielectric membranes and heterostructures, knowledge which until that point had found application mainly in the production of metallic point contacts [6].

Early work with charged particle beams led by Ciaran McMullan, Derek Stein, and Jiali Li started in the late 1990s, in the lab of Jene Golovchenko at Harvard. It was first shown that one could use an Argon ion beam to controllably shrink to single-digit-nanometer existing nanopores by locally melting silicon dioxide membranes [7], followed by a demonstration that the Argon beam could also directly open a nanopore [8, 9]. These solitary pores made by ion beam sculpting were used to detect, for the first time, single molecules of DNA using the fluctuations in the ionic current passing through a solid-state nanopore. Soon after, it was demonstrated by the group of Cees Dekker at Delft that not only was an analogous controlled shrinking effect accessible using the electron beam from a transmission electron microscope (TEM), but that the electron beam could also be used to directly drill single nanopores below 10 nm in size [10] before controllably shrinking them even further. Since TEMs are found in most well-equipped institutions and are therefore relatively accessible compared to the custom-built ion-beam sculpting instrument, TEM-based drilling allowed other groups to establish research programs employing solid-state nanopores. Notably, the groups of Amit Meller, Meni Wanunu, and Marija Drndić have all perfected the art of TEM pore drilling for biosensing applications [11, 12, 13]. State-of-the-art TEM-based drilling techniques can make pores on the order of a single nanometer [14], and despite its complexity and cost, it has been the workhorse of solid-state nanopore fabrication until recently. The last decade has also seen the development of commercial ion beam milling machines capable of reliably making pores < 20 nm with single nm precision, such as the Helium Ion Microscope (HIM) developed by Zeiss, which Adam Hall *et al.* has shown can make sub-4 nm pores [15]. Nonetheless, the requirement for such expensive infrastructure has been a severe limitation to making solid-state nanopore research accessible outside of the top-funded research universities globally, and has slowed the transition of solid-state nanopore technologies to industry.

A second class of nanopore is based on pulling glass pipettes to define nanometer-scale tips at the end of a capillary. Such nanopipettes were a natural extension of efforts to improve the spatial resolution of scanning probe microscopy techniques which employed micron-scale pipette tips to study ion activity in cells [16]. One way to fabricate nanopipettes with openings on the order of 20–40 nm quite simply is to pull quartz capillaries with a laser-equipped pipette puller with a careful choice of parameters such as glass type, thickness, temperature, and pulling force [17–23]. The groups of Ulrich Keyser [24–26] and Joshua Edel [27, 28] have, in recent years, perfected this technique for DNA sensing and other biomarker detection schemes by producing pores in the 10–20 nm range. Nanopipettes are in some ways different from solid-state nanopores on flat membranes in that they are inherently asymmetric, and have an effective pore length (i.e. the sensing region) that depends on the cone angle at the tip. These features have important consequences for capture and translocation dynamics [29].

Most recently, nanopore fabrication based on the dielectric breakdown of membranes in solutions under high electric field stress was reported by the lab of Vincent Tabard-Cossa. It was first shown that a voltage applied to generate a large ionic current through the pore could be used to controllably enlarge existing nanopores immersed in high concentration salt solutions [30], followed by the demonstration that a high electrical field could be imposed over an intact membrane immersed in liquid to drive the formation of a single pore when the value of the sustained electric field approached that of the dielectric breakdown strength of the membrane material [31]. *C*ontrolled *b*reak*d*own (CBD) can fabricate pores directly in an electrolyte solution through an insulating membrane [31–33] or a membrane stack including dielectrics and conductors [34]. The application of a high electric field results in the localized breakdown of the membrane, and the highly controllable formation of single nanopores [31, 35] at a random position on the membrane, though localization of the nanopore is possible with additional intervention [36–39]. Sub-nm nanopores can be fabricated using this technique [40], with high yield reported for 2-nm nanopores [35], and pores can be controllably enlarged with near-atomic precision using the same apparatus that was used for fabrication [35, 41, 42]. While the mechanistic details of nanopore fabrication are complex [43, 44], implementation of the method is simple [42]. Due to the inherent precision, low cost, and simplicity of the method, CBD is replacing beam-based drilling techniques as the method of choice for nanopore research. Most importantly, however, is that the method can be automated and parallelized, allowing solid-state nanopores a path out of the academic lab that was previously inaccessible [42, 45].

While there have been other worthy contributions to the growing list of nanopore fabrication techniques, including track-etching of polymeric membranes [46, 47], stretchable polymeric membranes with tuneable pore sizes [48–50], numerous wet-etching techniques [51–58], and more recent work using laser illumination to open single nanopores [36, 59], we will focus on the handful of techniques discussed above, which are or have been the most commonly used and that are capable of making sub-20 nm nanopores in solid-state membranes, a particularly important size range for single-molecule sensing. Techniques also exist with which to fabricate arrays of nanopores [56, 60–63], important contributions which are beyond the scope of this work.

1.2 Energetic Particle Beam Drilling

1.2.1 Fabrication

Nanopore fabrication by particle beam drilling is achieved by direct sputtering of membrane material with a tightly focused beam of high-energy ions or electrons. By converging a particle beam down to a spot size of a few nanometers, it is possible to remove material at the focus and thereby open a nanoscale hole in the membrane [7, 64–66]. There are two main subclasses of beam-based methods: those based on electron beams implemented in a TEM, and those implemented using ion beams using a variety of specialized microscopy equipment, as depicted in Figs. 1.1 and 1.2. Both methods are broadly applicable across membrane materials, with TEM drilling having been demonstrated on different dielectric thin films and 2D materials including SiN, SiO_2 [67], Al_2O_3 [68], HfO_2 [69], graphene [70–72], MoS_2 [73], WS_2 [59], and ZnO [74], and metalized membranes [10]. In addition to the patterning of dielectric membranes, focused ion beam (FIB) techniques have been applied to metalized membranes and graphene [75, 76].

Among ion-beam drilling techniques, light helium ions [14, 15, 78] are generally preferred for several reasons. Perhaps the most significant is that other heavy ions do not penetrate very far into the material they are drilling, and tend to get embedded in

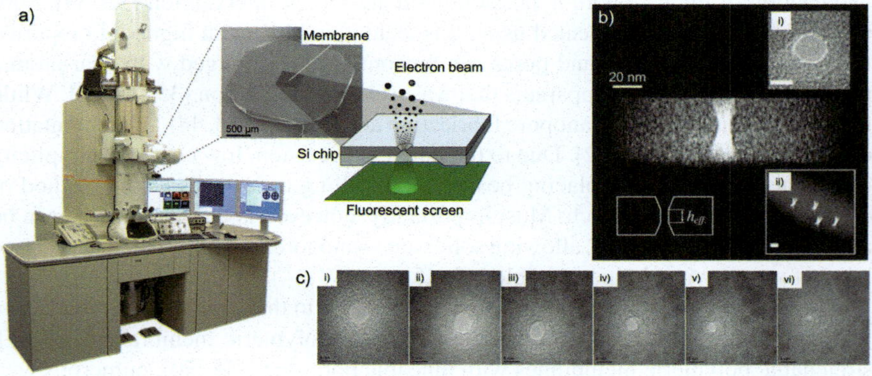

Fig. 1.1 **a** Image of a JEOL 2100 TEM. Insets: electron microscopy image of a Si chip supporting a SiN_x window (free-standing membrane) in its center and schematic of a tightly focused high energy electron beam sputtering the membrane and monitored by eye by looking at flickers on the fluorescent screen to indicate pore opening. **b** Electron beams from a TEM typically have a Gaussian intensity profile with hour-glass shaped contours, which leads to hourglass-shaped pores, as verified using TEM. Adapted with permission from [12]. Copyright 2006 WILEY-VCH Verlag GmbH & Co. KGaA, Weinheim. (i) and (ii) show top and angled views of the pores, respectively. **c** Continuous shrinking of an initially 5 nm nanopore down to 1 nm using 200 kV electron beam radiation from a TEM over the course of 1 h. Adapted with permission from [77]. Copyright 2006 IOP Publishing Ltd

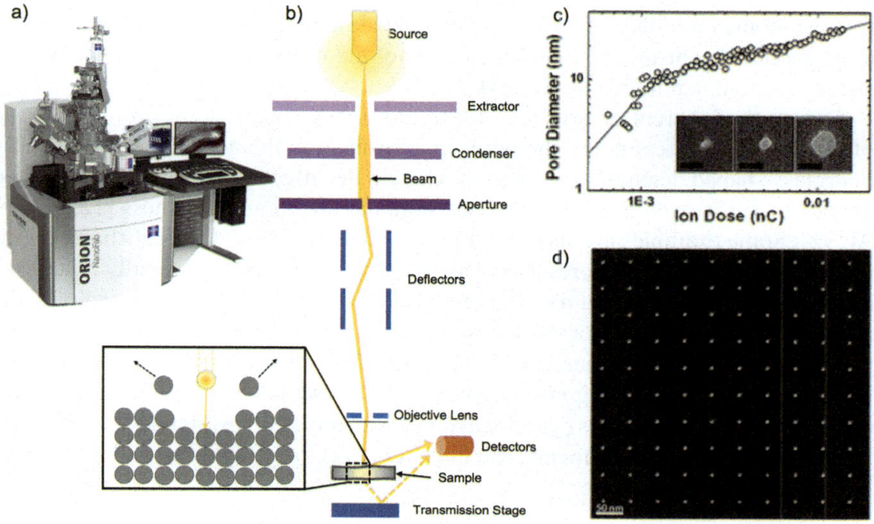

Fig. 1.2 **a** The Orion Nanofab, a helium ion microscope commercialized by Zeiss (image by Carl Zeiss Microscopy reprinted under cc-by-sa-2.0). **b** A schematic diagram of helium ion microscope components. **c** Nanopore diameter as a function of ion dose by drilling in 30 nm thick SiN_X membranes, showing that pore growth slows down once the pore exceeds 10 nm. **d** An array of pores fabricated by sequential helium-ion milling of a single membrane. Adapted with permission from [78]. Copyright 2016 Springer International Publishing Switzerland

the bulk material, resulting in buildup of charges which in turn defocus the incoming beam and reduce precision, limiting pores to larger sizes. While this effect can be mitigated by adding a conductive layer to the membrane to dissipate the charge, this is often undesirable due to the added fabrication steps. Helium ions, being very light, have a larger penetration depth, on the order of 100 nm [79], which often exceeds the thickness of the material being sputtered, limiting charge buildup and improving precision while simplifying the fabrication requirements [80–84]. A schematic of helium-ion drilling and examples of pores fabricated with this technique are shown in Fig. 1.2.

The kinetics of material removal by particle beams are complex, with different membrane elements being removed at different rates around the beam focus [85]. For example, when SiN_X membranes are locally thinned using an electron beam down to sub-2-nm, all that remains is amorphous silicon [86], which is chemically unstable in aqueous salt solutions. Only a handful of studies have reported the nanoscopic details of changes to the membrane material due to electron beam drilling at the point of pore formation [65, 87, 88]. When ions are used at low flux, simulations indicate that direct sputtering dominates, while at high flux it is a more thermally-driven process that involves local melting and even boiling of membrane material [89]. Helium ions are particularly interesting, since depending on the membrane thickness, helium ions with sufficient energy can penetrate deeply into the membrane material and remove atoms on both sides of the membrane simultaneously [90], or create voids and nanobubbles inside the membrane bulk for membranes thicker than the helium ion penetration depth [39, 81, 82, 84].

Due to the inherent divergence of a tightly focused particle beam on the scale of the size of the focal point, the pores that result from this technique are typically hourglass-shaped, leading to a situation wherein the effective thickness of the actual nanopore constriction is thinner than the nominal membrane thickness [12, 13, 85, 91, 92]. Some examples are shown in Fig. 1.1b. Despite this, electrical characterization of TEM-drilled nanopores and many ion-drilled nanopores typically assumes a cylindrical geometry with a reduced effective membrane thickness instead of using a model which accounts for curvature in the nanopore geometry. This is because the additional parameters needed for the fit are difficult to obtain without somehow varying the pore size during measurement [93–96]. Most work with particle-beam-drilled pores models them as cylindrical pores with an effective thickness of one-third to one-half of the nominal membrane thickness [92].

1.2.2 Size Control

Particle beams can be used to controllably grow or shrink existing nanopores in solid-state membranes through local melting of the membrane material using a beam with a wider focus, depending on the initial nanopore size [77]. The free energy difference between an intact membrane with surface tension γ of thickness h and one containing a pore of radius r is

$$\Delta F \propto \gamma r h\left(1-\frac{r}{h}\right), \tag{1.1}$$

which has a maximum at $r = h$, which indicates that if the membrane is locally liquified the pore will grow if $r > h$, and shrink otherwise, a condition which agrees well with experimental evidence [9]. An example of this effect is shown in Fig. 1.1c.

While heavier ions have been shown to be more efficient at controlling the nanopore size by shrinking them, the mechanism of closure is a competition between the mass flow of locally melted membrane material and the sputtering of existing material by incident ions. Indeed, these competing effects can be tuned to either grow or shrink the pore through the choice of ion and beam characteristics [97]. Control of the initial pore size can be improved even to below the FWHM of the incident beam by monitoring the transmission intensity of ions through the membrane [98]. Of the various ion beams used for fabrication, helium has been shown to be the most consistent at making small pores. Once a pore is formed, its size can be tuned using a technique similar to that used with TEM drilling, exposing the existing pore to a lower-intensity ion beam, or an electron beam from a TEM [7, 8, 97]. Pores can either grow or shrink using this method, depending on the beam intensity and the temperature [7].

1.2.3 Outlook

Nanopore drilling with a beam of high-energy particles directly in a microscope system has the advantage of providing a direct visual measurement of the resulting nanopore immediately after fabrication. However, care should be taken when correlating images obtained this way to conductance measurements of the resulting nanopore in liquid, since imaging them for too long can also alter their size. Moreover, pores fabricated by these methods are formed in vacuum and typically require treatment in Piranha solution to make them sufficiently hydrophilic for subsequent full wetting for molecular sensing. These additional processing steps can change the pore size from what was measured visually, so the pore that is eventually wetted in solution no longer necessarily retains its initial size measured when it was under the microscope. Any attempt to correlate pore size between visual imaging and electrical measurements must perform a book-end measurement to ensure that the pore was not altered by the interim processing steps in order to be valid.

Because many ion-beam systems are dedicated nanofabrication tools, unlike TEM, ion beam drilling is more amenable to automation, allowing programmable fabrication of small nanopore arrays [99], or pre-patterning (for example, local thinning) of membrane surfaces prior to nanopore fabrication by other methods [39].

While helium ions can achieve pore diameters in the 5–10 nm range, ion beam drilling is generally less widely used than electron beam drilling as a nanopore fabrication technique. This is mainly due to the fact that the equipment is less commonly found in nanofabrication facilities or needs to be home-built [7].

In the end, despite its many shortcomings, nanopore drilling by particle beams, in particular electron beams, has been the main solid-state nanopore fabrication method used by a majority of nanopore research labs since its initial development, largely because most major universities already have TEM infrastructure. These techniques have been an enabling tool which took solid-state nanopores from an idea to a dynamic research field and demonstrated numerous exciting applications [3, 100–102]. While the prohibitive equipment cost, extensive training required, and difficulties with scalability will most likely not allow this technique to be used outside of research settings, TEM and HIM drilling of nanopores will likely continue to be used in many labs for applications requiring precisely located nanopores or visual feedback of the details of pore fabrication.

1.3 Nanopipettes

1.3.1 Fabrication

Unlike most other methods of solid-state nanopore production, nanopipettes do away with the flat membrane entirely in favor of a nanoscale opening at the end of a glass or quartz capillary, most often produced using standard laser-based pulling techniques and instrumentation. Nanopipettes are fabricated using commercially available benchtop pipette pullers capable of sequential heating and pulling, like the P2000 from Sutter Instruments, and which, with careful choice of pulling parameters and starting pipette material, can generally produce inner tip diameters in the range of tens of nanometers [16, 18, 22, 23, 25–29]. The material of the capillary plays an important role in defining the minimum achievable pore size during pulling and in determining the electrical noise performance during ionic current recording. In particular, quartz's high mechanical strength and low dissipation factor and dielectric constant, are better suited to fabricate small pores with lower noise, thus improving the SNR for biosensing applications. The ratio of the outer diameter (OD) to the inner diameter (ID) of the capillary is also an important parameter that affects the ultimate achievable diameter [17, 103]. Typically, quartz capillaries with a starting inner diameter of 0.2 mm and an outer diameter of 0.5 mm and containing a fused filament are used for single-molecule biosensing experiments [24]. Figure 1.3 depicts the fabrication of nanopipettes with the laser pulling technique. Similar to HIM or TEM-drilled pores, capillaries need to be thoroughly cleaned using piranha solution to ensure proper wetting prior to use.

Nanopipette fabrication in double- and quad-barrelled pipettes has also been demonstrated using identical techniques to regular nanopipette pulling, allowing for the use of multiple pores either in parallel or in series, mediated by a meniscus of liquid at the tip [107–109], or functionalization into an electrically gated nanopore transistor [110]. The ability to fabricate two solid-state nanopores within close proximity to one another is compelling for a number of applications involving long-term capture of polymers between two pores, and nanopipettes remove the requirement of line-of-sight access that is necessary for beam-based drilling.

1.3.2 Size Control

The fabrication of pores < 20 nm, and even < 10 nm is possible using pipette pulling, but it requires the use of rigid quartz capillaries and a careful control of the environmental conditions of the instrument, as ambient temperature and humidity can affect the heating of the capillary by the laser. Indeed, the same set of pulling parameters may not always provide the same nanopore characteristics. Producing nanopores of consistent size can therefore be a challenge from run to run, and pores must either

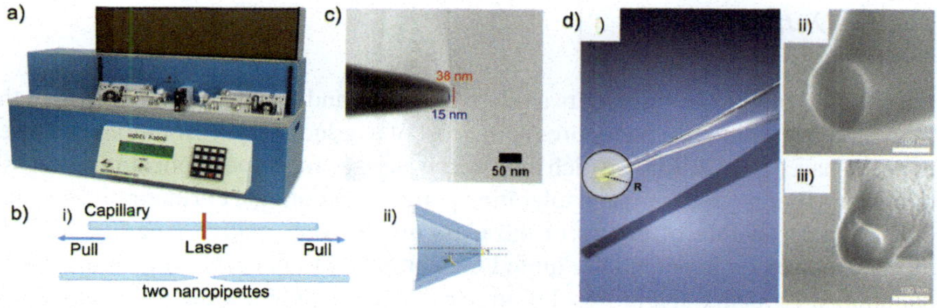

Fig. 1.3 **a** Image of the P-2000 laser-based (CO_2) pipette puller by Sutter Instrument (image by Sutter Instrument. Copyright Sutter Instrument Company 2022). **b** (i) Schematic of the heating-and-pulling method for the fabrication of a nanopore from a glass capillary and (ii) a nanopipette tip, where r and β are the nanopipette radius and cone half-angle. Adapted with permission from [104]. Copyright 2022 The Electrochemical Society. **c** SEM image of a nanopore made in a quartz capillary where ID = 0.2 mm and OD = 0.5 mm. The contrast allows users to differentiate between the walls and the inside of the nanopore. Assuming a constant wall thickness ratio before and after the pulling, the inner diameter can be inferred by measuring the outer diameter, when the contrast is not sufficient to measure directly (image by Bell and Keyser [105] reprinted under CC-BY license). **d** (i) Picture of the end of a nanopipette. (ii) SEM micrograph of a nanopipette tip prior to electron beam irradiation. Scale bar is 100 nm. (iii) The same nanopipette tip after 14 min of electron beam exposure, showing partial shrinking and likely carbon deposition. Adapted with permission from [106]. Copyright 2013 American Chemical Society

be imaged under a TEM, or mounted in salt solution to characterize their conductance, in order to confirm their size. Nanopipettes fabricated by laser-based pulling typically have a wide variation (many nanometers) in pore diameter upon fabrication even when following identical pulling parameters. In practice, users pull many nanopipettes and mount them in a liquid cell to measure their conductance to decide if they are in the appropriate size range for their application. Therefore, the precision of this fabrication method is poor, but this is offset by the speed of fabrication and cost and availability of capillaries. Furthermore, without significant additional work to shrink pores, it is not possible to reliably produce nanopipettes with openings much smaller than 10 nm. Although, once a nanopipette is formed, the size can be controllably reduced down to single nm diameters by irradiation in an electron beam [106, 109], like pores in flat membranes, but having to do so mitigates the advantages of being able to make them with a benchtop instrument, and most commercially available TEM systems require customization to accommodate a pipette. An example is shown in Fig. 1.3d. Since the fabrication of a new nanopipette is fast and cheap, shrinking is generally eschewed in favor of simply making a new one until the desired dimensions are achieved [111].

1.3.3 Outlook

Laser-based pipette pulling offers a relatively quick and inexpensive approach to the fabrication of solid-state nanopores at the cost of a reduced size resolution compared to beam-based pore drilling, which may not be an issue for some applications necessitating large pores (> 10 nm) and tolerating pore size variations. Furthermore, because of the tapered nature of the tip of the glass pipette after pulling (see Fig. 1.3b.ii), the sensing region is potentially quite long compared to flat SiN_X membrane-based nanopores, reducing spatial resolution and blockage depth, inducing a significant EOF, and making both ion transport and molecular translocation through nanopipettes inherently asymmetric and direction-dependent [29], and causing ion current rectification [18]. These features can be taken advantage of to enhance polymer transport characteristics (e.g., capture rate and polymer conformation) [112]. The excellent dielectric properties of some glass used for nanopipettes also leads to very low capacitance and reduced high-frequency noise in the ionic current signal compared to more conventional flat membrane-based nanopores [113], although recently developed chip architectures in glass substrates, instead of Si, allow flat membrane-based pores to have similar high-frequency noise performance [114]. Nanopipettes are therefore versatile experimental single-molecule tools, necessitating minimal equipment for fabrication, that offer excellent high-frequency noise performance, although studies to date have not taken advantage of this to perform high-bandwidth (~1 MHz) recording. Nanopipettes provide beneficial transport characteristics at the cost of limited size control, and will likely continue to be widely used in the research field for the foreseeable future.

1.4 Controlled Breakdown

1.4.1 Fabrication

*B*reak*d*own (BD) in dielectric materials is a ubiquitous process in solid-state electronics and is one of the most common causes of failure in most transistors. It results when the electric field applied across an electrically insulating material exceeds its dielectric strength, making it conductive while causing physical damage to the material. For most applications this is an uncontrolled and detrimental event, resulting in failure of the device in question. In the context of gate dielectric degradation in microelectronics, BD is a very well-studied problem, and most research efforts over the years have been applied toward understanding it and developing ways to avoid it. An excellent review of the physics underlying the dielectric breakdown phenomena in solid-state electronics is given by Lombardo *et al.* [115]. It was recently demonstrated, however, that this process can be controlled at the nanoscale to produce nanopores. This fabrication approach is quickly becoming the method of choice in the nanopore field [31, 42].

When an electric field on the order of the dielectric strength of the membrane material is applied across a dielectric, it partially loses its electrical insulating properties and a leakage current flow through it. This current is generally attributed to a form of trap-assisted tunneling of electrons or holes (i.e. hopping between defects). It causes localized damage to the bulk material (e.g. defects, i.e. broken chemical bonds), which is assumed to take the form of neutral charge traps between which electrons can also tunnel if they are sufficiently close. The breakdown process then proceeds via a percolation effect, mediated by the fact that as these traps build up, they locally reinforce the tunneling current at that location. Once a connected path of these traps spans the material, a highly conductive path through is formed, and dielectric breakdown occurs, irreversibly damaging the material at that location. This process is an example of a weakest link problem, which is well studied in research into the failure modes of electronics [43, 115]. Nanopore fabrication by controlled breakdown of membranes immersed in solution harnesses this normally destructive and uncontrollable process to achieve precise nanofabrication of nanopores in solid-state membranes [31, 42, 116].

In the most basic realization of this approach, an insulating membrane is immersed between two fluidic reservoirs containing a conductive solution, usually an aqueous electrolyte, such that the only path for current flow between the reservoirs is through the nominally insulating membrane itself. A voltage is applied between the two reservoirs such that the electric field is smaller than, but on the order of, the dielectric strength of the membrane material. Sustaining this electric field then drives a leakage current through several randomly distributed hot spots on the membrane, which do not need to be pre-existing, extrinsic defects or processing errors during film deposition. Eventually, one such spot generates enough defects to form a localized conductive path which spans the thickness of the membrane, and a discrete BD event occurs. The material at the damaged path is quickly washed away in the solution, leaving behind a nanoscale hole through the membrane [31, 43]. The conductive solution then fills the resulting nanopore and contributes an ionic current through the pore, which can be used as a feedback mechanism to terminate the voltage before further damage is done to other locations on the membrane. This process, termed controlled breakdown (i.e. controlled BD, leading to the acronym, CBD), is shown in Fig. 1.4.

Because nanopores fabricated by CBD can be made directly in the same aqueous solution and fluidic cell that are later used for single-molecule sensing experiments, this method of nanopore fabrication greatly increases the yield of precisely sized, low-noise nanopores compared to ex-situ fabrication methods such as TEM and FIB. This is because pores produced by CBD are immediately wetted as soon as they are fabricated, avoiding additional handling steps to subsequently facilitate wetting which would otherwise risk damage to the pores between fabrication and use.

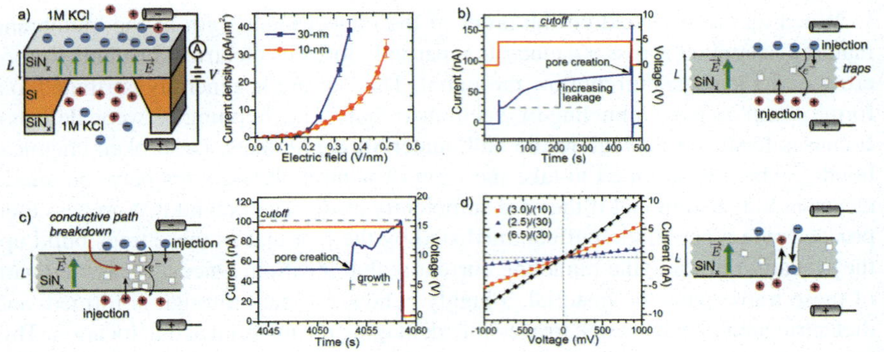

Fig. 1.4 The mechanism of dielectric breakdown in solution. **a** An intact membrane is placed in solution. At high E-field (> 0.2 V/nm; E = V/L), the insulating SiN_X membrane leaks (reversible process). Trap-assisted tunneling dominates the current and increases exponentially with voltage. The voltage applied across the 10 nm membrane (red circles) is from 0 to 5 V, and across the 30 nm thick membrane (blue squares) 0 V to 11 V. **b** Under a sustained 0.5 V/nm (here, 5 V for a 10-nm thick SiN_X—red line), the leakage current (blue line) slowly varies with time as localized damages in the form of locally conductive charge traps accumulate. The leakage current is monitored in real-time. **c** Once a connected path of defects spans the membrane, breakdown occurs (irreversible process). This breakdown event is detected as a sudden spike in the measured current (here, for a 30 nm SiN_X membrane). **d** Material is quickly removed, leaving behind a nanometer-scale hole. The nanopore size can immediately be estimated in the same setup by assessing the conductance of the pore by performing an I-V curve. Adapted with permission from [31]

1.4.2 Controlling Membrane Breakdown

There are numerous ways to achieve control of dielectric breakdown. The first published realization of this scheme used active feedback and a DC bias voltage. By monitoring the current through the membrane, one can detect an abrupt increase in the current when breakdown occurs which corresponds to the onset of ionic current passing through the newly opened pore. This can be used as a signal to terminate the voltage as soon as it is detected (Fig. 1.4b and c) [31]. Feedback speed requirements are modest, needing only ~ 0.1s response time to reliably achieve pores as small as 2 nm [35]. As discussed in detail in Sect. 1.4.4, the time to breakdown (i.e. the time to pore fabrication) is exponentially related to the electric field strength so as long as the fabrication time is orders of magnitude larger than the response time to cut off the voltage, the likelihood of forming more than one pore should be low. Note that the statistics of the first breakdown cannot be reliably used to predict a second BD event, which should be assumed to be faster because of defects already generated at other hot spots.

Several other fabrication control approaches have been developed. To break free from the tyranny of the kinetics of pore fabrication (see Sect. 1.4.4) and in order to use a minimal voltage to open the nanopore in a reasonable time frame, one can employ a slowly increasing voltage until the moment of breakdown, as depicted in

Fig. 5a [42]. A current clamp approach can also be employed, varying the voltage to maintain a fixed tunneling current [38], which naturally reduces the applied voltage when a pore is opened at which point the membrane resistance is consequently reduced (Fig. 1.5b). Alternatively, short DC voltage pulses can be applied to avoid setting a feedback mechanism, either of constant values or increasing in steps as shown in Fig. 1.5c and d [31–33]. These fabrication pulses can also be interleaved with low voltage monitoring to check if a pore formed during the high voltage step, similar to how the pore enlargement is precisely controlled (see Sect. 1.4.3). One of the simplest options, which obviates the need for active feedback entirely, is to use a resistor or otherwise a passive resistive element (such as a long fluidic channel, for example) with resistance on the same order as the desired nanopore itself in series with the membrane. Immediately after pore fabrication, this resistive element would provide passive reduction in the voltage across the membrane, since the resistive element then forms a voltage divider with the nanopore [116]. As the pore grows, the voltage across it is divided between the pore and the resistor element, limiting its size and limiting further growth once the pore voltage drops below a threshold. While these nanopore fabrication control schemes differ in their implementation and have been assigned a variety of acronyms and names, the mechanism by which the nanopore is formed is functionally the same in all cases and is based on the CBD principles [31–33, 43].

When using active feedback and a DC bias it is particularly important to limit the voltage to biases such that the time-to-pore-fabrication is much longer than the feedback time, since the time-to-breakdown is exponentially dependent on the applied voltage (see Sect. 1.4.4). Because of the highly concentrated salt solutions (> 1 M) used during fabrication, the formation of the first pore only perturbs the electric field in its immediate vicinity, meaning that the generation of defects on the rest of the membrane continues unabated even after the pore is opened if a sufficiently high electric field continues to be applied over the entire membrane. Since defects build up at different random locations over the membrane during electric field application, the time to fabrication of a second pore is much shorter than the time to the first pore, and using voltages too high for the feedback time of the system can lead to the formation of multiple pores before the voltage can be terminated [39, 117]. Because damage occurs to the membrane globally prior to the first fabrication event, the statistics of the first breakdown cannot be reliably used to predict the second, which should be assumed to be faster generally.

Nanopore fabrication by CBD has been demonstrated on a variety of membrane materials and architectures. In addition to thin films of SiN_x, SiO_2 [31], TiO_2 [118], and HfO_2 [119], it has been used to make pores in 2D materials, including graphene [33, 120] and transition metal dichalcogenides such as MoS_2 [32, 121]. In the 2D case the kinetics are qualitatively different, since there is no bulk material in which to build up charge traps; rather, the first atom damaged by tunneling current, and/or the redox reaction to transfer charges across the membrane, becomes the location of the pore. However, this is still a weakest-link problem governed by the kinetics underlying

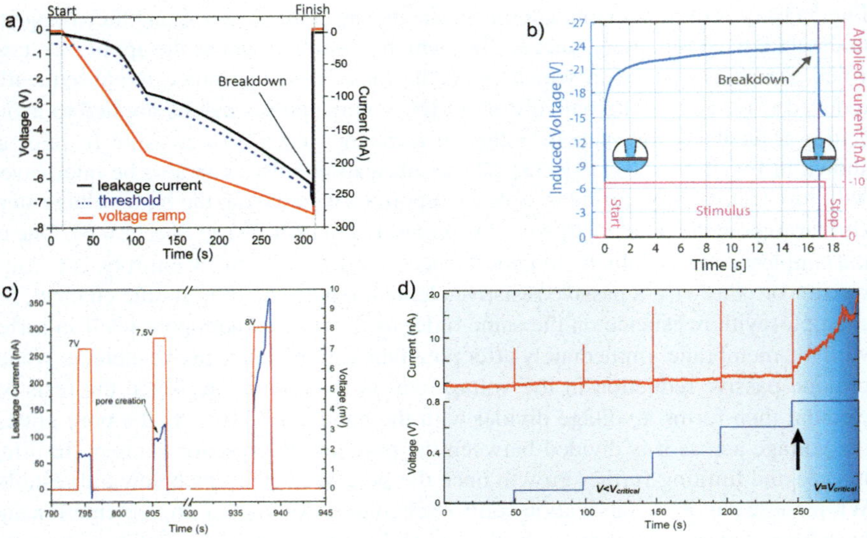

Fig. 1.5 **a** Realization of a voltage ramp control scheme, in which the voltage (red) is increased at a constant rate and terminates when the current (black) spikes above a determined threshold. Here two ramp speeds are used to more quickly raise the voltage to 5 V, the low end for breakdown in tens of seconds in these 10 nm thick SiN membranes. Adapted with permission from [42]. Copyright 2019, Waugh et al., under exclusive licence to Springer Nature Limited. **b** A realization of a current-clamp fabrication scheme, in which the voltage is varied under fixed current until the opening of a pore is heralded by a sudden decrease in the voltage. Adapted with permission from [38]. Copyright 2017 American Chemical Society. **c** A voltage pulse scheme, where the voltage is increased in steps until the pore formation is observed in a 20 nm thick SiO_2 membrane. In this scheme low voltage (200 mV) steps can also be used to periodically check the pore conductance in the ohmic regime to know the size. Adapted with permission from [31]. **d** Another implementation of the voltage pulse scheme in which the voltage across a MoS_2 membrane is periodically increased until pore formation is heralded by an increase in the current. Adapted with permission from [32]. Copyright 2015 American Chemical Society

dielectric breakdown in the limit of membrane thickness approaching zero. To the best of our knowledge, no studies have reported the details of time-to-breakdown distributions on 2D membranes.

CBD has also been used to realize pores through composite multi-layered structures involving both dielectrics and metals (SiN_x/Au, HfO_2/SiN_x/Au/HfO_2, shown in Fig. 1.6), as well as stacked layers of different dielectrics (SiN_x/HfO_2 [122]). In the case of a membrane consisting of a gold layer on top of silicon nitride, the breakdown process was shown to remove a circular section of gold concentric with the fabricated nanopore [34], hinting at the possibility of single-step and automated fabrication of a nanopore centered inside a zero-mode waveguide (ZMW). Depositing a few atomic layers of Hafnia on top of silicon nitride has been shown (unpublished data [123]) to stabilize solid-state nanopores, limiting pore growth to less than 0.1 nm over 6 days when stored in LiCl solution [122]. Moreover, CBD was used to fabricate a pore

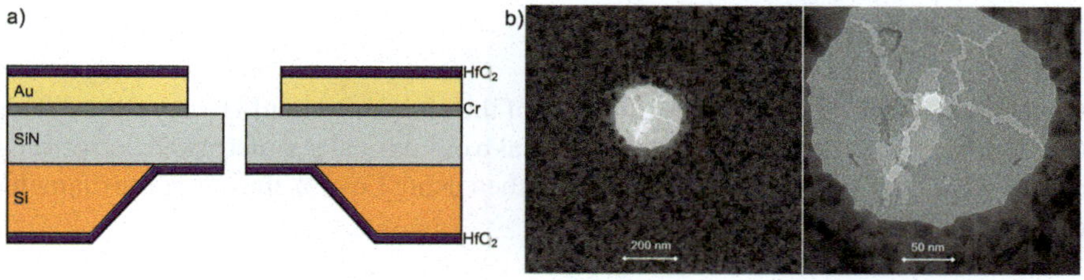

Fig. 1.6 **a** Schematic of the chip with dielectric/metal stack on the membrane. The HfO_2 layer (purple), deposited by atomic layer deposition (ALD), is 5 nm thick, the SiN layer (light grey) is 10 nm and the Au layer (yellow) is 15 nm (on top of a 5 nm Cr adhesion layer—dark grey). CBD is used to fabricate a pore through this multilayer structure in one step. **b** TEM images of the resulting nanopore. A single nanopore concentric with a larger circular area (~ 200 nm in diameter) in which the gold layer has been removed. Image from Zachary Roelen, reprinted with author's permission

through a silicon nitride membrane interfaced with conductive gels [124]. As long as the membrane is in electrical contact with the voltage source, CBD can be used to fabricate a nanopore in any configuration and in almost any device architecture [125, 126]. Thus far, CBD has been able to fabricate a nanopore in every combination of membranes and membrane stacks that we have attempted and has provided no reason to believe it will not continue to work as more complex structures are tested. This opens the possibility for simple fabrication of pores with embedded electrodes or plasmonic structures (see Sect. 1.4.5), which will likely find an important role in applications of nanopores moving forward.

It should be noted that care needs to be taken when fabricating nanopores in membranes that have very low capacitance. While low capacitance is desirable to reduce high-frequency noise in the nanopore ionic current signal, it can also cause a premature and unwanted breakdown if any small charge imbalance Q exists across the membrane prior to pore fabrication. Because the voltage across a capacitor with capacitance C is $V = \frac{Q}{C}$, even a small charge imbalance can induce voltages above the dielectric strength of the material in the low capacitance limit, causing the formation of multiple defects in the membrane. For typical SiN_x membranes, the capacitance is high enough (greater than 10 pF) to ignore this effect, but this can cause issues for capacitance-optimized membrane structures. Luckily this can easily be avoided by briefly shorting both sides of the membrane, with, for example, an Ag/AgCl electrode connecting the two opposing fluidic reservoirs, to balance any charges during the membrane wetting step [127].

1.4.3 Size Control

Once fabrication is complete, the diameter of the nanopore should be measured. This can be achieved by assuming the nanopore has a cylindrical geometry and performing an I-V sweep to extract its conductance. By then using Eq. 1.2, the nanopore diameter d can readily be obtained [128, 129]:

$$d = \frac{G}{2\sigma}\left(1 + \sqrt{1 + \frac{16\sigma L}{\pi G}}\right) \tag{1.2}$$

where G is the nanopore conductance, σ is the conductivity of the solution, and L is the length of the nanopore. While beam-based drilling methods often produce nanopores with an hourglass shape whose effective length is usually assumed to be 1/3 of the membrane thickness [130] (see Sect. 1.2.1), nanopores fabricated by CBD do not have this same characteristic as their formation is not the result of sputtering material via a focused beam. As a result, the effective pore length for CBD-fabricated nanopores is closer to the nominal membrane thickness, an assumption that has been observed indirectly through the level of current blockage produced by DNA translocations, though more quantitative data, including TEM tomography, is needed to characterize this fully.

Typically, nanopores fabricated by CBD are < 2 nm after creation (when using a typical 10 nm thick SiN_X membrane and regular fabrication conditions), as seen in Fig. 1.7a, which is too small for most applications. As a result, unlike beam-based drilling techniques which can tune pore size by starting with a large pore and shrinking it, nanopores fabricated by CBD typically require enlarging. The most convenient way of achieving nanopore enlargement is through a method termed "conditioning" that is procedurally similar, though mechanistically distinct, from the CBD method itself; while CBD is performed with one DC voltage or a voltage ramp that is applied across an intact membrane until a nanopore is fabricated, the conditioning step is performed by applying a series of DC pulses of alternating polarity across the nanopore-containing membrane, as shown schematically in Fig. 1.7b.

The conditioning process typically consists of voltage pulse trains that are separated by a lower-voltage measurement step that checks the nanopore size to determine whether further enlargement is necessary. An example of this is shown in Fig. 1.7c, which has a 5-pulse enlargement train at $\pm$ 4.5 V followed by a $\pm$ 0.2 V measurement step. While the current is measured during the high-voltage enlargement step, the typical nanopore enlargement voltage range is 3–5 V for 10–20 nm thick SiN_X membranes, which is high enough to produce non-ohmic current, meaning the nanopore size cannot be easily extracted from its magnitude [128, 129], hence the intermittent size measurement voltage step.

While CBD is a process governed by the magnitude of the electric field across the membrane, nanopore enlargement is a function of the amount of current passing through the nanopore [30, 41]. Therefore, when discussing the speed with which enlargement occurs, both the voltage being applied across the membrane and the

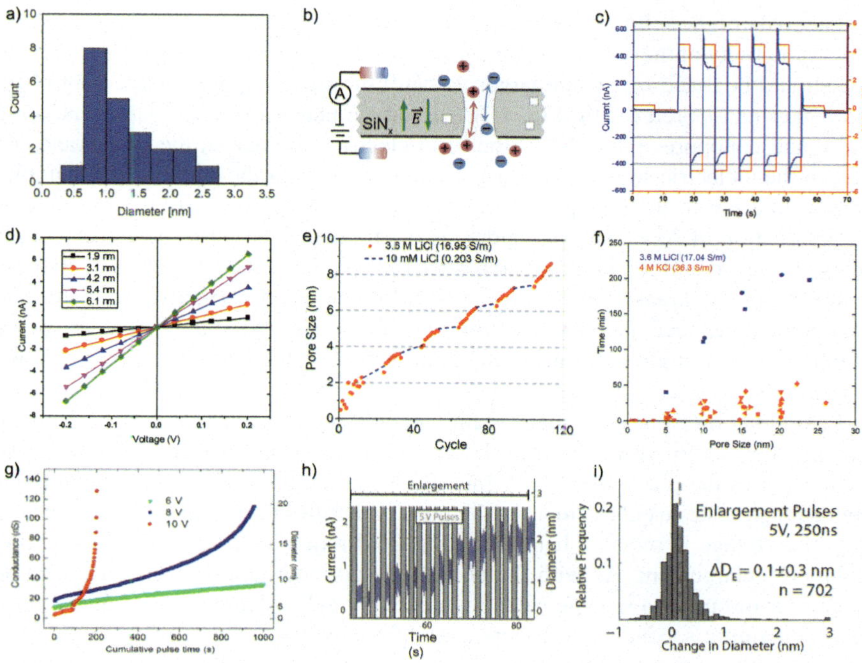

Fig. 1.7 **a** A histogram of nanopore sizes post CBD fabrication. Nanopores fabricated by CBD can start out at sub-nm in size, after which they can be controllably enlarged to any desired size. The distribution represents the pore size for nanopores made in 10 nm thick SiN membranes fabricated in 1 M KCl pH 10, after aging in 3.6 M LiCl pH 8 for up to 24 h. Adapted with permission from [35]. Copyright 2014 WILEY-VCH Verlag GmbH & Co. KGaA, Weinheim. **b** A schematic of the conditioning step where alternating voltages are used. Adapted with permission from [42]. Copyright 2019, Waugh et al., under exclusive licence to Springer Nature Limited. **c** A time series of the conditioning of a nanopore, including a 5-pulse train of ± 4.5 V followed by a ± 0.2 V measurement step (voltage shown in red) and the corresponding current (blue). Unpublished data. **d** Nanopores fabricated by CBD and enlarged using conditioning show linear I-V responses for different nanopore sizes. Unpublished data. **e** A comparison of the growth rate of a nanopore undergoing ± 4.5 V conditioning for 40 cycles in high molar LiCl (red) and low molar LiCl(blue). Note the blue dashes are a guide to the eye as the conductivity is too low for proper size measurements using Eq. 1. Adapted with permission from [41]. Copyright 2020 IOP Publishing Ltd. **f** Comparison of conditioning times in 4 M KCl (high conductivity) and traditional 3.6 M LiCl (lower conductivity). Adapted with permission from [41]. Copyright 2020 IOP Publishing Ltd. **g** Comparison of the impact voltages have on nanopore enlargement. Note that in this case, it is unclear whether a single nanopore is enlarging or multiple nanopores are being fabricated by CBD. Adapted with permission from [30]. Copyright 2020 IOP Publishing Ltd. **h** A series of 5 V pulses is used to enlarge a nanopore fabricated in graphene after pulsed fabrication was initially used to open the nanopore, adapted from [33], with the permission of AIP Publishing. **i** The change in graphene nanopore size per pulse indicates that the enlargement method can control pore size with atomic precision in 2D materials, adapted from [33], with the permission of AIP Publishing

conductivity of the solution are relevant parameters. The significant impact changing the conductivity of the solution can have on the growth rate is shown in Fig. 1.7e and f. It should be noted that while larger voltages during nanopore enlargement result in quicker nanopore growth (Fig. 1.7g), a voltage that is too large risks generating additional nanopores via CBD as the local electric field across the membrane can exceed the dielectric strength of the material in regions of the membrane not immediately adjacent to the nanopore [30, 41]. This can be particularly troublesome as the fabrication of secondary nanopores is often undetected until molecular data is analyzed. It is therefore advisable to reduce the voltage of the DC pulses to no more than 40% of the breakdown voltage. When implemented correctly, the controlled enlargement can take nanopores from fabrication to a range of diameters, as shown in Fig. 1.7d. It can allow for near-angstrom control of the nanopore size, as shown in Fig. 1.7h and i.

In slight contrast to the voltage clamp method described above, a current clamp can also be used to slowly reduce the voltage required to maintain a set value of the current as the pore grows [38]. Initially when the pore is small, this voltage is high enough to drive enlargement of the pore, and as the pore approaches the desired size, the voltage is reduced below the threshold required for growth, thus halting nanopore enlargement. As with the voltage clamp method, however, care must be taken to avoid using a voltage value high enough and sustained long enough that it risks generating secondary nanopores. Additionally, it can be difficult to interpret the magnitude of the current when the voltage is high enough to produce non-ohmic current behaviour [128, 129].

In addition to enlarging the nanopore for appropriate pairing with the target molecule of choice, the conditioning method has also proved to be capable of reducing the 1/f noise of solid-state nanopores, unclogging nanopores, and aiding improperly wetted nanopores [11, 30, 42].

In summary, by coupling the nanopore fabrication method of CBD with this controlled enlargement method, one can automate [42] the generation of nanopores with near-atomic precision [30, 41], allowing for excellent nanopore-analyte matching for improved biosensing performance.

1.4.4 Kinetics

The kinetics of pore formation and growth are complex, having numerous dependencies, not all of which are present in the case of a breakdown in semiconductor devices. There are dependencies on the membrane material, the electric field, and the composition of the electrolyte solution in which fabrication occurs. Additionally, breakdown kinetics have even been shown to vary between batches of nominally identical silicon nitride membranes, pointing to the effects of variations in thickness or stoichiometry [42, 43]. Due to the complexity of the physics involved in dielectric breakdown, very few experimentally useful theoretical models exist [115], so we instead discuss an empirical characterization of these dependencies. Interestingly, although both fabrication and growth are realized using the same setup to apply external electric fields across the pore, the kinetics of the two processes are completely different. These differences provide useful ways to optimize pore fabrication and growth for target pore sizes in a timely manner [131].

For the initial nanopore fabrication, since dielectric breakdown occurs at the first location on the membrane which achieves a connected path of defects spanning the two membrane-electrolyte interfaces, the time required to open the first pore in the membrane is a weakest-link problem, which is described by a Weibull distribution of time-to-breakdown. This property has been demonstrated for both voltage clamp and current clamp fabrication schemes [38, 43], as shown in Fig. 1.8a and b.

The microscopic details of the process are not well-characterized at this point. In the dry state, dielectric breakdown proceeds through an accumulation of locally conductive defects caused by leakage current through the membrane, which in turn reinforces leakage current and the generation of further defects until a connected

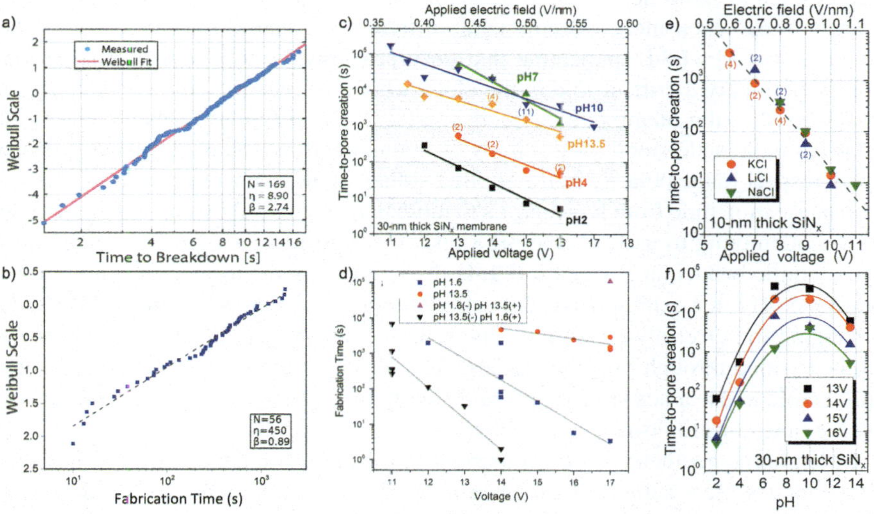

Fig. 1.8 **a** Time to breakdown follows a Weibull distribution for a current clamp applying 10 nA in 1 M LiCl at pH 8. Adapted with permission from [38]. Copyright 2017 American Chemical Society. **b** Weibull distribution for a voltage clamp applying 8 V in 1 M KCl pH at 10. The Weibull scale is $\ln(-\ln(1-F(t)))$ where F(t) is the cumulative fraction of membranes that have undergone CBD by time t. η and β are the Weibull timescale and slope, respectively. Adapted with permission from [43]. Copyright 2015 IOP Publishing Ltd. **c** Time to breakdown exponentially depends on the applied voltage (or equivalently, the applied electric field). Adapted with permission from [31]. **d** Time to breakdown is extremely sensitive to the local pH near the membrane. By modifying the local concentrations of OH^- and H^+ near the membrane, fabrication times can be modulated from seconds to days for a given electric field strength. Adapted with permission from [43]. Copyright 2015 IOP Publishing Ltd. **e** For a given electric field, fabrication times are insensitive to the choice of cation. Adapted with permission from [31]. **f** For a given electric field, fabrication times are highly sensitive to the pH of the electrolyte solution. Adapted with permission from [31]

path exists that makes the membrane locally conductive [115]. In the liquid state, material removal follows quickly, and a pore forms [43].

The mean time to breakdown depends exponentially on the applied electric field, as shown in Fig. 1.8c–e. This extreme electric field dependence can be used to change the mean time by many orders of magnitude for a given set of parameters using only a small variation in voltage. For this reason, care must be taken when choosing a voltage to ensure that the time it takes to fabricate the pore is much longer than the feedback time at which the pore can be detected and the voltage terminated, since as already noted, one otherwise risks opening several pores at the same time [39, 117]. On the other hand, if the electric field strength is too low, the fabrication time will be prohibitively long.

Breakdown kinetics also depend on the nature of the electrolyte used to carry the current, a dependence which is obviously not present in the case of dry dielectric breakdown [43]. In particular, the time to breakdown was found to be very sensitive to pH for a given electric field strength (Fig. 1.8f), reducing by orders of magnitude at both extremes of the pH scale, while being largely insensitive to the nature of the cation used to carry the current through the solution (Fig. 1.8e) [31]. This sensitivity can be made even more extreme using different pH values on either side of the membrane (Fig. 1.8d), indicating that unsurprisingly, protons and hydroxide ions likely both play a critical role in the electrochemical reactions by which charge is transferred at the membrane-solution interface [43], reaction pathways which are starting to be understood [44]. Simply changing the direction of the pH asymmetry with respect to the polarity of the applied voltage in the most extreme case changes the fabrication time from less than 1 s to more than 3 days [43]. Care should be taken when attempting to use extreme pH values to speed up nanopore formation, since pore ohmic behaviour is often lost at both extremes: pores formed in conditions of extreme pH often exhibit persistent current rectification that is not always solved by changing the electrolyte solution, suggesting geometric asymmetry which has yet to be verified through imaging studies. As a means to reduce fabrication time, it is preferable to use a near-neutral pH and simply increase the voltage instead, subject to the time constraints discussed above.

Membrane geometry is also important in breakdown kinetics. Membranes with larger active areas achieve breakdown faster since there is more area for defects to form [43], though the area dependence is logarithmic and therefore quite modestly affects the time to breakdown [115]. Thinner membranes also have faster fabrication for a given voltage due to the higher electric field applied [43, 132].

Dielectric breakdown is a universal process and can be controlled by a large selection of possible conditions. The universality of the fabrication process, combined with the wide range of parameters which have a significant effect, provides many knobs to turn to optimize nanopore fabrication for a given application.

1.4.5 Localization

Since CBD is usually performed using purely electrical feedback, there is no a priori way to know where on the membrane the nanopore was formed without additional steps. In the simplest form of CBD, because the electric field used to fabricate the nanopore is applied uniformly across the entire membrane, the actual location at which the pore is fabricated is random. Identifying the location of the pore has been achieved using a TEM to image the membrane after pore formation [31, 32, 133], and optically using a calcium-activated indicator dye on one side of the membrane, coupled with a calcium-based electrolyte on the other [117] as shown in Fig. 1.9a. Since the only place where significant concentrations of both the dye and calcium are both present is in the pore itself, the pore lights up and can be identified under a microscope when a voltage is applied to drive calcium ions into the region with the dye [134].

Despite this randomness in nanopore location, it is possible to take advantage of the extreme electric-field dependence of the time to breakdown to deliberately position pores with varying degrees of precision, either by locally increasing the electric field strength, or by limiting the membrane area which is exposed to an electrolyte. Precisely positioned nanopore fabrication by controlled breakdown has been successfully demonstrated using various embodiments of both approaches.

Locally thinning a membrane has the effect of amplifying the electric field in the thinned regions for a given applied voltage [39, 117], and can localize the nanopore to the same precision with which the membrane can be thinned by other means. The most precise localization by thinning demonstrated so far achieved accuracy on the order of 100 nm, using a helium ion microscope to perform partial milling down to 10 nm of 100 nm × 100 nm regions in 100 nm thick membranes prior to CBD (Fig. 1.9b). Of note in this approach is that while the probability per unit area of forming the pore at a given location is much higher in the thinned region, since the higher electric field there generates more defects per unit area per unit time, the thicker region is often orders of magnitude larger. While the area contribution breakdown in Weibull statistics is modest, care must be taken to tune the electric field such that the time to breakdown in the much smaller thinned region is much faster than the time to breakdown in the thick region [39].

Another way to achieve local electric field enhancements is to use the optical field from a laser. This technique involves locally changing the electric field by shining a focused laser on the desired location of the nanopore. The laser is hypothesized to increase the local conductivity by interacting with the photosensitive material, reducing the band gap for conduction in the membrane material near the laser focus and promoting controlled breakdown at that spot [139], a method which achieves diffraction-limited localization. The diffraction limit can be overcome for even more precise localization when the laser field is further modulated by plasmonic structures, which permits reliable fabrication of solid-state nanopores within nanometer distances of existing nanostructures on a membrane [37], as shown in Fig. 1.9c.

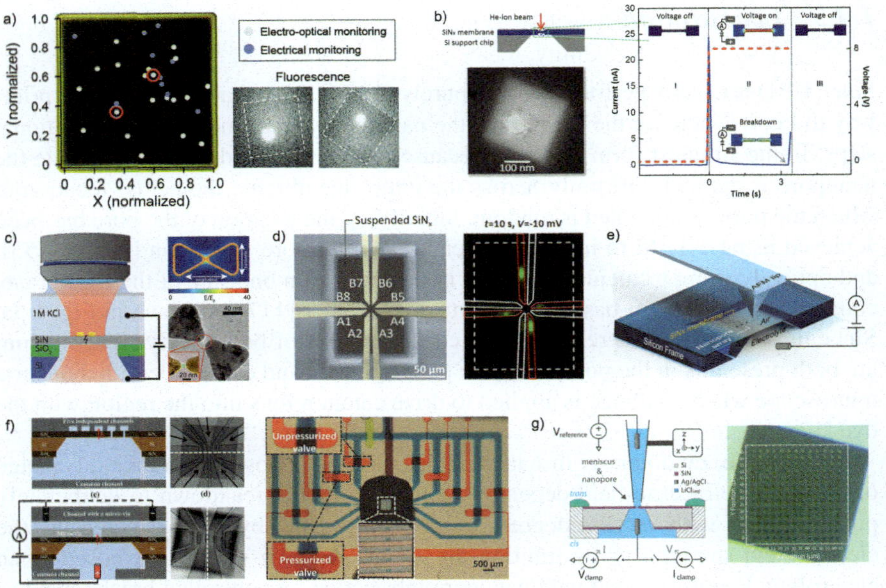

Fig. 1.9 **a** The locations of many nanopores fabricated on different membranes and overlaid, showing a roughly uniform distribution of pore locations on the membrane (~ 40 μm × 40 μm) in the absence of any additional effort to localize the electric field. The circled points correspond to the two nanopores shown in the fluorescent image. Pore locations are determined by Gaussian fitting of fluorescence spots produced by calcium dye (dotted square represents membrane outline). Adapted from [117] with permission from the Royal Society of Chemistry. **b** Local enhancement of the electric field through local thinning of the membrane by a helium ion beam to localize pore fabrication. Adapted with permission from [39]. Copyright 2017 IOP Publishing Ltd. **c** Electric field enhancement around a plasmonic bowtie antenna allows pore localization within the hot spot. Adapted with permission from [37]. Copyright 2015 American Chemical Society. **d** Localization of the pore by using CBD with one of the electrodes directly on the membrane surface. Adapted from [135] under CC BY 4.0. **e** Localization of the pore using a conductive AFM tip as one of the fabrication electrodes to limit the pore formation to the contact region. Adapted with permission from [136]. Copyright 2019 WILEY-VCH Verlag GmbH & Co. KGaA, Weinheim. **f** Microfluidic isolation of separate membrane segments for individual pore fabrication in each channel (left). Reproduced from [137] with permission from the Royal Society of Chemistry. Top view of the same system, with pressure valves for flow control (right). Adapted with permission from [138]. Copyright 2017 WILEY-VCH Verlag GmbH & Co. KGaA, Weinheim. **g** A schematic of the pore localization strategy using a micropipette tip to make microscale liquid contact with a target membrane area. Scanning the tip of the glass pipette can produce an array of pores tens of microns apart (although not individually addressable in parallel). Adapted with permission from [38]. Copyright 2017 American Chemical Society

However, this comes at the cost of complexity in nanofabricating the plasmonic structures and aligning the laser spot. This method works well for SiN_x membranes, and more work remains to ascertain its efficacy with other membrane materials.

Localization can also be achieved by focusing the electric field by having one of the electrodes in close proximity or directly in contact with an area of the membrane. Micropatterned metal electrodes on the membrane surface have been used to drive nanopore fabrication somewhere directly along the electrode (Fig. 1.9d) [44, 135]. This process also perforates the metal layer, similar to when a nanopore is fabricated in a metalized membrane [34] (see Sect. 1.4.2). One of the electrodes used to control the electric field during CBD can be in the form of a conductive AFM tip, which limits pore formation to the point of contact between the tip and the membrane where the field is focused (Fig. 1.9e) [136].

It is possible to limit the application area of the electric field using microfluidics to confine the fluidically (and thus the electrically) active areas to a pre-specified portion of an intact membrane. One realization of this approach has been demonstrated by partitioning the membrane into various microfluidic channels [137, 138, 140], as shown in Fig. 1.9f. Due to the exponential dependence of fabrication time on the local electric fields, care must be taken to ensure that the applied electric field is uniform across the target area.

Localization of the pore has also been achieved by limiting the area over which liquid contacts the membrane using the meniscus at the tip of a micropipette [38]. Using a micropositioner to scan the pipette tip terminating in microscale liquid contact across the membrane permits serial fabrication of hundreds of nanopores on a single microscale membrane, as shown in Fig. 1.9g. Localization is achievable with precision equal to the diameter of the liquid meniscus at the tip of the pipette. While micrometer precision has been demonstrated, it could in principle be improved with smaller pipette tips.

1.4.6 Outlook

Owing to the small initial size of CBD-fabricated nanopores (< 2 nm) and its sub-nm precision in controlling the final desired size [35], the CBD method is very well suited to fabricate pores in the sub-20 nm pore size range which is critically important for single-molecule sensing applications where size matching to the analyte of interest is key. Additionally, as the nanopores are fabricated and grown in-situ (i.e. directly in an aqueous salt solution), the precision of the fabrication and enlarging process is better suited to forming nanopores of a specific target size than other techniques that make pores in air or vacuum and require harsh cleaning chemicals prior to use. In-situ fabrication leads to fewer delicate manipulation steps of the chips, which facilitates the utilization of thinner, more mechanically fragile, membranes (≤ 10 nm) that are desirable given their increased spatial sensitivity. It is interesting to note that the fabrication does not need to be performed in aqueous solutions and can be performed in organic solvents. In fact, functional pores have been fabricated in EtOH [31] using the same protocol as otherwise used with water-based solutions.

One of the other breakthroughs CBD provides is that line-of-sight access to the eventual location of the pore is not necessary. All that is required is a liquid contact connecting the electrodes to either side of the membrane to be perforated. This permits pores to be fabricated deep inside complex geometries after they are fully assembled, greatly increasing yield, and putting mass-manufacturing of devices with embedded nanopores within reach. Such an integrated device, comprising a nanopore equipped with a nanofilter within molecular distance was recently demonstrated, which shows the possibility of fabricating a single nanopore through an existing porous structure to produce two solid-state nanopores in series within nanoscale distances of one another [125, 126, 141].

On the other hand, CBD is limited to relatively thin membrane materials and membrane stacks practically < 30 nm in thickness because the electric fields required to make pores in thicker stacks are too high for reliable size control. Additionally, it is difficult to make pores much larger than 30 nm in diameter due to the excessive time required to controllably enlarge nanopores past that point. On that topic, some groups have proposed adaptations of the CBD protocol, principally by changing the chemical composition of the fabrication solution, to improve the yield of fabricated pores past a certain size and above a certain membrane thickness (> 20 nm). It has also been reported that soft breakdown [115] can occur in thicker membranes, which creates an ionically conductive membrane area, but leaves behind solid material that can mimic the electrical characteristics of a nanopore without enabling the passage of molecular analytes [133]. In particular, it's been suggested that working with alkaline solutions (pH > 11) can avoid a soft breakdown during fabrication [142], and that adding bleach can help produce pores less prone to clogging by affecting the surface chemistry of the post-breakdown nanopore [143]. All these limitations are the subject of ongoing research, and new advances are reported regularly.

1.5 Conclusions

Electron-beam nanopore drilling has sustained the needs of a small number of academic labs performing solid-state nanopore research until recent years. Despite its remarkable early impact, new groups entering the field now tend to employ laser-pulled nanopipettes or nanopores fabricated by CBD to support their research programs due to the reduction in cost and complexity that these methods offer. Furthermore, while quartz nanopipettes offer interesting built-in advantages such as nanoconfinement prior to polymer passage and low noise performance, fabricating small pores (< 10 nm) remains a challenge for most users and relies on the fabrication of a large number of pipettes to obtain a single usable one of the desired size. Moreover, pulling pipettes is not a scalable nanofabrication method and is not conducive to large-scale array fabrication, which will be needed to achieve the application of solid-state nanopores outside of research settings. In contrast, the CBD method can be fully automated and parallelized, while maintaining the pore size resolution and precision required for many single-molecule biosensing applications making the commercialization of solid-state nanotechnology within reach. In addition, it has the potential to enable solid-state nanopores to make it beyond specialized labs to allow non-experts to experiment with solid-state nanopores and focus on the development of applications. Where other methods involve serial fabrication of nanopores using expensive equipment, CBD can be realized in an affordable benchtop format [42]. While instrumentation that achieves this automated nanopore fabrication has been fully open-sourced in previous work on the subject [42], a turn-key instrument, called the Spark-E2, has been made available commercially by the authors.

Nanopore fabrication techniques are rapidly maturing toward commercial readiness, yet there are still segments of the parameter space of fabrication variables that have not yet been explored. Most nanopore research is conducted using silicon nitride membranes, which is largely due to familiarity and convenience, since robust thin membrane fabrication protocols exist for this material. A few groups are exploring alternative membranes, however. Hafnium oxide shows promise as a more stable and robust membrane material which could be made much thinner than silicon nitride while retaining long-term stability [69]. The unique and infinitely tunable electrical properties of transition metal dichalcogenides make them attractive as membrane materials, especially as components of multilayer devices for more precise control of local electric fields in and around the pore, but so far only a few have been explored [144]. In short, while considerable strides have been made in fabrication techniques for solid-state nanopores over the past two decades, there can be no doubt that the pace of discovery will continue unabated for the foreseeable future.

Competing Financial Interest KB, MW, and VTC declare competing financial interests in the form of ownership of Northern Nanopore Instruments Inc., a for-profit company commercializing the controlled breakdown method of solid-state nanopore fabrication.

References

1. Bezrukov SM, Vodyanoy I, Parsegian VA (1994) Counting polymers moving through a single ion channel. Nature 370(6487):279–281. https://doi.org/10.1038/370279a0
2. Kasianowicz JJ, Brandin E, Branton D, Deamer DW (1996) Characterization of individual polynucleotide molecules using a membrane channel. Proc Natl Acad Sci USA 93(24):13770–13773. https://doi.org/10.1073/pnas.93.24.13770
3. Dekker C, Article R, Dekker C (2007) Solid-state nanopores. Nat Nanotechnol 2(4):209–215. https://doi.org/10.1038/nnano.2007.27
4. Xue L, Yamazaki H, Ren R, Wanunu M, Ivanov AP, Edel JB (2020) Solid-state nanopore sensors. Nat Rev Mater 5(12):931–951. https://doi.org/10.1038/s41578-020-0229-6
5. Branton D, Deamer DW, Marziali A, Bayley H, Benner SA, Butler T, Di Ventra M, Garaj S, Hibbs A, Huang X, Jovanovich SB, Krstic PS, Lindsay S, Ling XS, Mastrangelo CH, Meller A, Oliver JS, Pershin YV, Ramsey JM, Riehn R, Soni GV, Tabard-Cossa V, Wanunu M, Wiggin M, Schloss JA (2008) The potential and challenges of nanopore sequencing. Nat Biotechnol 26(10):1146–1153. https://doi.org/10.1038/nbt.1495
6. Gribov NN, Theeuwen SJCH, Caro J, Radelaar S (1997) A new fabrication process for metallic point contacts. Microelectron Eng 35:317–320

7. Li J, Stein D, McMullan C, Branton D, Aziz MJ, Golovchenko JA (2001) Ion-beam sculpting at nanometre length scales. Nature 412(6843):166–169. https://doi.org/10.1038/35084037
8. Stein D, Li J, Golovchenko JA (2002) Ion Beam sculpting time scales. Phys Rev Lett 89:276106. https://doi.org/10.1103/PhysRevLett.89.276106
9. Stein DM, McMullan CJ, Li J, Golovchenko JA (2004) Feedback-controlled ion beam sculpting apparatus. Review of Scientific Instruments 75(4):900–905. https://doi.org/10.1063/1.1666986
10. Wanunu M, Meller A (2008) Single-molecule techniques: a laboratory manual. Cold Spring Harbor Laboratory, New York
11. Fischbein MD, Drndić M (2007) Sub-10 nm device fabrication in a transmission electron microscope. Nano Lett 7(5):1329–1337. https://doi.org/10.1021/nl0703626
12. Kim MJ, Wanunu M, Bell DC, Meller A (2006) Rapid fabrication of uniformly sized nanopores and nanopore arrays for parallel DNA analysis. Adv Mater 18(23):3149–3153. https://doi.org/10.1002/adma.200601191
13. Kim MJ, McNally B, Murata K, Meller A (2007) Characteristics of solid-state nanometre pores fabricated using a transmission electron microscope. Nanotechnology 18(20):205302. https://doi.org/10.1088/0957-4484/18/20/205302
14. Emmrich D, Beyer A, Nadzeyka A, Bauerdick S, Meyer JC, Kotakoski J, Gölzhäuser A (2016) Nanopore fabrication and characterization by helium ion microscopy. Appl Phys Lett 108(16). https://doi.org/10.1063/1.4947277
15. Yang J, Ferranti DC, Stern LA, Sanford CA, Huang J, Ren Z, Qin L-C, Hall AR (2011) Rapid and precise scanning helium ion microscope milling of solid-state nanopores for biomolecule detection. Nanotechnology 22(28):285310. https://doi.org/10.1088/0957-4484/22/28/285310
16. Lieberman K, Lewis A, Fish G, Shalom S, Jovin TM, Schaper A, Cohen SR (1994) Multi-functional, micropipette based force cantilevers for scanned probe microscopy. Appl Phys Lett 65(5):648–650. https://doi.org/10.1063/1.112259
17. Stanley J, Pourmand N (2020) Nanopipettes—the past and the present. APL Mater 8(10). https://doi.org/10.1063/5.0020011
18. Wei C, Bard AJ, Feldberg SW (1997) Current rectification at quartz nanopipet electrodes. Anal Chem 69(22):4627–4633. https://doi.org/10.1021/ac970551g
19. Shao Y, Mirkin MV (1997) Fast kinetic measurements with nanometer-sized pipets. transfer of potassium ion from water into dichloroethane facilitated by dibenzo-18-crown-6. J Am Chem Soc 119(34):8103–8104. https://doi.org/10.1021/ja971824s
20. Ying L, White SS, Bruckbauer A, Meadows L, Korchev YE, Klenerman D (2004) Frequency and voltage dependence of the dielectrophoretic trapping of short lengths of DNA and DCTP in a nanopipette. Biophys J 86(2):1018–1027. https://doi.org/10.1016/S0006-3495(04)74177-6
21. Karhanek M, Kemp JT, Pourmand N, Davis RW, Webb CD (2005) Single DNA molecule detection using nanopipettes and nanoparticles. Nano Lett 5(2):403–407. https://doi.org/10.1021/nl0480464
22. Actis P, Mak AC, Pourmand N (2010) Functionalized nanopipettes: toward label-free, single cell biosensors. Bioanal Rev 1(2):177–185. https://doi.org/10.1007/s12566-010-0013-y
23. Morris CA, Friedman AK, Baker LA (2010) Applications of nanopipettes in the analytical sciences. Analyst 135(9):2190–2202. https://doi.org/10.1039/c0an00156b
24. Bell NAW, Thacker VV, Hernández-Ainsa S, Fuentes-Perez ME, Moreno-Herrero F, Liedl T, Keyser UF (2013) Multiplexed ionic current sensing with glass nanopores. Lab Chip 13(10):1859–1862. https://doi.org/10.1039/c3lc50069a
25. Steinbock LJ, Otto O, Chimerel C, Gornall J, Keyser UF (2010) Detecting DNA folding with nanocapillaries. Nano Lett 10(7):2493–2497. https://doi.org/10.1021/nl100997s
26. Bell NAW, Keyser UF (2016) Digitally encoded DNA nanostructures for multiplexed, single-molecule protein sensing with nanopores. Nat Nanotechnol 11(7):1–28. https://doi.org/10.1038/NNANO.2016.50
27. Sze JYY, Ivanov AP, Cass AEG, Edel JB (2017) Single molecule multiplexed nanopore protein screening in human serum using aptamer modified DNA carriers. Nat Commun 8(1):1–10. https://doi.org/10.1038/s41467-017-01584-3

28. Gong X, Patil AV, Ivanov AP, Kong Q, Gibb T, Dogan F, Demello AJ, Edel JB (2014) Label-free in-flow detection of single DNA molecules using glass nanopipettes. Anal Chem 86(1):835–841. https://doi.org/10.1021/ac403391q
29. Bell NAW, Chen K, Ghosal S, Ricci M, Keyser UF (2017) Asymmetric dynamics of DNA entering and exiting a strongly confining nanopore. Nat Commun 8(1):380. https://doi.org/10.1038/s41467-017-00423-9
30. Beamish E, Kwok H, Tabard-Cossa V, Godin M (2012) Precise control of the size and noise of solid-state nanopores using high electric fields. Nanotechnology 23(40):405301. https://doi.org/10.1088/0957-4484/23/40/405301
31. Kwok H, Briggs K, Tabard-Cossa V (2014) Nanopore fabrication by controlled dielectric breakdown. PLoS ONE 9(3):e92880. https://doi.org/10.1371/journal.pone.0092880
32. Feng J, Liu K, Graf M, Lihter M, Bulushev RD, Dumcenco D, Alexander DTL, Krasnozhon D, Vuletic T, Kis A, Radenovic A (2015) Electrochemical reaction in single layer MoS_2: nanopores opened atom by atom. Nano Lett 15(5):3431–3438. https://doi.org/10.1021/acs.nanolett.5b00768
33. Kuan AT, Lu B, Xie P, Szalay T, Golovchenko JA (2015) Electrical pulse fabrication of graphene nanopores in electrolyte solution. Appl Phys Lett 106(May):203109. https://doi.org/10.1063/1.4921620
34. Kwok H, Waugh M, Bustamante J, Briggs K, Tabard-Cossa V (2014) Long passage times of short SsDNA molecules through metallized nanopores fabricated by controlled breakdown. Adv Funct Mater 24(48):7745–7753. https://doi.org/10.1002/adfm.201402468
35. Briggs K, Kwok H, Tabard-Cossa V (2014) Automated fabrication of 2-nm solid-state nanopores for nucleic acid analysis. Small 10(10):2077–2086. https://doi.org/10.1002/smll.201303602
36. Gilboa T, Zrehen A, Girsault A, Meller A (2018) Optically-monitored nanopore fabrication using a focused laser beam. Sci Rep 8(1):9765. https://doi.org/10.1038/s41598-018-28136-z
37. Pud S, Verschueren D, Vukovic N, Plesa C, Jonsson MP, Dekker C (2015) Self-aligned plasmonic nanopores by optically controlled dielectric breakdown. Nano Lett 15(10):7112–7117. https://doi.org/10.1021/acs.nanolett.5b03239
38. Arcadia CE, Reyes CC, Rosenstein JK (2017) In situ nanopore fabrication and single-molecule sensing with microscale liquid contacts. ACS Nano 11(5):4907–4915. https://doi.org/10.1021/acsnano.7b01519
39. Carlsen AT, Briggs K, Hall AR, Tabard-Cossa V (2017) Solid-state nanopore localization by controlled breakdown of selectively thinned membranes. Nanotechnology 28(8):085304. https://doi.org/10.1088/1361-6528/aa564d
40. Yanagi I, Akahori R, Hatano T, Takeda K (2014) Fabricating nanopores with diameters of sub-1 nm to 3 nm using multilevel pulse-voltage injection. Sci Rep 4:5000. https://doi.org/10.1038/srep05000
41. Leung C, Briggs K, Laberge M-P, Peng S, Waugh M, Tabard-Cossa V (2020) Mechanisms of solid-state nanopore enlargement under electrical stress. Nanotechnology, 31(44). https://doi.org/10.1088/1361-6528/aba86e
42. Waugh M, Briggs K, Gunn D, Gibeault M, King S, Ingram Q, Jimenez AM, Berryman S, Lomovtsev D, Andrzejewski L, Tabard-Cossa V (2020) Solid-state nanopore fabrication by automated controlled breakdown. Nat Protoc 15(1):122–143. https://doi.org/10.1038/s41596-019-0255-2
43. Briggs K, Charron M, Kwok H, Le T, Chahal S, Bustamante J, Waugh M, Tabard-Cossa V (2015) Kinetics of nanopore fabrication during controlled breakdown of dielectric membranes in solution. Nanotechnology 26(8):084004. https://doi.org/10.1088/0957-4484/26/8/084004
44. Fried JP, Swett JL, Nadappuram BP, Fedosyuk A, Sousa PM, Briggs DP, Ivanov AP, Edel JB, Mol JA, Yates JR (2021) Understanding electrical conduction and nanopore formation during controlled breakdown. Small 17(37):2102543. https://doi.org/10.1002/smll.202102543
45. Fried JP, Swett JL, Nadappuram BP, Mol JA, Edel JB, Ivanov AP, Yates JR (2021) In situ solid-state nanopore fabrication. Chem Soc Rev 50(8):4974–4992. https://doi.org/10.1039/d0cs00924e

46. Yu AP, Blonskaya IV, Lizunov NE, Olejniczak K, Orelovitch OL, Sartowska BA, Dmitriev SN (2017) Asymmetrical nanopores in track membranes: fabrication, the effect of nanopore shape and electric charge of pore walls, promising applications. Russ J Electrochem 53(1):58–69. https://doi.org/10.1134/S1023193517010037
47. Siwy Z, Apel P, Dobrev D, Neumann R, Spohr R, Trautmann C, Voss K (2003) ion transport through asymmetric nanopores prepared by ion track etching. Nucl Instrum Methods Phys Res B 208(1–4):143–148. https://doi.org/10.1016/S0168-583X(03)00884-X
48. Willmott GR, Moore PW (2008) Reversible mechanical actuation of elastomeric nanopores. Nanotechnology 19(47):475504. https://doi.org/10.1088/0957-4484/19/47/475504
49. Roberts GS, Kozak D, Anderson W, Broom MF, Vogel R, Trau M (2010) Tunable nano/micropores for particle detection and discrimination: scanning ion occlusion spectroscopy. Small 6(23):2653–2658. https://doi.org/10.1002/smll.201001129
50. Sowerby SJ, Broom MF, Petersen GB (2007) Dynamically resizable nanometre-scale apertures for molecular sensing. Sens Actuators B Chem 123(1):325–330. https://doi.org/10.1016/j.snb.2006.08.031
51. Zhang B, Galusha J, Shiozawa PG, Wang G, Bergren AJ, Jones RM, White RJ, Ervin EN, Cauley CC, White HS (2007) Bench-top method for fabricating glass-sealed nanodisk electrodes, glass nanopore electrodes, and glass nanopore membranes of controlled size. Anal Chem 79(13):4778–4787. https://doi.org/10.1021/ac070609j
52. Zhang B, Zhang Y, White HS (2004) The nanopore electrode. Anal Chem 76(21):6229–6238. https://doi.org/10.1021/ac049288r
53. Lan WJ, Holden DA, White HS (2011) Pressure-dependent ion current rectification in conical-shaped glass nanopores. J Am Chem Soc 133(34):13300–13303. https://doi.org/10.1021/ja205773a
54. White RJ, Ervin EN, Yang T, Chen X, Daniel S, Cremer PS, White HS (2007) Single ion-channel recordings using glass nanopore membranes single ion-channel recordings using glass nanopore membranes. J Am Chem Soc 129(15):11766–11775. https://doi.org/10.1021/ja073174q
55. Schibel AEP, Edwards T, Kawano R, Lan W, White HS (2010) Quartz nanopore membranes for suspended bilayer ion channel recordings. Anal Chem 82(17):7259–7266. https://doi.org/10.1021/ac101183j
56. Deng T, Chen J, Si W, Yin M, Ma W, Liu Z (2012) Fabrication of silicon nanopore arrays using a combination of dry and wet etching. J Vac Sci Technol, B: Nanotechnol Microelectron Mater Process Meas Phenom 30(6):061804. https://doi.org/10.1116/1.4766322
57. Park SR, Peng H, Ling XS (2007) Fabrication of nanopores in silicon chips using feedback chemical etching. Small 3(1):116–119. https://doi.org/10.1002/smll.200600268
58. Deng T, Chen J, Wu CN, Liu ZW (2013) Fabrication of inverted-pyramid silicon nanopore arrays with three-step wet etching. ECS J Solid State Sci Technol 2(11):P419–P422. https://doi.org/10.1149/2.005311jss
59. Danda G, Masih Das P, Chou YC, Mlack JT, Parkin WM, Naylor CH, Fujisawa K, Zhang T, Fulton LB, Terrones M, Johnson ATC, Drndić M (2017) Monolayer WS2 nanopores for DNA translocation with light-adjustable sizes. ACS Nano 11(2):1937–1945. https://doi.org/10.1021/acsnano.6b08028
60. Tong HD, Jansen HV, Gadgil VJ, Bostan CG, Berenschot E, van Rijn CJM, Elwenspoek M (2004) Silicon nitride nanosieve membrane. Nano Lett 4(2):283–287. https://doi.org/10.1021/nl0350175
61. DesOrmeaux JPS, Winans JD, Wayson SE, Gaborski TR, Khire TS, Striemer CC, McGrath JL (2014) Nanoporous silicon nitride membranes fabricated from porous nanocrystalline silicon templates. Nanoscale 6(18):10798. https://doi.org/10.1039/C4NR03070B
62. Streimer CC, Gaborski TR, McGrath JL, Fauchet PM (2007) Charge- and size-based separation of maeromolecules using novel ultrathin silicon membranes. Nature 445(February):177. https://doi.org/10.1109/LEOS.2007.4382334
63. Qi C, Striemer CC, Gaborski TR, McGrath JL, Fauchet PM (2014) Highly porous silicon membranes fabricated from silicon nitride/silicon stacks. Small 10(14):2946–2953. https://doi.org/10.1002/smll.201303447

64. Kim H-M, Lee M-H, Kim K-B (2011) Theoretical and experimental study of nanopore drilling by a focused electron beam in transmission electron microscopy. Nanotechnology 22(27):275303. https://doi.org/10.1088/0957-4484/22/27/275303
65. Wu MY, Krapf D, Zandbergen M, Zandbergen H, Batson PE (2005) Formation of nanopores in a SiN/SiO_2 membrane with an electron beam. Appl Phys Lett 87(11):1–4. https://doi.org/10.1063/1.2043247
66. Stein D (2003) Ion beam sculpting molecular scale devices. Ph.D., Harvard
67. Storm AJ, Chen JH, Ling XS, Zandbergen HW, Dekker C (2005) Electron-beam-induced deformations of SiO_2 Nanostructures. J Appl Phys 98(1). https://doi.org/10.1063/1.1947391
68. Venkatesan BM, Dorvel B, Yemenicioglu S, Watkins N, Petrov I, Bashir R (2009) Highly sensitive, mechanically stable nanopore sensors for DNA analysis. Adv Mater 21(27):2771–2776. https://doi.org/10.1002/adma.200803786
69. Larkin J, Henley R, Bell DC, Cohen-Karni T, Rosenstein JK, Wanunu M (2013) Slow DNA transport through nanopores in hafnium oxide membranes. ACS Nano 7(11):10121–10128. https://doi.org/10.1021/nn404326f
70. Russo CJ, Golovchenko JA (2012) Atom-by-atom nucleation and growth of graphene nanopores. Proc Natl Acad Sci 109(16):5953–5957. https://doi.org/10.1073/pnas.1119827109
71. Liu S, Zhao Q, Xu J, Yan K, Peng H, Yang F, You L, Yu D (2012) Fast and controllable fabrication of suspended graphene nanopore devices. Nanotechnology 23(8):085301. https://doi.org/10.1088/0957-4484/23/8/085301
72. Goyal G, Lee YB, Darvish A, Ahn CW, Kim MJ (2016) Hydrophilic and size-controlled graphene nanopores for protein detection. Nanotechnology 27:1–30. https://doi.org/10.1088/0957-4484/27/49/495301
73. Liu K, Feng J, Kis A, Radenovic A (2014) Atomically thin molybdenum disulfide nanopores with high sensitivity for DNA translocation. ACS Nano 8(3):2504–2511. https://doi.org/10.1021/nn406102h
74. Park KB, Kim HJ, Kang YH, Yu JS, Chae H, Lee K, Kim HM, Kim KB (2017) Highly reliable and low-noise solid-state nanopores with an atomic layer deposited ZnO membrane on a quartz substrate. Nanoscale 9(47):18772–18780. https://doi.org/10.1039/c7nr05755e
75. Deng Y, Huang Q, Zhao Y, Zhou D, Ying C, Wang D (2017) Precise fabrication of a 5 nm graphene nanopore with a helium ion microscope for biomolecule detection. Nanotechnology 28(4):045302. https://doi.org/10.1088/1361-6528/28/4/045302
76. Hemamouche A, Morin A, Bourhis E, Toury B, Tarnaud E, Mathé J, Guégan P, Madouri A, Lafosse X, Ulysse C, Guilet S, Patriarche G, Auvray L, Montel F, Wilmart Q, Plaçais B, Yates J, Gierak J (2014) FIB patterning of dielectric, metallized and graphene membranes: a comparative study. Microelectron Eng 121:87–91. https://doi.org/10.1016/j.mee.2014.03.020
77. Lo CJ, Aref T, Bezryadin A (2006) Fabrication of symmetric sub-5 nm nanopores using focused ion and electron beams. Nanotechnology 17(13):3264. https://doi.org/10.1088/0957-4484/17/13/031
78. Zahid OK, Hall AR (2016) Helium ion microscope fabrication of solid-state nanopore devices for biomolecule analysis. In: Helium ion microscopy. Springer International Publishing, Cham, pp 447–470. https://doi.org/10.1007/978-3-319-41990-9_18
79. Ziegler J (2013) Stopping and range of ions in matter (SRIM) computer software
80. Tan S, Klein K, Shima D, Livengood R, Mutunga E, Vladár A (2014) Mechanism and applications of helium transmission milling in thin membranes. J Vacuum Sci Technol B Nanotechnol Microelectr Mater Process Meas Phenomena 32(6):06FA01. https://doi.org/10.1116/1.4900728
81. Livengood R, Tan S, Greenzweig Y, Notte J, McVey S (2009) Subsurface damage from helium ions as a function of dose, beam energy, and dose rate. J Vacuum Sci Technol B Microelectr Nanometer Struct 27(6):3244. https://doi.org/10.1116/1.3237101
82. Stanford MG, Lewis BB, Iberi V, Fowlkes JD, Tan S, Livengood R, Rack PD (2016) In situ mitigation of subsurface and peripheral focused ion beam damage via simultaneous pulsed laser heating. Small 12(13):1779–1787. https://doi.org/10.1002/smll.201503680

83. Nguyen MA, Ruault MO, Fortuna F (2012) Formation and growth of nanocavities and cavities induced by He+ implantation in silicon. Adv Nat Sci Nanosc Nanotechnol 3(1). https://doi.org/10.1088/2043-6262/3/1/015015
84. Siegele R, Weatherly GC, Haugen HK, Lockwood DJ, Howe LM (1995) Helium bubbles in silicon: structure and optical properties. Appl Phys Lett 66(11):1319–1321. https://doi.org/10.1063/1.113228
85. Wu M, Smeets R, Zandbergen M, Zlese U, Krapf D, Batson PE, Dekker NH, Dekker C, Zandbergen H (2009) Control of shape and material composition of solid-state nanopores. Nano Lett 9(1):479–484
86. Rodriguez-Manzo JA, Puster M, Nicolai A, Meunier V, Drndic M, Rodríguez-Manzo JA, Puster M, Nicolaï A, Meunier V, Drndić M (2015) DNA Translocation in nanometer thick silicon nanopores. ACS Nano 9(6):6555–6564. https://doi.org/10.1021/acsnano.5b02531
87. Egerton RF, Li P, Malac M (2004) Radiation damage in the TEM and SEM. Micron 35(6):399–409. https://doi.org/10.1016/j.micron.2004.02.003
88. Howitt DG, Chen SJ, Gierhart BC, Smith RL, Collins SD (2008) The electron beam hole drilling of silicon nitride thin films. J Appl Phys 103(2). https://doi.org/10.1063/1.2828157
89. Das K, Freund JB, Johnson HT (2015) Mechanisms of material removal and mass transport in focused ion beam nanopore formation. J Appl Phys 117(8). https://doi.org/10.1063/1.4913449
90. Marshall MM, Yang J, Hall AR (2012) Direct and transmission milling of suspended silicon nitride membranes with a focused helium ion beam. Scanning 34(2):101–106. https://doi.org/10.1002/sca.21003
91. van den Hout M, Hall AR, Wu MY, Zandbergen HW, Dekker C, Dekker NH (2010) Controlling nanopore size, shape and stability. Nanotechnology 21(11):115304. https://doi.org/10.1088/0957-4484/21/11/115304
92. Kowalczyk SW, Grosberg AY, Rabin Y, Dekker C (2011) Modeling the conductance and DNA blockade of solid-state nanopores. Nanotechnology 22(31):315101. https://doi.org/10.1088/0957-4484/22/31/315101
93. Frament CM, Bandara N, Dwyer JR (2013) Nanopore surface coating delivers nanopore size and shape through conductance-based sizing. ACS Appl Mater Interfaces 5(19):9330–9337. https://doi.org/10.1021/am4026455
94. Bandara YMNDY, Karawdeniya BI, Dwyer JR (2016) Real-time profiling of solid-state nanopores during solution-phase nanofabrication. ACS Appl Mater Interfaces 8(44):30583–30589. https://doi.org/10.1021/acsami.6b10045
95. Bandara YMNDY, Nichols JW, Iroshika Karawdeniya B, Dwyer JR (2018) Conductance-based profiling of nanopores: accommodating fabrication irregularities. Electrophoresis 39(4):626–634. https://doi.org/10.1002/elps.201700299
96. Frament CM, Dwyer JR (2012) Conductance-based determination of solid-state nanopore size and shape: an exploration of performance limits. J Phys Chem C 116(44):23315–23321. https://doi.org/10.1021/jp305381j
97. Cai Q, Ledden B, Krueger E, Golovchenko JA, Li J (2006) Nanopore sculpting with noble gas ions. J Appl Phys 100(2). https://doi.org/10.1063/1.2216880
98. Patterson N, Adams DP, Hodges VC, Vasile MJ, Michael JR, Kotula PG (2008) Controlled fabrication of nanopores using a direct focused ion beam approach with back face particle detection. Nanotechnology 19(23):235304. https://doi.org/10.1088/0957-4484/19/23/235304
99. Sawafta F, Clancy B, Carlsen AT, Huber M, Hall AR (2014) Solid-state nanopores and nanopore arrays optimized for optical detection. Nanoscale 6:6991–6996. https://doi.org/10.1039/c4nr00305e
100. Wanunu M (2012) Nanopores: a journey towards DNA sequencing. Phys Life Rev 9(2):125–158. https://doi.org/10.1016/j.plrev.2012.05.010
101. Branton D, Deamer DWDW, Marziali A, Bayley H, Benner SA, Butler T, di Ventra M, Garaj S, Hibbs A, Huang X, Jovanovich SB, Krstic PS, Lindsay S, Ling XS, Mastrangelo CH, Meller A, Oliver JS, Pershin YV, Ramsey JMM, Riehn R, Soni GV, Tabard-Cossa V, Wanunu M, Wiggin M, Schloss JA et al (2008) The potential and challenges of nanopore sequencing. Nat Biotechnol 26(10):1146–1153. https://doi.org/10.1038/nbt.1495

102. Atas E, Singer A, Meller A (2012) DNA sequencing and bar-coding using solid-state nanopores. Electrophoresis 33(23):3437–3447. https://doi.org/10.1002/elps.201200266
103. Gao R, Cui LF, Ruan LQ, Ying YL, Long YT (2019) A closed-type wireless nanopore electrode for analyzing single nanoparticles. J Vis Exp 2019(145):1–9. https://doi.org/10.3791/59003
104. Kececi K, Dinler A, Kaya D (2022) Review—nanopipette applications as sensors, electrodes, and probes: a study on recent developments. J Electrochem Soc 169(2):027502.https://doi.org/10.1149/1945-7111/AC4E58
105. Bell NAW, Keyser UF (2015) Specific protein detection using designed DNA carriers and nanopores. J Am Chem Soc 137(5):2035–2041. https://doi.org/10.1021/ja512521w
106. Steinbock LJ, Steinbock JF, Radenovic A (2013) Controllable shrinking and shaping of glass nanocapillaries under electron irradiation. Nano Lett 13(4):1717–1723. https://doi.org/10.1021/nl400304y
107. Cadinu P, Paulose Nadappuram B, Lee D, Sze JYY, Campolo G, Zhang Y, Shevchuk A, Ladame S, Albrecht T, Korchev YE, Ivanov AP, Edel JB (2017) Single molecule trapping and sensing using dual nanopores separated by a zeptolitre nanobridge. Nano Lett 17(10). https://doi.org/10.1021/acs.nanolett.7b03196
108. Cadinu P, Kang M, Nadappuram BP, Ivanov AP, Edel JB (2020) Individually addressable multi-nanopores for single-molecule targeted operations. Nano Lett 20(3):2012–2019. https://doi.org/10.1021/acs.nanolett.9b05307
109. Nadappuram PB, McKelvey K, Byers JC, Güell AG, Colburn AW, Lazenby RA, Unwin PR (2015) Quad-barrel multifunctional electrochemical and ion conductance probe for voltammetric analysis and imaging. Anal Chem 87(7):3566–3573. https://doi.org/10.1021/acs.analchem.5b00379
110. Ren R, Zhang Y, Nadappuram BP, Akpinar B, Klenerman D, Ivanov AP, Edel JB, Korchev Y (2017) Nanopore extended field-effect transistor for selective single-molecule biosensing. Nat Commun 8(1):586. https://doi.org/10.1038/s41467-017-00549-w
111. Alawami MF, Bošković F, Zhu J, Chen K, Sandler SE, Keyser UF (2022) Lifetime of glass nanopores in a PDMS chip for single-molecule sensing. iScience 25(5). https://doi.org/10.1016/j.isci.2022.104191
112. Chen K, Jou I, Ermann N, Muthukumar M, Keyser UF, Bell NAW (2021) Dynamics of driven polymer transport through a nanopore. Nat Phys 17(9):1043–1049. https://doi.org/10.1038/s41567-021-01268-2
113. Steinbock LJ, Bulushev RD, Krishnan S, Raillon C, Radenovic A (2013) DNA translocation through low-noise glass nanopores. ACS Nano 7(12):11255–11262. https://doi.org/10.1021/nn405029j
114. Chien CC, Shekar S, Niedzwiecki DJ, Shepard KL, Drndić M (2019) Single-stranded DNA translocation recordings through solid-state nanopores on glass chips at 10 MHz measurement bandwidth. ACS Nano 13(9):10545–10554. https://doi.org/10.1021/acsnano.9b04626
115. Lombardo S, Stathis JH, Linder BP, Pey KL, Palumbo F, Tung CH (2005) Dielectric breakdown mechanisms in gate oxides. J Appl Phys 98(12):121301. https://doi.org/10.1063/1.2147714
116. Kwok WHH (2015) New approach in fabrication of solid-state nanopore for bio-sensing applications. Ph.D., University of Ottawa. https://doi.org/10.20381/ruor-3888
117. Zrehen A, Gilboa T, Meller A (2017) Real-time visualization and sub-diffraction limit localization of nanometer-scale pore formation by dielectric breakdown. Nanoscale 9(42):16437–16445. https://doi.org/10.1039/C7NR02629C
118. Wang R, Gilboa T, Song J, Huttner D, Grinstaff MW, Meller A (2018) Single-molecule discrimination of labeled DNAs and polypeptides using photoluminescent-free TiO_2 nanopores. ACS Nano 12(11):11648–11656. https://doi.org/10.1021/acsnano.8b07055
119. Wang Y, Chen Q, Deng T, Liu Z (2017) Nanopore fabricated in pyramidal HfO_2 film by dielectric breakdown method. Appl Phys Lett 111(14):143103. https://doi.org/10.1063/1.4990717
120. Zhang X, van Deursen PMG, Fu W, Schneider GF (2020) Facile and ultraclean graphene-on-glass nanopores by controlled electrochemical etching. ACS Sens 5(8):2317–2325. https://doi.org/10.1021/acssensors.0c00883

121. Graf M, Lihter M, Thakur M, Georgiou V, Topolancik J, Ilic BR, Liu K, Feng J, Astier Y, Radenovic A (2019) Fabrication and practical applications of molybdenum disulfide nanopores. Nat Protoc 14(4):1130–1168. https://doi.org/10.1038/s41596-019-0131-0
122. Carlsen A, Tabard-Cossa V (2015) Enhanced stability of 2-nm nanopores fabricated by controlled breakdown. In: Biophysical society of Canada 1st annual meeting. Biophysical Society of Canada
123. Briggs K (2015) Applications of nanopore fabrication by controlled breakdown. In: Swissnanopore conference. Lenzerheide
124. Waugh M, Carlsen A, Sean D, Slater GWGWGW, Briggs K, Kwok H, Tabard-Cossa V (2015) Interfacing solid-state nanopores with gel media to slow DNA translocations. Electrophoresis 36(15):1759–1767. https://doi.org/10.1002/elps.201400488
125. Madejski GR, Briggs K, DesOrmeaux J-P, Miller JJ, Roussie JA, Tabard-Cossa V, McGrath JL (2019) Monolithic fabrication of NPN/SiN_x dual membrane cavity for nanopore-based DNA sensing. Adv Mater Interfaces 6(14). https://doi.org/10.1002/admi.201900684
126. Briggs K, Madejski G, Magill M, Kastritis K, de Haan HW, McGrath JL, Tabard-Cossa V (2017) DNA translocations through nanopores under nanoscale pre-confinement. Nano Lett 18(2):660–668. https://doi.org/10.1021/acs.nanolett.7b03987
127. Matsui K, Yanagi I, Goto Y, Takeda KI (2015) Prevention of dielectric breakdown of nanopore membranes by charge neutralization. Sci Rep 5:1–9. https://doi.org/10.1038/srep17819
128. Höltzel A, Tallarek U (2007) Ionic conductance of nanopores in microscale analysis systems: where microfluidics meets nanofluidics. J Sep Sci 30(10):1398–1419. https://doi.org/10.1002/jssc.200600427
129. Liu Y, Huber DE, Dutton RW (2010) Limiting and overlimiting conductance in field-effect gated nanopores. Appl Phys Lett 96(25):2010–2013. https://doi.org/10.1063/1.3457350
130. Wanunu M, Sutin J, McNally B, Chow A, Meller A (2008) DNA translocation governed by interactions with solid-state nanopores. Biophys J 95(10):4716–4725. https://doi.org/10.1529/biophysj.108.140475
131. Leung C, Briggs K, Laberge M-P, Peng S, Waugh M, Tabard-Cossa V (2020) Mechanisms of solid-state nanopore enlargement under electrical stress. Nanotechnology 31(44). https://doi.org/10.1088/1361-6528/aba86e
132. Yanagi I, Fujisaki K, Hamamura H, Takeda KI (2017) Thickness-dependent dielectric breakdown and nanopore creation on sub-10-nm-thick SiN membranes in solution. J Appl Phys 121(4). https://doi.org/10.1063/1.4974286
133. Yanagi I, Hamamura H, Akahori R, Takeda KI (2018) Two-step breakdown of a SiN membrane for nanopore fabrication: formation of thin portion and penetration. Sci Rep 8(1):1–13. https://doi.org/10.1038/s41598-018-28524-5
134. Ivankin A, Henley RY, Larkin J, Carson S, Toscano ML, Wanunu M (2014) Label-free optical detection of biomolecular translocation through nanopore arrays. ACS Nano 8(10):10774–10781. https://doi.org/10.1021/nn504551d
135. Fried JP, Swett JL, Nadappuram BP, Fedosyuk A, Gee A, Dyck OE, Yates JR, Ivanov AP, Edel JB, Mol JA (2022) Localised solid-state nanopore fabrication via controlled breakdown using on-chip electrodes. Nano Res 15(11):9881–9889. https://doi.org/10.1007/S12274-022-4535-8/METRICS
136. Zhang Y, Miyahara Y, Derriche N, Yang W, Yazda K, Capaldi X, Liu Z, Grutter P, Reisner W (2019) Nanopore formation via tip-controlled local breakdown using an atomic force microscope. Small Methods 3(7):1900147. https://doi.org/10.1002/smtd.201900147
137. Tahvildari R, Beamish E, Tabard-Cossa V, Godin M (2015) Integrating nanopore sensors within microfluidic channel arrays using controlled breakdown. Lab Chip 15(6):1407–1411. https://doi.org/10.1039/C4LC01366B
138. Tahvildari R, Beamish E, Briggs K, Chagnon-Lessard S, Sohi AN, Han S, Watts B, Tabard-Cossa V, Godin M (2017) Manipulating electrical and fluidic access in integrated nanopore-microfluidic arrays using microvalves. Small 13(10):1–7. https://doi.org/10.1002/smll.201602601

139. Bustamante J (2014) Synchronous optical and electrical measurements of single DNA molecules translocating through a solid-state nanopore. M.Sc., University of Ottawa
140. Jain T, Rasera BC, Guerrero RJS, Lim JM, Karnik R (2017). Microfluidic multiplexing of solid-state nanopores. J Phys Condensed Matter 29(48). https://doi.org/10.1088/1361-648X/aa9455
141. Lam MH, Briggs K, Kastritis K, Magill M, Madejski GR, McGrath JL, de Haan HW, Tabard-Cossa V (2019) Entropic trapping of DNA with a nanofiltered nanopore. ACS Appl Nano Mater 2(8). https://doi.org/10.1021/acsanm.9b00606
142. Yanagi I, Akahori R, Takeda K (2019) Stable fabrication of a large nanopore by controlled dielectric breakdown in a high-PH solution for the detection of various-sized molecules. Sci Rep 9(1):1–15. https://doi.org/10.1038/s41598-019-49622-y
143. Bandara YMNDY, Saharia J, Karawdeniya BI, Hagan JT, Dwyer JR, Kim MJ (2020) Beyond nanopore sizing: improving solid-state single-molecule sensing performance, lifetime, and analyte scope for omics by targeting surface chemistry during fabrication. Nanotechnology 31(33). https://doi.org/10.1088/1361-6528/ab8f4d
144. Liu K, Feng J, Kis A, Radenovic A (2014) Atomically thin molybdenum disulfide nanopores with high sensitivity for DNA translocation. ACS Nano 8(3):2504–2511. https://doi.org/10.1021/nn406102h
145. Storm AJ, Chen JH, Ling XS, Zandbergen HW, Dekker C (2003) Fabrication of Solid-state nanopores with single-nanometre precision. Nat Mater 2:537–540. https://doi.org/10.1038/nmat941

Chapter 2
Emerging Abnormal Phenomena in Confined Nanofluidics

Mengdi Niu, Fanfan Chen, and Jiandong Feng

Abstract Nanofluidics is an emerging field that investigates the properties and transport behavior of fluids confined at the nanometer or even sub-nanometer level. Its development is of great importance for interfacial catalysis, biological science, energy conversion, and other fields. Recent developments in micro-nano processing have allowed the scale of confined channels to be down to the molecular or even atomic level, in which the structure of the molecule or the ion is not negligible. Therefore the confined fluid exhibits properties and behaviors that are different from those of the bulk phase, such as nonlinear transport, Coulomb blockade and ultra-fast water transport. In this chapter, we summarize the anomalies of ion, water, and biological molecules under confinement. The first part introduces the current research techniques for confined fluids, followed by discussing the abnormal phenomena of ions, water, and molecules in confined fluids, respectively. Finally, we outline the applications and future challenges of confined nanofluidics.

Keywords Nanofluidics · Confined fluids · Nanochannels · Ion behavior · Water transport · molecular sensing · Micro-nano processing

Recently nanofluidics has shown its great importance in various fields such as interfacial science [1, 2], catalytic chemistry [3, 4], physical biology and genetics [5–7]. Nanofluidics was first regarded as an extension of microfluidics towards smaller scale, though fluids at nanoscale have been explored for decades [8]. Until diverse anomalous phenomena found in confined nanofluids, researchers started to realize that the specific effects in nanofluidics are differing from the bulk phase, and this field finally earned its own name [9, 10]. Thanks to the development of micro-nano processing technology and the improvement of methods for regulating and detecting nanofluids, emerging new phenomena involving static properties and dynamic transport of particles (ion/water/molecule) in confined nanofluids has been observed.

M. Niu · F. Chen · J. Feng (✉)
Laboratory of Experimental Physical Biology, Department of Chemistry, Zhejiang University, Hangzhou 310027, China
e-mail: jiandong.feng@zju.edu.cn

J.-P. Leburton (ed.), *Solid State Nanopores*, Nanostructure Science and Technology,
https://doi.org/10.1007/978-3-031-42336-9_2

These anomalous phenomena include non-linear ionic transport, gating, ionic rectification, abnormal permittivity, phase transition, layering and ultrafast transport of water and so on. Here we summarize a number of novel phenomena in the field of confined nanofluidics including the related developments of theoretical models. This chapter is mainly divided into four parts: basic understanding, research technique, new phenomena in ion/water/molecule behavior, as well as the conclusion and perspective. Firstly, we introduce several important concepts in nanofluidic measurement systems including electrical double layer (EDL), noise and resistance. Then the main part describes new phenomena of ion/water/molecule behavior. For ion transport, we summarize abnormal ionic behavior including ionic Coulomb blockade (ICB), gating, mechanically activated ion transport, ion current rectification and solid–liquid interaction. For water behavior, we start from static properties like ultra-low dielectric constant, structural phase transition and layering to dynamic transport effects like ultra-fast water flow. For molecular transport, we summarize the transport behavior of nucleic acids, proteins, and other neutral molecules, discussing their applications in sensing, molecular sieve, seawater desalination and energy harvesting. Finally, we make the conclusion and prospect of the full chapter. We outline the practical applications of these emerging phenomena, and point out the shortcomings of current research and future challenges in the field of confined nanofluidics.

2.1 Basic Concepts

2.1.1 Electrical Double Layer (EDL) and Debye Length

Under strong confinement, the ion transport is completely different from the free ion diffusion. The spatial distribution of confined transport and free diffusion is dependent on electrical double layer (EDL) thickness. EDL appears on the solid–liquid interface and can be divided into two parallel layers of ion distribution on the surface [11]. The first layer called absorption layer is consisted of counter ions which are absorbed on the interface closely due to Coulomb interaction with the surface charge. The second layer named diffusion layer is composed of ions attracted to first layer which screens the charge of first layer and is not strongly integrated with interface. Electrical double layer is influenced by surface charge, ion valence, ionic concentration and so on. Debye length λ_D is a parameter that could reflect the effects of these conditions on EDL and could be estimated by means of the Debye–Hückel parameter κ,

$$\kappa = \frac{1}{\lambda_D} = \sqrt{\frac{e^2 \sum_i n_i z_i^2}{\varepsilon \varepsilon_0 k_B T}}$$

where n is ionic concentration, k_B is Boltzmann constant, e is electron charge, z is ion valence, T is absolute temperature, ε is solvent dielectric constant and ε_0 is vacuum dielectric constant.

Debye length is a concept describing the valid distance of electrostatic effect in electrolytes or plasmas. In electrolytes, Coulomb interaction between opposite charged ions will lead to ionic charge screening and ion hydration. Debye length is defined as the distance from the surface to where the potential reduces to 1/e of the original value. At room temperature and water solvent, the function of Debye length with monovalent ions could be simplified as $\lambda_D(nm) = \frac{0.304}{\sqrt{c}}$, where c is ionic concentration [12].

By adjusting ionic concentration or pH value of electrolytes, surface charge and EDL thickness could be modified which have great effects on ion transport under extreme confinement. These effects including strong noncovalent hydrated cation-π interactions and size exclusion effect make the application of controllable ion transport possible. The strong noncovalent hydrated cation-π interactions between the empty orbital of cation and aromatic ring structure of graphene oxide membranes (GOMs) can control the interlayer spacing [13]. The interlayer spacing can be confined down to 11.4 Å with K^+ fixed which is shorter than the reported narrowest spacing of GOMs and the controlling precision can be minimized to 1 Å. The GOMs with strong confined structure reject cations but allow water penetration. The great reduction of ion permeation rate has shown promising applications in water desalination and molecular sieving. Moreover, the surface electric field of nanochannels can also adjust the polarization of neutral molecules' p or π orbital, which allows the realization of neutral-molecular controllable transport [14]. Esfandiar et al. fabricated two-dimensional graphite channels with only 6.6 Å height and the hydration shells of ions would be crushed and rearranged under the angstrom-scale confinement. This reconfiguration affected the mobility of ions. For example, the mobility of Cl^- could decrease three times compared to bulk solution and the Al^{3+} also show ten times less mobility. This was due to extra friction between the OH group of Cl^- hydration shells and the polarized water molecules [15].

2.1.2 *Noise*

Low noise and high signal-to-noise ratio are essential for effective measurements. Different noise sources could affect different frequency of signal. The total noise could be transferred to power spectral density (PSD) and fitted by the function [16]:

$$S = \frac{a}{f} + b + cf + df^2\left(\frac{A^2}{Hz}\right)$$

where f is the frequency. The parameters a, b, c, d arise from flicker noise (pink noise), Johnson noise, capacitive noise and voltage noise, respectively. In ion transport, the low-frequency noise characterizes current fluctuations and the high-frequency noise represents capacitance effect of device materials [16]. By fitting noise with function, one cound obtian the effective factors that influence the measurement system and optimize the noise.

Low frequency noise is a vital performance metric for devices and determines the detection limitation in numerous applications [17]. Low-frequency 1/f noise could be fitted and compared by the low-frequency noise amplitude A:

$$A = S_I \frac{f}{I^2},$$

where f is the frequency, I is the average current, S_I is the current power spectral density [18]. The low-frequency noise originates from surface charge fluctuations and mobility of charge carriers which could be used to estimate charge sites [19, 20]. By varying pH value and concentration of electrolytes, the low-frequency noise could be adjusted [21]. Besides, after treated with piranha, the silicon nitride devices could show lower flicker noise [17].

High-frequency noise is also an important factor that determines device performance. High-frequency noise is dependent on dielectric properties which could be adjusted by improvement of the device structure or employing new type materials. Usually, biological nanopores such as α-hemolysin and MspA show lower noise level than artificial nanochannels [22]. However, artificial nanochannels can give greater assembly diversity and stability than protein nanopores that could only work at mild environment with unstable performance. Therefore, a lot of efforts to decrease noise level of artificial nanochannel devices have been made. As shown in Fig. 2.1a, the method of PDMS sealing on solid-state nanopores to reduce noise have been demonstrated [17, 23]. The PDMS coating could reduce the exposed area of support chip surface and improve the dielectric constant of systems. Similarly, deposition of high dielectric constant materials on the devices could also lower high-frequency noise, such as metal oxides semiconductor (CMOS) [24] such as Al_2O_3 [23, 25] (Fig. 2.1b and c). Glass nanopore is another type of high dielectric constant devices which shows low noise level and it has been widely used in molecular detection due to its low noise performance [26, 27] (Fig. 2.1d).

2.1.3 Resistance

The resistance of nanochannels has been modeled based on continuum fluid mechanics which is consist of resistance of nanochannels, chambers and field focusing [28, 29],

$$R = \overline{R_{nano}} + 2R_{micro} + 2R_{ff}$$

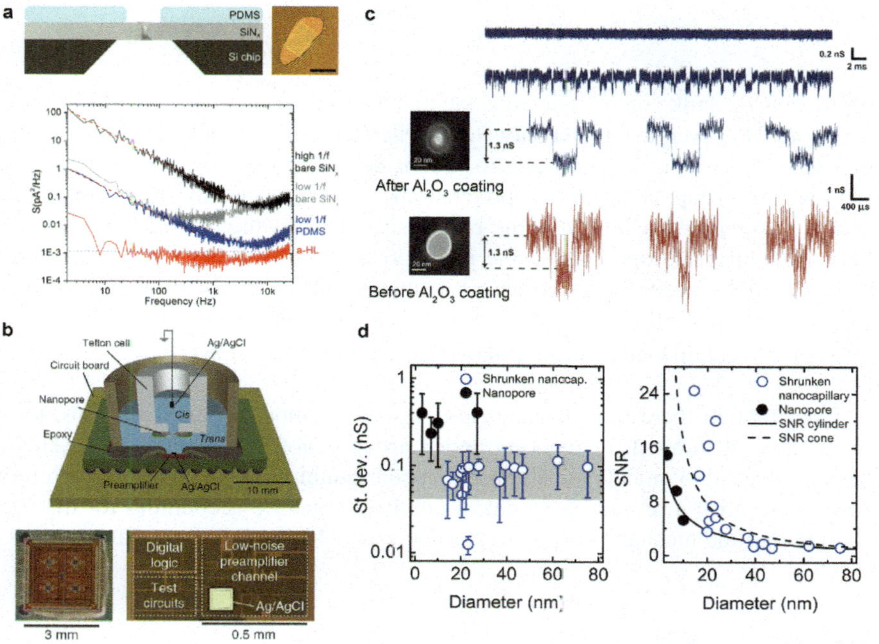

Fig. 2.1 The reduction of high-frequency noise to optimize device performance. **a** The reduction of dielectric noise with PDMS coated on the free-standing membrane [17]. The schematic of the cross section and microscope image of a chip with PDMS coated are shown. The 1/f noise is decreased by PDMS painting. Adapted with permission from Ref. [17] Copyright (2023) (IOP Publishing). **b** The low-noise measurement nanopore device integrated with complementary metal-oxide semiconductor (CMOS) preamplifier [24]. The measurement setup, optical microscope image and magnified image of devives are shown. Adapted with permission from Ref. [24] Copyright (2023) (Springer Nature) **c** The typical low noise current signal with Al_2O_3 coating [23]. Adapted with permission from Ref. [23] Copyright (2023) (American Chemical Society) **d** The low-noise glass nanopores with diameter variation [26]. Adapted with permission from Ref. [26] Copyright (2023) (American Chemical Society)

$$R_{nano} = \rho_{res}\frac{d}{hw},\ \overline{R_{nano}} = R_{nano}\frac{c_0}{N},\ R_{micro} = \rho_{res}\frac{L}{HW},$$
$$R_{ff} = \rho_{res}\frac{\overline{f}}{L},\ \rho_{res} = \frac{RT}{DF^2c_0},$$

where R is the gas constant, T is the absolute temperature, D is the diffusion coefficient, F is the Faraday constant, c_0 is the bulk concentration, d is the channel length, h is the channel height, L is the chamber length, H is the chamber height, W is the chamber width, $\overline{f}$ is the nondimensional function.

The conductance of nanopore which is a vital indicator in molecular detection and ion transport has also been modeled. Smeets et al. modeled the conductance of nanopores based on cylindrical structure [30]. It is described as the function:

$$G = \sigma \frac{\pi d^2}{4l}$$

where σ is the bulk conductivity, d is the diameter of nanopore, l is the thickness of nanopore. However, the model only works when the diameter of nanopore is smaller than 15 nm. Access resistance needs to be considered into the system. Hille assumed a semi-spherical cupola above nanopore entrance and described the access resistance as $R_{access} = \rho/\pi d$, where ρ is the bulk resistivity, d is the diameter of electrode [31]. Hall further proposed that the nanopore structure could be considered as a plane-parallel capacitor and the access resistance could be obtained with the function $R_{access} = \rho/2d$ [32]. Kowalczyk et al. combined the channel resistance and access resistance to obtain total resistance of nanopore with the function $G = \sigma\left[\frac{4l}{\pi d^2} + \frac{1}{d}\right]^{-1}$ [33].

Even though there have been various resistance models putted forward, these model are still limited for working under strong nanofluidic confinement. This is because that these models are all based on continuum fluid mechanism. Therefore, how to establish a resistance model of confined structure accounting for the sub-continuum fluid mechanism is an important issue which needs more efforts.

2.2 Research Technique

The lack of understanding in fluids under extreme confinement originated from several aspects, and the most important one was the behindhand development of research technique including confinement construction, fabrication process as well as regulation, observation, and quantification methods. Since graphene was first isolated in 2004 by Novoselov and Geim [34], carbon materials started to play an important role in nanofluidic research [35, 36]. Thanks to the atomically thickness but strong mechanical rigidity, excellent photoelectric properties and most importantly the facility of graphene to construct various of confinement structure, nanofluidics start to be able to delineate its contour in the sub-continuum realm. In addition to carbon materials, other 2D materials such as h-BN [37], MoS_2 [38], as well as the emerging MXenes [39], have also made advances to nanofluidic research. The advent of nanomaterials solved partial problem to a certain extent, but the progress in nanofluidics depends largely on the development of more precise processing technology for addressing artificial fluid channels [40–43]. At present, the artificial devices can be fabricated at confined scale down to the size of water molecules-3 Å, which allows exploring the fluid behavior at molecular or even atomic level [44]. In this scale range, many anomalous fluid behaviors with exciting results have been found, like nonlinear transport mechanism [45], single-file arrangement of water molecular [46], coupling of ion-water transport [47] and so on. Last but not least, the exploitation of light, electricity and mechanical pressure allows us to accurately regulate, observe, and quantify fluid behavior. For instance, Comtet et al. [48] used single-molecule localization microscopy to resolve optically the transport of individual excess protons at the interface of hexagonal boron nitride crystals and aqueous solutions at room temperature. Light, electricity, and mechanical pressure, as typical techniques for studying nanofluids, have played a significant role in revealing the peculiar phenomenon of nanofluids in confinement.

Here, we summarize the current research methods of nanofluidics, which are mainly divided into three parts: the dimension of the confinement, the fabrication process as well as the regulation, observation, and quantification of nanofluids. Firstly, the fluidic control can be achieved by constructing different confinement dimensions and length scales of the nanochannel. And the next discussion involves the precise processing technology which allows the confinement construction to be realized. Finally, liquid flow can be driven by light, electricity and mechanical pressure which can be defined as regulation technique. And these three tools can also be used as measurement methods and visualization techniques to characterize nanofluids.

2.2.1 Confinement Dimension

The confinement structures range from three-dimensional (3D) porous materials, two-dimensional (2D) slits, one-dimensional (1D) nanotube and nanochannel, finally down to zero-dimensional (0D) nanopores [49] (Fig. 2.2).

We start with the simplest structure—0D nanopore with atomic thickness. The 'zero-dimension' is viewed from the structure of the pore itself, and we usually name it nanopores in 2D materials (Fig. 2.2a). And it needs to be distinguished from tranditional solid-state nanopore, which was classified as one-dimensional nanochannel. Due to their simple confined structure, extremely thin thickness, modifiable pore edges, and adjustable surface charge density, nanopores in 2D materials have extraordinary advantages in the study of sub-continuum nanofluids [50], single-molecule sensing [7], and high efficiency energy harvesting [38, 51]. When the fluid is confined in a single-nanopore system, models commonly explained in continuous fluids such as the Navier–Stokes equation no longer apply [52]. The many-body ion-ion interactions and the dehydration of ions or molecules in the pore [53], could lead to various unusual fluid transport phenomena, such as enhanced ion/water permeability [54], nonlinear transport [45], ionic Coulomb blockade [50] and so on. Feng et al. [50] firstly in 2016 experimentally observed and measured ionic Coulomb blockade in nanopores. Recently they found nonlinear ion transport phenomena with mechanical response in graphene nanopore [45]. Except for researching anomalous ionic transport, nanopore is also exploited to molecular sensing of nucleic acids and proteins [7]. Compared to biological and solid-state nanopore, nanopores in 2D materials have their own unique advantages, such as stronger mechanical stability, shorter sensing length and 'atom-by-atom' adjustable pore size [55]. Besides, the surface charge density of the nanopore can be regulated by external stimulations such as light [56] and pH [38], resulting in a number of interesting effects including gating

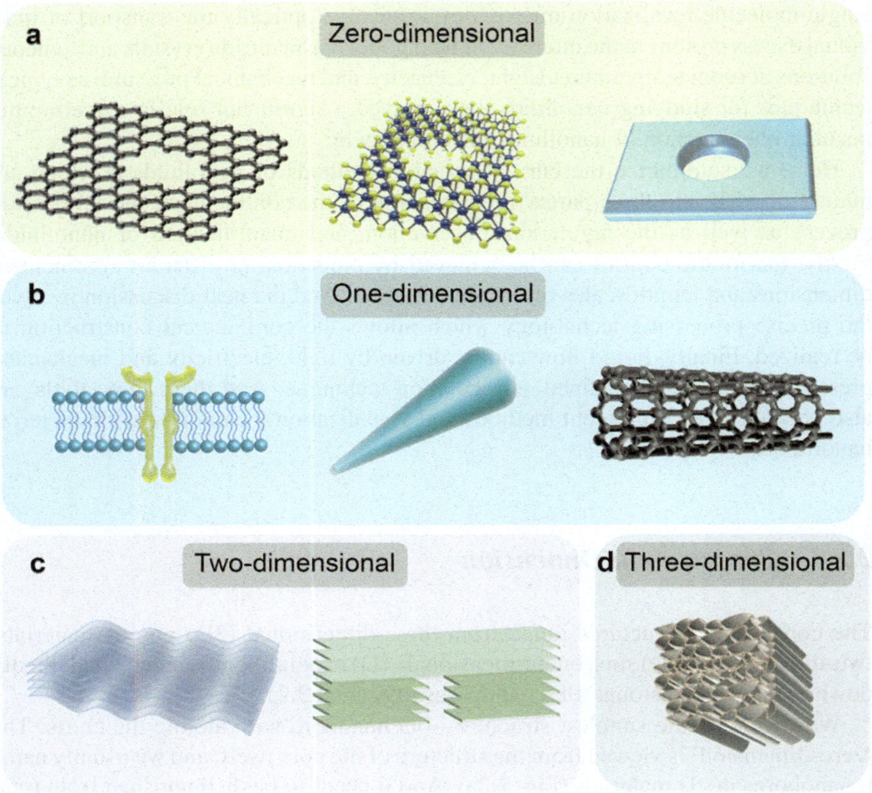

Fig. 2.2 Different confinement dimensions for regulating nanofluids. **a** Zero-dimensional nanopore like graphene and MoS_2 naopores and so on. **b** One dimensional biological channels (MspA), nanopipette and nanotubes (carbon nanotubes). **c** Two-dimensional nanochannels and nanoslits. **d** Three-dimensional porous nanochannels and bundles aligned nanotubes

phenomena, enhanced ion selectivity, and higher energy harvesting capacity. Graf et al. [56] used light to improve to surface charge density of a single-layer MoS_2 nanopore and enhanced the osmotic power output.

1D nanochannels and nanotubes are analogous to 0D nanopores. The difference between them is their diameter/length ratio. For 0D nanopore, the diameter is larger than its length but 1D nanochannel is on the opposite (Fig. 2.2b). Materials constructing these one-dimensional nanochannels typically contain carbon materials [57], hexagonal boron nitride (h-BN) [37], titania [58], titanate [59], silicon [60], and cadmium sulfide [61]. Carbon nanotubes (CNTs), for example, are often used to study exotic phenomena in nanofluidics, such as ultrafast water flow, single-file arrangement of water molecules, nonlinear ion transport, ion Coulomb drag, and water freezing at higher temperature [46, 47, 54, 62, 63]. That is attributed to its adjustable nanoscale sizes and helical symmetry of carbon atoms that enable nearly frictionless interfaces. Tunuguntla et al. [54] focused on the CNTs inserted into lipid membranes, which confined water molecules down to a single-file chain, and the permeability of water molecules in CNTs exceeded that in biological nanochannels and wider CNT pores by an order of magnitude. In general, CNTs can be constructed with size down to the molecule or atomic level and provide a simple and effective model system for nanofluidic research under extreme confinement. Besides, boron nitride nanotubes (BNNTs) are a preliminary attempt to develop new avenue to high efficiency concentration difference energy harvesting. Siria et al. [37] fabricated a hierarchical nanofluidic device, which is made of a boron nitride nanotube pierced into an ultrathin membrane and connected two fluid reservoirs. They observed very large osmotically induced electric currents generated by salinity gradients, exceeds its pressure-driven counterpart by two orders of magnitude.

Naturally, the confined structure of nanofluids can extend to two-dimensional nanochannels from one-dimensional nanotubes. These two-dimensional channels include planar channel formed by layered 2D membranes like graphene oxide (GO) [64, 65], and van der Waals (vdW) assembled nanoslits [15, 66] with confined length scale down to Ångstrom level (Fig. 2.2c). Due to richer functionalization and adjustable interlayer spacing, planar channels made from layered graphene oxide membrane were widely used in sieving, energy harvesting, ion transport regulating and so on. For planar channels, a superior advantage is the adjustable interlayer distance. For example, Abraham et al. [67] realized to control interlayer distance by physical confinement and achieve accurate and tunable ion sieving. Because the layer spacing is adjustable, we can even narrow the channel further, even down to molecular or atomic scale, to explore the unknown transport mechanism under extreme confinement. Analogously, Xue et al. [68] fabricated Ångstrom-level channels formed by interlayer of layered graphene and realized atomic-scale ion transistor by controlling the gate potential applied in the ion channel. Also they proved that the ion permeability in nanochannels was enhanced to two orders of magnitude higher than that in bulk water [68]. In recent years, thanks to the development of stacked van der Waals assembly materials [66, 69], the abnormal ion transport behavior in two-dimensional slits has been gradually explored. These van der Waals assembly materials usually include graphene, hBN and MoS_2. The development of angstrom slits makes the monatomic space possible. The hydrated ions were found to be dehydrated to enter the monatomic confinement, which is smaller than the hydrated ions themselves. For example, Esfandiar et al. [15] fabricated Ångstrom-scale channels by van der Waals assembly allowing deformation of ionic hydration shell and ions squeezing into the channel.

Due to the more complex many-body interaction originated from the structural configuration of 3D porous materials, it is seldom to be used to explore basic mechanism of fluids behavior in the field of confined nanofluidics (Fig. 2.2d). The fluid flow and heat transfer through porous media draw great attention, which were generally exploited to develop applications in desalination and energy harvesting [70, 71]. 3D porous materials can be made of carbides, nitrides, borides, and silicide of various metals, alloys, and refractory metals [72]. 3D graphene oxide was commonly used among them, because it has more oxygen-containing groups and stronger hydrophilicity, which is conducive to the transport of water molecules. Thus graphene oxide has greater application potential in harvesting water energy. Zhao et al. [70] fabricated a high-performance power generator based on a super hydrophilic three-dimensional assembly of graphene oxide. After electrochemical polarization treatment, it has a unique concentration gradient of oxygen-containing groups, which makes the water molecules and protons migrate in a directional way and output voltage. Besides, 3D porous GO membrane was also used to study ion transport mechanism influenced by electrical double layer (EDL) induced by surface potential. For example, Cheng et al. [73] regulated the potential sign, ion species and concentration to reverse or change the ion permeability confined in sub 2 nm.

2.2.2 Fabrication

The fabrication process of nanopores in 2D materials includes the preparation of supporting solid-state pores or polymer materials, the growth and transfer of two-dimensional membranes as well as the drilling of nanopores. Here we focus on the drilling methods of nanopores in. Common nanopore drilling techniques include heavy ion bombardment [74], focused ion beam (FIB) [75], transmission electron microscopy (TEM) [76] and electrochemical reaction drilling (ECR) [55] (Fig. 2.3). TEM drilling utilizes controllable energy and adjustable electron beam spot size. The advantage of TEM drilling is its ability to precisely control the drilling process, to open the pore atom by atom, and most importantly, and to observe the process in situ. However, the disadvantages are that single layer two-dimensional material is radiation-sensitive and prone to cause defects and carbon deposition. In addition, TEM drilling is time-consuming, costly, and hard to be scaled up or automated. Ion bombardment can make up for the disadvantages of TEM in scaling up, but it cannot finely control the position, diameter, density, spacing and geometry of nanopores (Fig. 2.3d). Compared to ion bombardment, FIB is highly accurate and operated in a lower energy range without causing direct radiation damage (Fig. 2.3f). Electrochemical reaction (ECR) is another reliable drilling method, in which the membrane is placed in ionic solution and an electrochemical reaction is triggered on the surface of the material by applying a voltage (Fig. 2.3i). The advantages of this method are as follows: (i) the formation of nanopores can be completed in situ, meanwhile avoiding TEM contamination; (ii) low cost, simple and easy way to operate (iii) large-scale processing of nanopores. In addition, reactive ion etching (RIE) [77] can be used to obtain porous arrays on 2D materials by using masks, electron beam lithography and dry etching, which requires extremely careful manipulation and control (Fig. 2.3h). It is also possible to treat the surface nucleation of defect with plasma or ozone first and then sputtering or chemical etching to obtain the desired aperture.

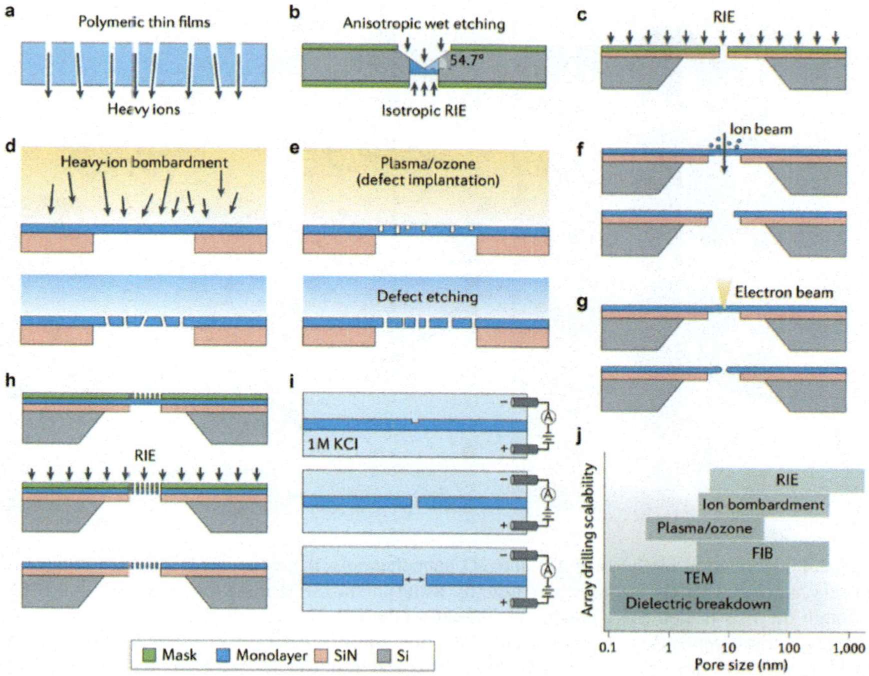

Fig. 2.3 Fabrication methods of nanopores. **a–c** Different methods for drilling solid state nanopores: (a) Heavy-ion irradiation, (b) Anisotropic etching combined with isotropic RIE and (c) RIE. **d–i** Different methods for drilling 2D nanopore: (d) Heavy-ion bombardment (e) Plasma/ ozone for defect implantation combined with subsequent etching, (f) Focused ion beam (FIB), (g) Transmission electron microscopy (TEM), (h) Reactive ion etching (RIE) and (i) Electrochemical reaction drilling (ECR). **j** Overall obtainable pore sizes and process upscaling capabilities [51]. Adapted with permission from Ref. [51] Copyright (2023) (Springer Nature)

For one-dimensional nanofluidic channels, we mainly focus on the typical carbon nanotubes and asymmetric one-dimensional channel. The growth of carbon nanotube including arc-discharge, laser ablation, chemical vapor deposition (CVD) has been developed for years [78] and we will not elaborate it here. Single carbon nanotubes are usually deposited on a substrate [66], or embedded in SiN_x [47], lipid bilayer [54], or nanocapillary [79], and then packaged with electrodes and ionic solution as nanochannel devices. Rabinowitz et al. [80] reported that Coulomb drag occurred between carriers in carbon nanotube and charges in ions. They deposited SWCNT on a SiO_2/Si chip and used electron beam lithography (EBL) to pattern and electron beam evaporation to deposit Au electrodes. Besides, in 2013, Siria et al. [37] has realized precisely control of inserting boron nitride nanotubes into SiN_x holes by using in situ SEM manipulation. Then in 2016, they deployed the same technique to

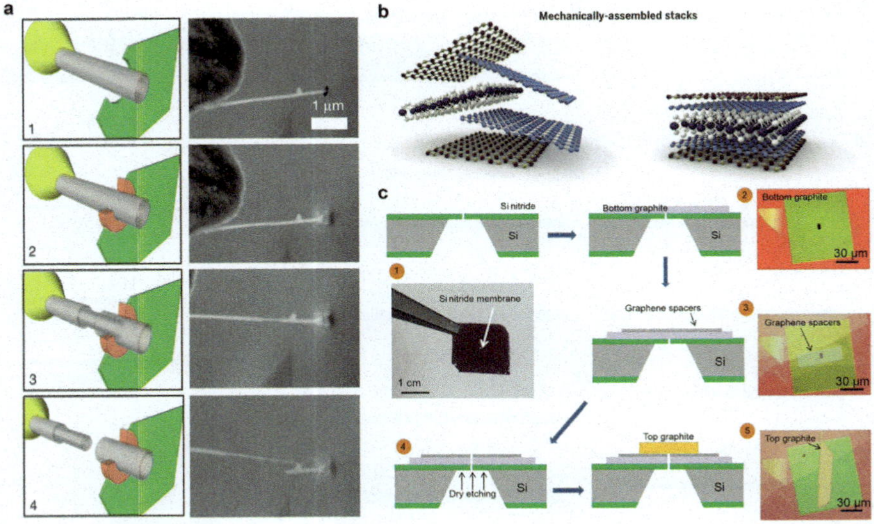

Fig. 2.4 Fabrication methods of 1D and 2D nanochannels. **a** Precise control of inserting boron nitride nanotubes (BNNTs) into SiN_x hole by using in situ SEM insertion. 1. Inserting a BNNT through the FIB-drilled SiN_x nanopore. 2. Sealing the FIB-drilled hole by local electron-beam-induced carbon deposition. 3. Retraction of internal walls of the BNNT. 4. Realization of trans-membrane BNNT device [37]. Adapted with permission from Ref. [37] Copyright (2023) (Springer Nature). **b** Van der Waals stacking of different two-dimensional materials [66]. Adapted with permission from Ref. [66] Copyright (2023) (The American Association for the Advancement of Science). **c** Fabricating 2D nanoslits by van der Waals stacking of top/bottom materials and spacers [81]. Adapted with permission from Ref. [81] Copyright (2023) (Springer Nature)

pierce a CNT into nanocapillary and found that slip length for CNT was dramatically large, but for boron nitride nanotube (BNT) similar in crystal structure to CNT was totally different [79] (Fig. 2.4a).

Asymmetric one-dimensional nanochannels exhibit special properties such as ion selectivity, ionic rectification, gating etc. [82] The core principles of asymmetric channel construction includes external environment control and channel structure design [83]. The control of the external environment, including the regulation of pH, ion species, concentration, and other factors that has been discussed thoroughly in other reviews [83] is not within the scope of our discussions here. The design of the internal structure of the channel includes geometry, surface charge distribution, chemical composition, and wettability. Conical pores are an common type of asymmetric nanochannels. In 2006, Cervera et al. used etching liquid KOH on one side and acidic neutralizing liquid on the other to create conical asymmetric channels and realize ion rectification and selectivity [84]. By controlling the etching process, other shapes of symmetrical or asymmetrical channels can also be fabricated, such as hour-glass shape, cigar-like shape and bullet-like shape [83]. If the channel itself is asymmetric, the charge on the inner surface can be regulated by uniform modifications to achieve the asymmetric ion transport through the nanochannel. Also, it is possible to create non-uniform modifications to construct asymmetric channels. These modification methods include ion sputtering [85], electron-beam evaporation [86], plasma [87] and self-assembly [88].

For two-dimensional channel, primitive planar channels made of layered materials such as layered graphene oxide were simple to be fabricated. In recent years, by adjusting the layer spacing of layered 2D materials, the scale of confined channels has been narrowed to the order of Ångstrom. Xue et al. [68] used modified Hummer's method [89] to prepare multilayer graphene oxide (GO) flakes, which was then transferred to a pre-etched SiN_x chip and finally encapsulated to form a 2D channel device. Cheng et al. [90] exploited a method to stack nanosheets into a layered graphene gel membrane by direct flow filtration of chemically converted graphene (CCG), which formed a kind of planar nanochannel. They regulated ion transport by modulating the interlayer spacing, revealing a quantitative relationship between the electrokinetic ion transport and the channel size. In addition, the development of van der Waals assembly technology has enabled nanoslits to play a role in nanofluidics. Van der Waals stacked interlayer of 2D material exhibits unexpected self-cleaning ability (Fig. 2.4b), and hBN-encapsulated graphene has been reported to feature high carrier mobility [91]. Esfandiar et al. [15] fabricated a angstrom-scale slit. They exfoliated 2D materials from bulk crystals and then transferred to a SiN_x/Si chip with an etched hole, as bottom and top layer. The spacer was patterned by electron beam lithography and etched into an array of stripes. The cavity left in the interlayer between the top and the bottom material became the fluid channel. Gopinadhan et al. [44] exploited the same fabrication process and prepared a 3.4 Ångstrom height nanoslit in which the spacer was single layer graphene (Fig. 2.4c) This slit barely allowed permeation of hydrated ions like Na^+ or K^+. However, 2D water layer formed in the slit allowed H^+ transport by hopping between water molecules.

For three dimensional porous materials, carbon-based materials have been mostly studied, including carbon nitride, carbon oxide, graphene, and their derivatives [92]. Due to the wide range of areas covered, different ways of preparing porous material were developed, including Incipient Wetness sheik, film formation, and colloidal imprinting [93]. In the field of nanofluidics, the most common three-dimensional porous materials are graphene oxide (GO) nanofluidic channels. Cheng et al. used supramolecular assembly strategy to produce layered nanoporous gel membranes [73]. Andreeva et al. [94] self-assembled polyamine macro-molecules (PA) and graphene to prepare a three-dimensional intercalated fluid channel. Besides, Jia et al. [95] prepared three-dimensional porous channels made of sulfide nanosheets by vacuum filtration, and heterojunction channels were assembly by step-by-step pumping and direct stacking. The preparation methods of three-dimensional materials is diverse and creative. In addition to materials such as GO and MoS_2, nanowire films are also a promising material. Liu et al. [96] cultured geobacter sulfurreducens and centrifuged to remove the cells to obtain the protein nanowire films.

2.2.3 Regulation, Observation, and Quantification

The confinement structure at different dimensions, including its surface properties (such as the surface charge distribution, etc.), plays a key role for regulating the transport of ions or water. In this section, we will discuss the regulation of transport by externally imposed conditions such as light, electricity, pressure, or pH, concentration gradient and the presence of specific ions in the solution environment, to gain a deep understanding of the transport mechanism of ions or water under confinement. We then discuss methods for quantifying and visualizing the abnormal behavior of ions or water, including current super resolution light microscopy, precision electrical measurements, and some indirect visualizations.

In solution environment, ion transport can be regulated by pH, concentration, and the presence of specific ions. pH control is a commonly used regulation method. It has been reported that when increasing the pH value of ionic solution in which MoS_2 nanopore was immersed, the charge density of the pore was accordingly increased and the osmotic energy power density was greatly enhanced [38]. Andreeva et al. [94] showed that pH regulated ion permeation through a 2D PA-GO self-assembled membrane material featuring strong stability in pH switching circulation. Subsequently, they found that the existence of K^+ would dramatically increase the permeability of Cs^+, suggesting that K^+ could serve as a signal molecule of gating effect.

Electricity, light, and mechanical pressure, as the well studied systems for nanofluids, have played significant roles in revealing the peculiar phenomenon of nanofluids under confinement. These three energy forms can be used to drive and control nanofluids (Fig. 2.5). Electrical signal is one of the important indicators for measuring ion transport. Almost all phenomena involving ion transport, such as electroosmotic power generation, Coulomb blockade, and nanopore DNA sequencing, require electrical current measurements. Electric field can be used to drive and control nanofluid flow [97]. Electrical measurement is also the core principle of nanopore sensing (Fig. 2.5d). When molecules transport though a single nanopore under electrical field, the ionic current through the pore will be partially blocked which visually shows an current drop [98]. One can get rich information from the electrical measurements, including the directly obtained conductance and further analyzed the ion permeability, selectivity, and pore size, etc. [51]. Besides, electricity involves electrostatic modulation that can regulate ion behavior by controlling the surface charge of solid-water interface. Xue et al. [68] fabricated a multilayer 2D nanochannel with 0.45 nm height, which prohibits ion permeation. When applying a voltage higher than 1.0 V, the channel was switched on and the ionic diffusion coefficient reached 2.0×10^{-7} m^2/s, which was two magnitude higher than that in bulk water. They successfully used voltage to control the ion transport and realized gating behavior.

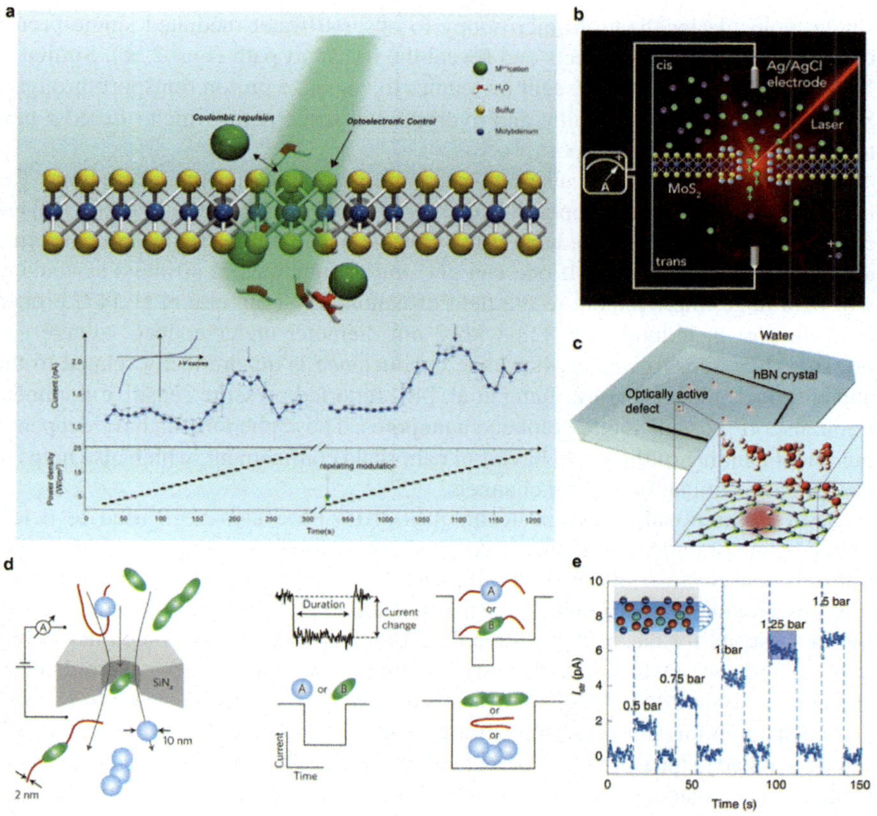

Fig. 2.5 Nanofluids regulation, observation, and quantification methods. **a** Optoelectronic control of ionic conductance oscillation phenomenon in sub-nanometer pores [99]. Adapted with permission from Ref. [99] Copyright (2023) (Elsevier). **b** Light is used to photogate the MoS_2 and modulate the surface charge of the nanopore [56]. Adapted with permission from Ref. [56] Copyright (2023) (Elsevier). **c** Single molecule localization microscopy to observe water-mediated single-proton transport between hBN defects [48]. Adapted with permission from Ref. [48] Copyright (2023) (Springer Nature). **d** Molecule transport through solid state nanoporedetected by electric measurement [100]. Adapted with permission from Ref. [100] Copyright (2023) (Springer Nature). **e** Mechanically activated ion transport through single carbon nanotube [47]. Adapted with permission from Ref. [47] Copyright (2023) (Springer Nature)

Light, as a control method, has recently been used by researchers to adjust the surface charge density of nanopores [56], thereby tuning ionic selectivity, and ultimately achieving the goal of controlling nanofluidic transport [101, 102] (Fig. 2.5a–c). Chen et al. [99] observed ion oscillations by photoelectrically controling the surface charge of single sub-nm nanopore (Fig. 2.5a). Compared with electrical gating, light control does not introduce direct crosstalk with ionic channel. The development of single-molecule microscopy has brought the optical technology for studying single molecules and nanofluids to a new level. Comtet et al. [48] used single-molecule localization microscopy to observe water-mediated single-proton transport between hBN defects and reveal the transport path (Fig. 2.5c). Similarly, Yang et al. [103] used fluorescent molecules to visualize proton transport through a SiN_x nanopore and hydrophilic and hydrophobic stripe path, which offered a new tool to uncover proton transport process.

Mechanical pressure can also be used as a means of control and detection. The mechanical response of nanopores is an important topic (Fig. 2.5e). In biological ion channels, pressure can be used as a way of mechanical gating to regulate the opening and closing of the channel. If one can get similar functions in artificial systems, it will be a huge improvement to the field of nanofluids. Marcotte et al. [47] probed ion transport in individual CNTs with 2 nm diameter under couped voltage and pressure drop. They revealed that ionic conductance is quadratically related to the applied mechanical pressure. Jiang et al. [45] reported pressure sensitive nonlinear ion transport through single graphene nanopores. These phenomena have deepened our understanding of the ion behavior in nanoscale confinement, which also helps to design new artificial biological channels.

Several traditional analytical techniques [104] have also been used to detect fluidic behavior under confinement, such as atomic force microscopy (AFM) [105], transmission electron microscopy (TEM) [106], gravimetry [107], Raman spectroscopy and mass spectrometry [108]. Fumagalli et al. [105] used scanning dielectric microscopy for electrostatic force detection based on atomic force microscopy to observe the anomalous low dielectric constant of water confined in a 1 nm height channel. Early in 2004 [106], TEM has been reported to observe water in a single carbon nanotube and the phase transition of water at nanoscale. As for gravimetry, it's often exploited to measure water permeability since it is difficult to drive and measrue neutral water molecules with electrical field. Recently, Agrawal et al. [109] observed water phase transitions in carbon nanotube by dynamic shift in the frequency of the Raman radial breathing mode (RBM). Besides, mass spectrometry is widely adopted to measure gas selectivity and permeability through nanochannel.

2.3 Ion Transport

2.3.1 *Ionic Coulomb Blockade*

In typical Coulomb blockade configuration, a voltage bias is applied to two electrodes (called as source and drain) which were separated by an insulating island [110, 111]. An electron traveling from source to drain must go through the insulating island. Under small voltage, when the electrostatic energy $\left(E = \frac{e^2}{2C}\right)$ of single traveling electron is much higher than electron thermal energy ($k_B T$), the hopping of electrons from source to drain will be blocked. And the occurance of electronic Coulomb blockade requires small enough island and low enough temperature so that the electrostatic energy could outdistance the thermal energy fluctuations.

Similar to electronic Coulomb blockade, ionic Coulomb blockade is appearing in the ion diffusion process through sub-nanometer artificial or biological pores. Continuum model is based on a mean-field approximation of ion interaction. However, in the sub-nanometer pores, the continuum models such as Poisson-Nernst-Planck (PNP) model do not work any more and many-body interaction must be taken into consideration.

In the all-atom molecular dynamics simulations of ion transport, Zwolak et al. took the effects of hydration layer broken into account (Fig. 2.6a) [112]. They found that when the diameter of nanopore is small enough, ions must shed the hydration layers to transport through the nanopore. The ion transport could show strong nonlinear relationship as a function of pore diameter which remind us of conductance quantization steps controlled by cross section of a quantum dot contact. They further showed the importance of ionic concentration and ionic valency in the ionic conductance quantization [113]. The ionic concentration could influence the Debye length and higher valency ions would bind hydration layers more strongly.

Ionic Coulomb blockade was first proposed by Krems and Di Ventra [114]. They found through simulations that the ions of two different polarities would accumulate around the two sides of the pore resulting in specific capacitances and hindering additional ions from transport as shown in Fig. 2.6b. They took the ion-ion interactions into consideration in the nanopore systems. In the system, conductance showed strong nonlinear response to ionic concentration and higher charged ions would show stronger effects of ionic Coulomb blockade. They further proposed that controlling the surface charge intensity of the nanopore could determine the occurrence of ionic Coulomb blockade.

Based on self-consistent electrostatic and Brownian dynamics simulations, Kaufman et al. described the ionic selectivity of ion channels originated from ionic Coulomb blockade and predicted a periodic conductance oscillation of calcium ions as a function of fixed surface charges (Fig. 2.6c) [115]. They considered a channel model with radius R = 0.3 nm and length L = 1.6 nm and ranged the charge value of the channel wall, and found that the current of Ca^{2+} showed oscillation with peak values (> 0) at $Q_f = -ze\left(n + \frac{1}{2}\right)$ and valley value (= 0) at $Q_f = -zen$ due to strong Coulomb blockade. They also observed the channel occupancy staircase during calcium ion transport. With the concentration increasing, the current showed stronger oscillation. However, in Na^{+} ion transport, the current only showed one peak and did not fall to zero. The channel occupancy showed almost linear response to charge values result from weak Coulomb blockade.

Analogy to electronic Coulomb blockade, the electrostatic energy of ionic transport system can be obtained.

$$U_n = \frac{Q_n^2 L}{8\pi\varepsilon_0\varepsilon_w R^2} = \frac{Q_n^2}{2C_s} = \frac{zen + Q_f}{2C_s}$$

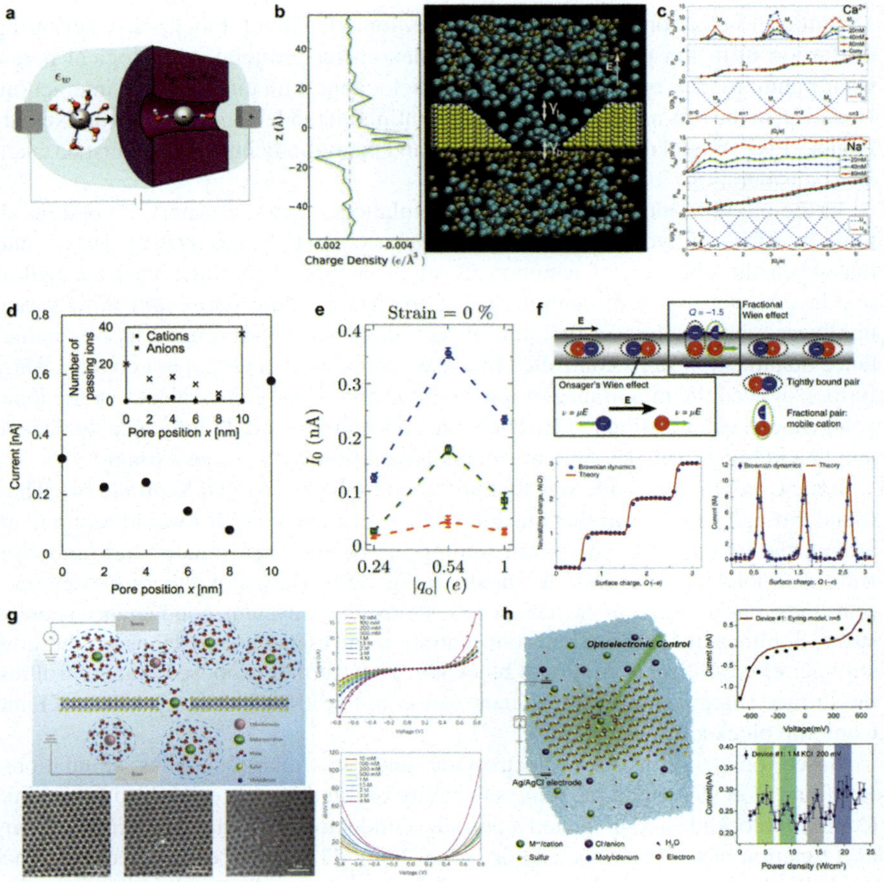

Fig. 2.6 Ionic Coulomb blockade under nanoconfinement. **a** The snapshot of single ion transport with first hydration moving in nanopore [113]. Adapted with permission from Ref. [113] Copyright (2023) (IOP Publishing). **b** The charge density and ion distribution in the cone-shaped Si_3N_4 nanopore. The Cl^- ions accumulated inside the pore but the K^+ gathered outside the pore [114]. Adapted with permission from Ref. [114] Copyright (2023) (IOP Publishing). **c** The different simulation results of ionic current oscillation with Ca^{2+} ions and Na^+. The plots of ion current J, channel occupancy P and electrostatic energy U versus the effective fixed charge Q_f are shown [115]. Adapted with permission from Ref. [115] Copyright (2023) (IOP Publishing). **d** The surface effects of pore diameter on ion selectivity [116]. Adapted with permission from Ref. [116] Copyright (2023) (IOP Publishing). **e** Ionic current oscillation with surface charge variation [117]. Adapted with permission from Ref. [117] Copyright (2023) (The American Association for the Advancement of Science). **f** The ionic Coulomb blockade in nanotubes with the fractional Wien effect mechanism. The sketch of Onsager's Wien effect and of fractional Wien effect are shown. The neutralizing charge N(Q) and ion current vary with surface charge changeing [118]. Adapted with permission from Ref. [118] Copyright (2023) (Springer Nature). **g** The experimental observation of ionic Coulomb blockade in sub-nm MoS_2 pores. The cureves of nonlinear *I-V* response and *dI/dV* versus voltage are shwon [50]. Adapted with permission from Ref. [50] Copyright (2023) (Springer Nature). **h** The experimental ionic conductance oscillation probed by optoelectronic control. The schematic of experimental setup, non-linear *I-V* curve and current oscillation are shown [99]. Adapted with permission from Ref. [99] Copyright (2023) (Elsevier)

Here, C_s is the self-capacitance of nanochannels, Q_n is the excess charge of ions, R and L are the radius and length of nanochannel respectively, ε_0 and ε_w are the dielectric constant of vacuum and water respectively, Q_f is the negative charge of nanochannel.

Kaufman et al. further proposed the anomalous mole fraction effect and ionic Coulomb blockade in biological nanochannels [119]. They measured the permeability of Na^+ and Ca^{2+} through NaChBac ion channels and its charged mutants, and clarified that nanochannel with different charges could influence the ion selectivity.

Tanaka et al. further explored the surface effects of ionic Coulomb blockade in the nanopore (Fig. 2.6d) [116]. They reported that the diameter of nanopore could influence the ion selectivity and ions would form crystal-like structure and transport with hopping-like mechanism under strong confinement.

By changing surface charge of crown ether pore, Sahu et al. presented current and potential well oscillations during ion transport and in the molecular dynamics (MD) simulation (Fig. 2.6e) [117]. Kavokine et al. unveiled the physical image of ionic Coulomb blockade in the nanotubes and interpreted it as a fractional Wien effect (Fig. 2.6f) [118]. They reported that under sufficient confinement, charged ions would form 'Bjerrum pairs'. The ion pairs would dissociate at high electric field which resulted in non-linear I-V curves. They observed the current and neutralizing charge oscillations as a function of surface charge and obtained phase diagram of Coulomb blockade. Feng et al. first reported the experimental observation of ionic Coulomb blockade in a sub-nm MoS_2 pore (Fig. 2.6g) [50]. In the experiments, the non-linear I-V characteristics with different concentrations and cations were studied. The current was strongly suppressed at the low voltage bias and started to increase forcefully when the voltage is higher than the threshold value. The typical I-V curves are strongly dependent on the diameter of the nanopore. They further found that conductance oscillations as a function of pH value which adjusted surface charge state of nanopore. Chen et al. reported the ionic conductance oscillations probed by optoelectronic control and found its ionic concentration and ionic valency dependence (Fig. 2.6h) [99]. They reported that ion transport in low ionic concentration solution could exhibit strong ionic oscillations due to the longer Debye length. Ions with higher valance such as Al^{3+} ions showed more obvious oscillation resultted from stronger Coulomb interaction in line with simulation expectation.

Realization of surface charge controlling technique including voltage gate [120, 121], pH gate [50] may help us to gain an insight into the mechanism. And with the rapid development of nanofabrication, one can expect to build artificial ion pumps or ion machines based on ionic Coulomb blockade.

2.3.2 *Gating*

Rapid switching of single channel ionic current between different levels in biological membranes was observed at denervated frog muscle fibers [122]. The behavior is a response of organism to environmental stimulation such as temperature [123], pH value [124] and ionic strength variation [125] to control live activities. The gating behavior involves two state switching including high-conductance state and low conductance state, called ‘open’ and ‘closed’ states respectively. At open state, the nanochannels show positive response and increased permeability. But at low-conductance state, the channels exhibit reversed performance. Various mechanisms have been proposed to explain the rapid conductance switching. These mechanisms include conformational transition, reversible protonation, hydrophobic influence, resonant conduction and invertible formation of nanoprecipitates.

Conformational transition including ligand reaction was mostly found in postsynaptic neuron of biological system and various conformation controlled nanochannels on the cell membrane have been investigated in the past several decades [126]. The reaction between ligand and receptors would change the conformation of nanochannels with the open and closed state and control water and ionic permeability as shown in Fig. 2.7a [127]. Proton receptor is a special type of ion channels that could response to environmental pH stimulatation. The observed rapid ionic current switching was due to reversible protonation in the nanochannels (Fig. 2.7b) [19]. Two-state Markov process [128] and Kramers analysis [129, 130] have been used to study the swiching process. Based on reversible protonation mechanism, numerous pH-sensitive artificial nanochannels were fabricated by micro-nano technique. For example, Xia et al. manufactured pH dependent Poly (ethyleneterephthalate) (PET) nanopore with DNA motors immobilized inside [124]. DNA motors would fold into nanopore and decrease the valid diameter in low-pH solution, thereby showing low conductance. In high-pH solution, the packed DNA motors would relax and enhance the ion permittivity through the nanopores. Besides, Jain et al. fabricated single-layer graphene pores with sub nanometer and observed gating behavior [131].

The hydrophobicity of nanopores plays an important role in ion transport. In MD simulations, Beckstein et al. proposed that hydrophobic nanopores could act as gate and control water entering [134]. The geometry and hydrophobicity of nanopores could determine the flux of water. And it was reported that adding dipoles to nanopore liner could open the closed hydrophobic nanochannels. Beckstein et al. further carried out simulations in short hydrophobic nanopores with various diameters and found that the gating behavior could be interpreted by capillary evaporation induced by pressure and density fluctuations in sub-nm pores [135]. They showed that the water flow is not continuous but in burst. Hydrophobicity effect was first demonstrated experimentally by Smirnov et al. [136]. They showed that the gating behavior could be induced by applying voltage across the hydrophobic nanochannels which also leads to non-linear *I-V* curves. The voltage stimulatation resulted in dry and wet state transition in the nanochannels and the state transition rate could be adjusted by channel length. Shorter channels showed faster transition rate. Powell et al. also showed that condensation and evaporation of water in the hydrophobic nanopores resulted in rapid ionic current switching (Fig. 2.7c) [132]. Polste et al. studied ion transport in the hydrophobic–hydrophilic junction and found transport bahavior could be adjusted by applied voltage and anion type [137]. Larger polarizable anions had weaker solvation shells and tended to move to nanopore lining. This caused anion accumulation which led to pore wetting and reduced applied stimulation voltage.

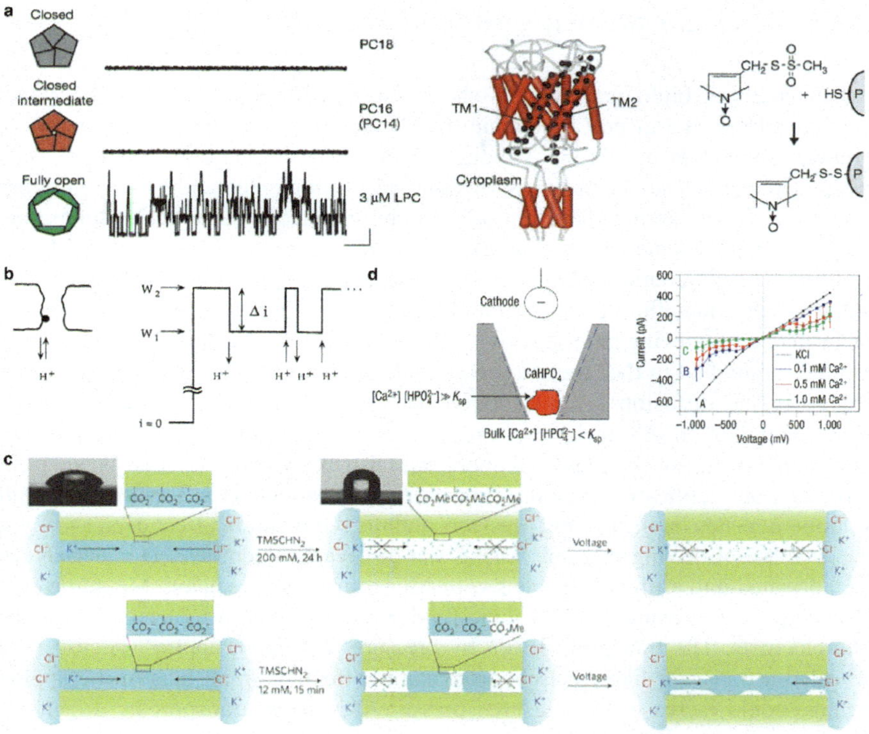

Fig. 2.7 Diverse mechanisms of gating behavior. **a** The on/off state switching of ion transport with nanopore conformational transition under different conditions [127]. Adapted with permission from Ref [127] Copyright (2023) (Springer Nature). **b** The schematic plot of reversible protonation and induced rapid ionic current switching [19]. Adapted with permission from Ref. [19] Copyright (2023) (the American Physical Society). **c** The electric-field-gated ion transport in hydrophobic nanopores [132]. Adapted with permission from Ref. [132] Copyright (2023) (Springer Nature). **d** The reversible formation and redissolution of nanoprecipitates ($CaHPO_4$) in nanopores [133]. Adapted with permission from Ref. [133] Copyright (2023) (Springer Nature)

Nanoprecipitation-assisted rapid ionic current switching was reported by Powell et al. and they found that adding divalent cations in the monovalent electrolyte could give rise to invertible formation and redissolution of nanoprecipitates in the nanopores (Fig. 2.7d) [133].

Although there are various mechanisms proposed, a definite mechanism of gating behavior in biological nanochannels still remains unknown. Research of artificial nanochannels with well-controlled gating behavior may provide alternative information for building such understandings.

2.3.3 Mechanically Activated Ion Transport

Ion channels in biological systems show sensitive mechanical responses, which exhibits advanced transport functionality like touch/pain perception, hearing, blood pressure regulation etc. [138] Although artificial systems are not comparable to biological nanochannels in terms of structural complexity and biological function diversity, researchers have reproduced this advanced mechanical response and explored how mechanical pressure affects ion transport in these channels [47].

In the case of coupling voltage and pressure drop, there is a considerable challenge to detect nanofluidic transport under confinement, especially for device manufacturing and achieving high sensitivity. Pressure-electricity coupled ion transport is initially embodied in the streaming potential. In confined nanochannel, the excess charge in the diffusion layer is driven by mechanical pressure and transported with water flow, resulting in a streaming current. Subsequently, the phenomenon of streaming current in glass nanopore has also been reported. Lan et al. [139] reported that rectification of ionic current in glass nanopores in low ionic strength solution depends on the driving pressure, and the rectification effect decreased with the increase of driving pressure. Using van der Waals assembly, 2D materials can be fabricated into artificial channels with Angstrom-scale precision. Mouterde et al. [140] reported measurements of ionic fluid transport through molecular-sized slit-like channels. The transport, driven by pressure and by an applied electric field, revealed a transistor-like electrohydrodynamic effect. A small bias of the applied voltage could increase the measured pressure-driven ion transport by a factor of 20. This effect is observed in both graphene and hexagonal boron nitride channels but shows obvious material-dependent discrepancies. Different from the previously reported results, Marcotte et al. [47] investigated nonlinear transport in individual 'single-digit' carbon nanotubes (CNTs) with radii of ~ 2 nm. Their findings indicated that the mechanically activated ion transport in the form of ion conductance depends quadratically on the applied pressure. This highly nonlinear nanofluidic transport under the confinement in molecular scale provided a new way to control the transport of molecules and ions and indicted that electro-mechanical coupling may play a role in understanding mechanically sensitive ion channels. Recently, Jiang et al. [45] reported pressure-modulated ionic conductance of graphene nanopore, and the conductance increased by a few percentage points to over 200% with increasing pressure. These above-mentioned studies may have potential applications in the characterization of channel surface characteristics, energy conversion, ion sieving, and so on.

2.3.4 Ion Current Rectification

Ion current rectification (ICR) is an electrodynamic phenomenon in confined environments with asymmetric current–voltage (*I*–*V*) characteristics. Condcutance recorded at one voltage polarity is higher than that at the opposite polarity. It is generally acknowledged that the phenomenon of ICR is usually observed in pores whose diameter is comparable to the EDL thickness. Rectification could be described quantitatively as the ion current rectification ratio RR: $RR = \left|\frac{I(-E)}{I(+E)}\right|$, where I(− E) and I(+ E) are the current values recorded at the negative and the positive potential respectively. Since the first experimental observation reported by Wei et al. [141], numerous experimental and theoretical researches have been carried out in order to figure out the underlying mechanism and main factors for ICR. There are two theoretical models of ICR mechanism proposed. The first one is electrokinetic trapping of mobile ions in the potential well [142]. A potential well is formed through the combination of the applied field and electric field due to local surface charge near the tip of a conical pore. The second one is that ion transport in cone-shaped nanopore is "actived with external field fluctuations" [143]. Counterions accumulate within the pore, leading to conductance enhancement. It also has been reported that the geometry of nanochannels, ionic strength of electrolyte, surface charge, convective flow, temperature and scan rate could all influence ICR behaviors.

ICR often occurrs in asymmetric pores such as glass nanopores or polymer nanopores (Fig. 2.8a–c). Jacobson et al. reported the influence of nanopore diameter on ICR behaviors in the poly (ethylene terephthalate) (PET) nanopores [144] and nanopipettes [145]. With the decrease of tip diameter, the ICR behaviors increased. Liu et al. studied the nanopore length effect and found that a long pore is necessary for ICR occurrence [146]. Ramírez et al. and Apel et al. theoretically and experimentally studied the influence of tip shape of the nanopore on ICR behavior, respectively [147, 148]. Zhang et al. reported the precise control of tip position along the channel and its influence on ICR [149].

Because the ionic strength of the electrolyte affects the thickness of EDL in the nanopore directly, the ICR behavior is greatly influenced by ionic concentration (Fig. 2.8d). Wei et al. found that ICR preferred to occur in low concentration solution (≤ 0.01 M KCl) and would disappear in high concentration electrolyte due to reduction of the EDL thickness [141]. Siwy et al. studied the influence of concentration gradient on the ICR [142]. In a synthetic PET (polyethylene terephthalate) nanodevice, potassium ions could transport against concentration gradient under activation with external field fluctuations. Jung et al. found that concentration gradient could induce ion electromigration current in a 175 nm pore whose size was ten times larger than the EDL thickness [156].

The surface charge of nanopore is recognized as an important factor for ICR which makes realization of nanofluidic logic circuit networks possible (Fig. 2.8e and f). The vital role of surface charge was first reported by Siwy et al. [150]. They deposited gold nanotubes into nanopores and found the occurrence of ICR behavior in KCl solution. However, in KF solution the I-V curve showed linear and Ohmic relationship. It was because that F^- could not absorb to Au nanotubes, while Cl^- could. The absorption of Cl^- would lead nanopore to be negatively charged. They also noticed that varying the pH values of solution could control the surface charge of the nanopore and influence ICR behavior. By introducing various chemical modifications, the surface charge

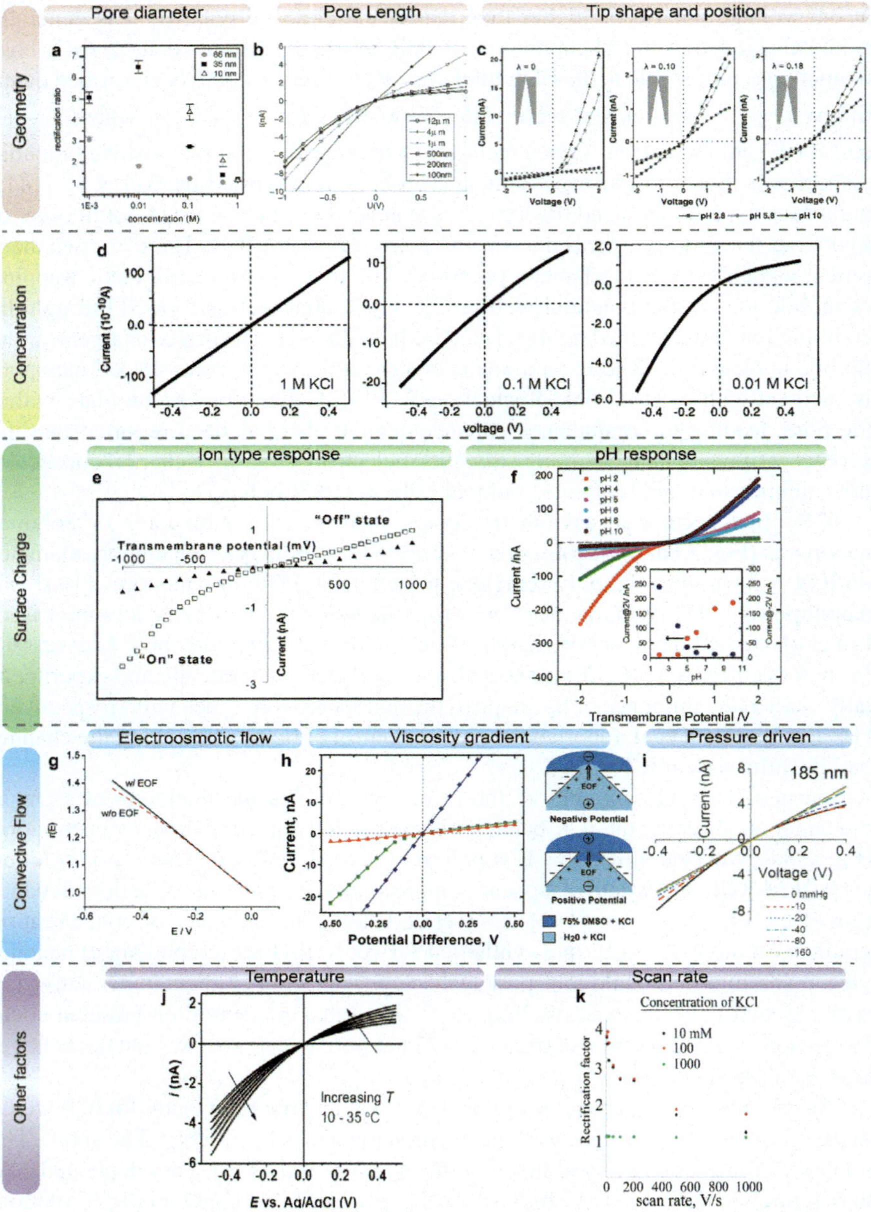
Geometry
Pore diameter
Pore Length
Tip shape and position
a
b
c
Concentration
d
1 M KCl
0.1 M KCl
0.01 M KCl
voltage (V)
Surface Charge
Ion type response
pH response
e
"Off" state
"On" state
Transmembrane potential (mV)
Current (nA)
f
Transmembrane Potential /V
Current /nA
Convective Flow
Electroosmotic flow
Viscosity gradient
Pressure driven
g
w/ EOF
w/o EOF
h
Potential Difference, V
Current, nA
Negative Potential
Positive Potential
75% DMSO + KCl
H2O + KCl
i
185 nm
Current (nA)
Voltage (V)
Other factors
Temperature
Scan rate
j
Increasing T
10 - 35 °C
E vs. Ag/AgCl (V)
k
Concentration of KCl
10 mM
100
1000
Rectification factor
scan rate, V/s

◂**Fig. 2.8** Various factors affecting ion current rectification. **a–c** The effects of the nanochannel geometry including pore diameter, pore length, tip shape and tip position [144, 146, 149]. Adapted with permission from Ref. [144] Copyright (2023) (American Chemical Society), Ref. [146] Copyright (2023) (the American Physical Society) and Ref. [149] Copyright (2023) (John Wiley and Sons). **d** The preferred occurrence of ion current rectification under low concentration condition [141]. Adapted with permission from Ref. [141] Copyright (2023) (American Chemical Society). **e–f** The influence of surface charge induced by surface modification and different response of ion current rectification on ion types and pH values [150, 151]. Adapted with permission from Ref. [150] Copyright (2023) (American Chemical Society) and Ref. [151] Copyright (2023) (American Chemical Society). **g–i** The ion current rectification driven by convective flow including electroosmotic flow, viscosity gradient and pressure driven [139, 152, 153]. Adapted with permission from Ref. [139] Copyright (2023) (American Chemical Society), Ref. [152] Copyright (2023) (American Chemical Society) and Ref. [153] Copyright (2023) (American Chemical Society). **j–k** Other factors of ion current rectification including temperature change and scan rate variation [154, 155]. Adapted with permission from Ref. [154] Copyright (2023) (American Chemical Society) and Ref. [155] Copyright (2023) (American Chemical Society)

density and distribution of nanopore can be tuned. Yameen et al. combined the pH control and polymer brush modification to realize the ion transport manipulation and ICR observation [151]. Lin et al. reported that high surface charge density and high pH of solution could extend the scale of pores with ICR behavior to sub micrometer [157]. Kalman et al. reported the controllable modification of nanopore inner wall and achieved the fabrication of first bipolar ionic transistor with p-n-p structure. The on/off state of nanopore could be tuned by changing ionic concentration and pH value of electrolyte [158].

Convective flow, including electroosmotic flow and pressure-driven flow, could also contribute to the ICR behavior (Fig. 2.8g–i). The influence of electroosmotic flow on ICR was reported by White et al. They found that the electroosmotic flow had a small effect on the current rectification but the effect would increase with the applied potential [152]. Making use of electroosmotic flow of different viscosity liquid, Yusko et al. observed the ICR in micropores whose diameter is 500 times larger than the EDL thickness and extended the range of fluidic diodes (A device where ion current can be rectified in a liquid, similar to electronic diodes.) to micro scale [153]. Lan et al. studied the influence of pressure-driven flow on ICR behavior [139]. They found that the pressure-driven flow could change the distributions of cation and anion and led to ICR reduction with the increasing of pressure-driven flow.

Temperature and scan rate could also influence ICR behavior in nanopores (Fig. 2.8j and k). Perera et al. reported that at higher temperature, the ion current shows less rectification and the electrostatic fields is reduced [154]. P. Guerrette et al. reported that ICR behavior could be strongly affected by scan rate [155]. At low scan rate, ICR would occur in the nanopore. but at high scan rate, I-V curves show almost linear relationship.

By controlling the ICR behavior in the nanopore, various applications have been realized including filtration and desalination [159], nanofluidic logic networks [158, 160–162] and energy conversion [38]. Further development of ICR requires improved nanofabrication techniques.

2.3.5 *Liquid–Solid Interaction*

Another new phenomenon in confined nano-fluids is reflected in energy conversion. Through ion-selective membrane, salinity difference causes the current of one type of ion (reverse electrodialysis, RED). Pressure induced ion migration through narrow channels called electrokinetic effects (streaming current). Besides, other new mechanisms involving solid–liquid interaction [163, 164] have been revealed. These mechanisms usually involve the transfer of protons, phonons, electrons, molecules, and ions in fluids and solids and their interactions among each other. In terms of obtaining fluid energy [165], researchers have tried to explain the solid–liquid interaction from different angles and with different theoretical mechanisms, such as streaming potential [166] (Fig. 2.9a), waving potential [167] (Fig. 2.9c), asymmetric chemical doping [168] (Fig. 2.9g–j), pseudocapacitive mechanism [169] (Fig. 2.9b), EDL-induced ion polarization [10, 170] (Fig. 2.9j), Coulomb drag [171], phonon coupling [172] etc.

Yin et al. [169] observed sub-millivolt voltage across a single layer of graphene by moving an ionic solution droplet over it (Fig. 2.9b). The authors showed that when the ionic solution was dropped on graphene, an electric double layer (EDL) formed only in the region of the dropped graphene. When the droplet moved on graphene, the formation of the front-end EDL attracted the charge carriers in the graphene, while the disappearance of the back-end EDL would release charges into the graphene. The pseudo capacitance before and after the droplet charges and discharges generated a voltage across the single layer graphene, which they called the drawing potential.

As for the waving potential, its principle was similar to the pseudo-capacitance mechanism proposed by Yin et al. [167]. They rapidly immersed and pulled out two-dimensional graphene in an aqueous NaCl solution, to generate a current at both ends of the graphene, which they attributed to the movement of the electric double layer (EDL) (Fig. 2.9c). Besides, Kim et al. [174] proposed a unique "twistron" structure. They placed the whole CNT yarn in solution, which acted as a surface area-controlled charge capacitor. When in electrochemical equilibrium with the surrounding fluid, the "twistron" generated electric current when stretched or twisted (Fig. 2.9f).

On this basis, researchers have reported a lot of related work, and tried to explain the generation of voltage by divise mechanisms. Yao et al. [185] reported a transpiration-driven dynamic generator (TEPG), which utilized capillary flow of water in asymmetrically wetted cotton fabrics coated with carbon black (Fig. 2.9j). Proton accumulation occurred at the solid (carbon black)/liquid (water) interface caused by the electric double layer and resulted in a potential difference between the wet side and the dry side. Similarly, water molecules migrate in the TiO_2 material due to capillary action, and the negatively charged surface cause more positive charges

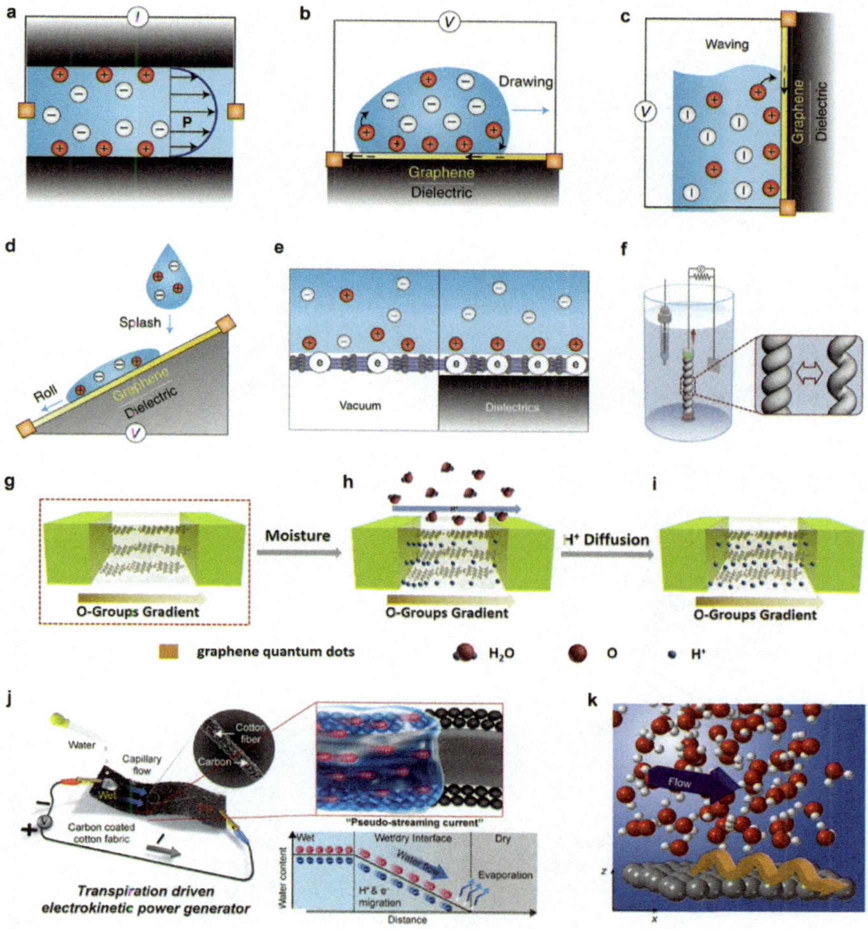

Fig. 2.9 Different forms of solid–liquid interactions for energy harvesting. **a** Schematic diagram of classical electrokinetic effects resulting in streaming current in nanochannels. **b** Voltage generated across a single layer of graphene by moving an ionic solution droplet over it. **c** Waving potential generated by rapidly immersing and pulling out two-dimensional graphene in an aqueous NaCl solution. **d** Power generation in the form of mimicking rainwater. **e** Solid–liquid interaction between free standing graphene or graphene overlaid on a dielectric substrate and ionic solution [173]. Adapted with permission from Ref. [173] Copyright (2023) (Springer Nature). **f** Electric current generated from a unique "twistron" structure when it is stretched or twisted [174]. Adapted with permission from Ref. [174] Copyright (2023) (The American Association for the Advancement of Science). **g–i** Energy harvesting from water vapor by designing asymmetric functional group materials [168]. Adapted with permission from Ref. [168] Copyright (2023) (American Chemical Society). **j** Transpiration-driven dynamic generator (TEPG) utilizing capillary flow of water in asymmetrically wetted cotton fabrics coated with carbon black [10]. Adapted with permission from Ref. [10] Copyright (2023) (American Chemical Society). **k** ab initio molecular dynamics simulations to research the mechanism of solid-water friction [175]. Adapted with permission from Ref. [175] Copyright (2023) (Springer Nature)

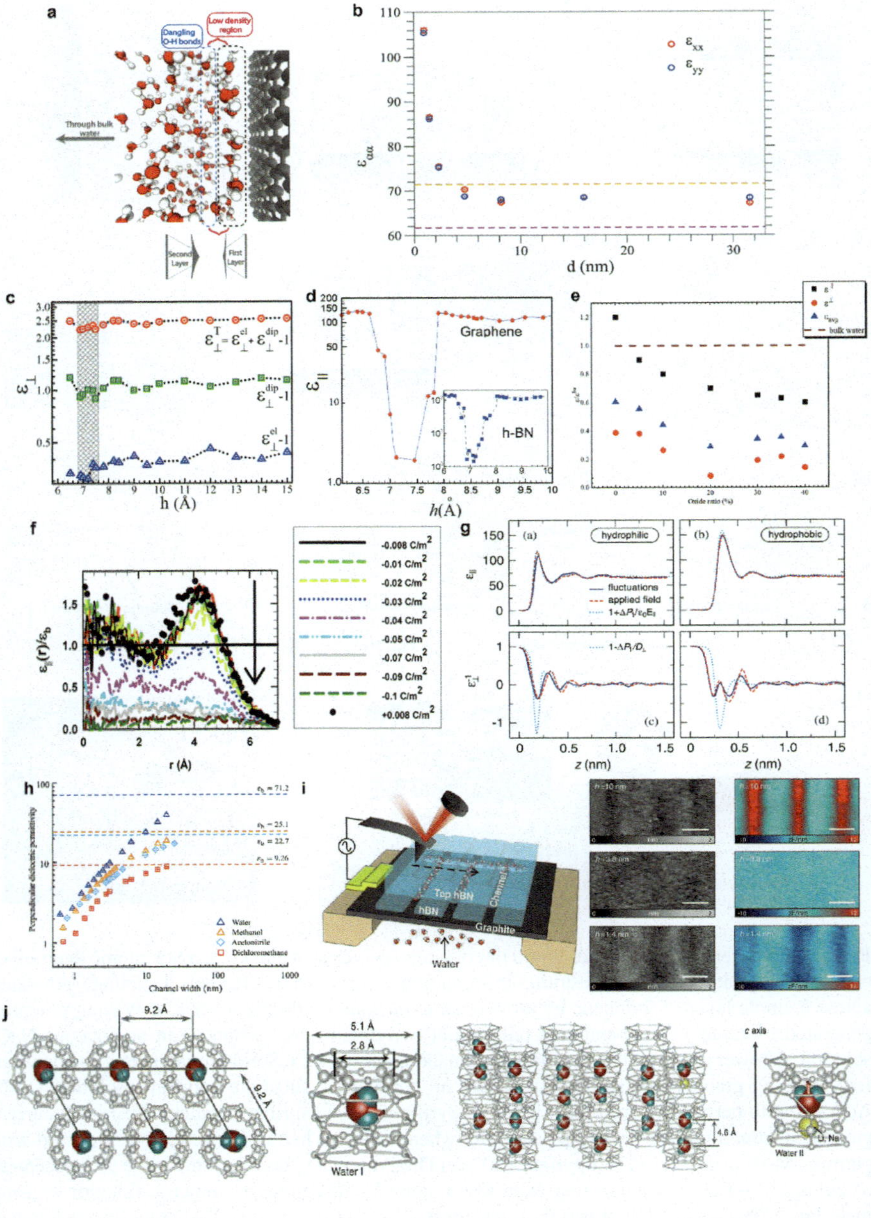
a
Dangling O-H bonds
Low density region
Through bulk water
Second Layer
First Layer
b
ε_{xx}
ε_{yy}
$\varepsilon_{\alpha\alpha}$
d (nm)
c
$\varepsilon_\perp^T = \varepsilon_\perp^{el} + \varepsilon_\perp^{dip} - 1$
$\varepsilon_\perp^{dip} - 1$
$\varepsilon_\perp^{el} - 1$
$\varepsilon_\perp$
h (Å)
d
Graphene
h-BN
$\varepsilon_\parallel$
h(Å)
e
bulk water
Oxide ratio (%)
f
-0.008 C/m^2
-0.01 C/m^2
-0.02 C/m^2
-0.03 C/m^2
-0.04 C/m^2
-0.05 C/m^2
-0.07 C/m^2
-0.09 C/m^2
-0.1 C/m^2
+0.008 C/m^2
$\varepsilon_\parallel(r)/\varepsilon_b$
r (Å)
g
(a) hydrophilic
(b) hydrophobic
fluctuations
applied field
$1+\Delta P_\parallel/\varepsilon_0 E_\parallel$
$1-\Delta P_\perp/D_\perp$
(c)
(d)
$\varepsilon_\parallel$
$\varepsilon_\perp^{-1}$
z (nm)
h
$\varepsilon_b = 71.2$
$\varepsilon_b = 25.1$
$\varepsilon_b = 22.7$
$\varepsilon_b = 9.26$
Perpendicular dielectric permittivity
Water
Methanol
Acetonitrile
Dichloromethane
Channel width (nm)
i
Top hBN
Channel
hBN
Graphite
Water
j
9.2 Å
9.2 Å
5.1 Å
2.8 Å
Water I
4.6 Å
c axis
Li, Na
Water II

◂**Fig. 2.10** Ultra-low dielectric constant of water molecules. **a** Water layer structure on the liquid–solid interface [176]. Adapted with permission from Ref. [176] Copyright (2023) (the American Physical Society). **b** Dielectric constant at parallel direction enhanced about 50% under strong confinement [177]. Adapted with permission from Ref. [177] Copyright (2023) (American Chemical Society). **c** Electronic and dipolar contribution to dielectric constant in confined water [178]. Adapted with permission from Ref. [178] Copyright (2023) (the American Physical Society). **d** The in-plane dielectric constant as a function of channel height in graphene and h-BN sheets [179]. Adapted with permission from Ref. [179] Copyright (2023) (AIP Publishing). **e** The influence of oxidation rates on dielectric properties of confined water [180]. Adapted with permission from Ref. [180] Copyright (2023) (Elsevier). **f** Dielectric permittivity of confined water molecules as a function of surface charge density [181]. Adapted with permission from Ref. [181] Copyright (2023) (the American Physical Society). **g** The parallel and inverse perpendicular dielectric function in hydrophilic and hydrophobic structure [182]. Adapted with permission from Ref. [182] Copyright (2023) (the American Physical Society). **h** The universal reduction of dielectric constant in confined fluids [183]. Adapted with permission from Ref. [183] Copyright (2023) (American Chemical Society). **i** The dielectric image of nano-confined water by AFM measurement [105]. Adapted with permission from Ref. [105] Copyright (2023) (The American Association for the Advancement of Science). **j** The water molecules confined in nanochannels of the beryl crystal lattice and showing incipient ferroelectricity [184]. Adapted with permission from Ref. [184] Copyright (2023) (Springer Nature)

to migrate along with it, resulting in the separation of positive and negative charges, thereby generating a potential difference between the two ends of the material [186].

Harvesting energy from water vapor by designing asymmetric functional group materials has also been reported. As shown in Fig. 2.9g–i, a concentration gradient of oxygen-containing functional groups was generated inside the electrochemically polarized graphene quantum dot device by Huang et al. [168]. When the device was exposed to water vapor environment, a large number of water molecules were adsorbed on the oxygen-rich end of the device. At this time, the OH bond was broken to release H^- ions, resulting in local enrichment of H^+ ions. Under the action of the concentration gradient, H^+ ions diffused to the other end of the device. Therefore, a voltage was generated, which was reflected in the reverse direction of the electrons in the external circuit. When the water molecules desorbed from the surface of the device, the free H^+ ions migrated back and recombined with the negatively charged oxygen-containing groups, resulting in an electrical signal in the opposite direction. Subsequently, the adsorbed remaining water molecules were released, and the graphene quantum dot device returned to the initial state, thereby completing a power generation cycle.

In addition to the above mechanisms for generating electricity, Coulomb drag is an emerging phenomenon about liquid–solid interaction. Coulomb drag originally describes that when two conductors are close to each other but insulated, the migration of carriers in one of the conductors 'drags' the carriers in the other, thereby generating a corresponding current in the driven conductor [187]. First it was extensively studied in one dimensional wire [188] and two dimensional "layers" like graphene [189–192]. In recent years, the Coulomb drag phenomenon was served as an emerging mechanism to explain fluid induced power generation [193, 194], which emphasized its importance in confined nanofluidics. The theoretical idea was first proposed by Král and Shapiro [194], who ascribed the liquid induced current in nanotube to two possible mechanisms. The first mechanism referred to the transfer of phonon momentum from the flowing liquid molecules to the carbon lattice, which induced a drag force on the free carriers in the nanotubes, and the second mechanism involved more direct Coulomb coupling between ions or polar molecules in the liquid and the medium carriers in carbon nanotubes. Then two years later, Ghosh et al. [62] first experimentally demonstrated the Coulomb drag mechanism in carbon nanotube as flow sensors. They reported voltage generation induced by liquid flow on SWNT bundles of 1 mm length, 2 mm wide and 0.2 mm thick. The voltage showed logarithmic dependence as a function of the flow velocity. However, Persson et al. [195] proposed a "Stick–slip" model to interpret the general Coulomb scattering mechanism,. Afterwards, Zhao et al. [196] reported a voltage generation induced by water flow inside metallic single-walled carbon nanotubes. In this study, a single SWNT with an average diameter of 1.6 nm was "filled" with water molecules, which generated a voltage difference of about 1 mV across the SWNT when the water was electrically driven to flow inside the tube. The researchers also simulated the energy harvesting process of hydrochloric acid–water solution flowing through CNTs [197], where the induced voltage along the axial direction was related to the corresponding ion drift velocity. However, experimental evidence for energy conversion on this scale remained scarce and elusive. In 2011, Dhiman et al. [198] reported energy harvesting from water flow over graphene. However, Yin et al. [199] found that the electric voltage was from the interactions between exposed electrodes and water, and once the electrodes were insulated, no measurable voltage was observed. Zhong et al. [193] generated a potential difference of about 10 mV across the graphene by moving the graphene oxide nanosheets on the graphene surface, which they interpreted as the Coulomb interaction between the charged GO nanosheets and graphene, because the slippage of h-BN on graphene surface did not produce a similar voltage. Recently, Rabinowitz et al. [80] demonstrated an electrically actuated biomimetic ion pump by combining these electronic and nanofluidic transport capabilities within an individual SWCNT device. They attributed this phenomenon to the Coulomb drag mechanism but gave no strong evidence. Besides, some other experimental systems like water-metal oxide have reported the Coulomb interaction at the solid–liquid interface [200]. These pioneering research about Coulomb drag gave us a broader view of the confined nanofluidics.

However, the exact mechanism used to explain the process of fluid-induced power generation has not been clearly clarified, which leads to the urge need of visualization, quantification, and theory development of confined nanofluids. Recently, Kavokine et al. [175] used ab initio molecular dynamics simulations to research the mechanism of solid-water friction (Fig. 2.9k). Furthermore, they revealed that charge fluctuations in liquids are coupled with electronic excitations in solids, thereby predicting quantum effects at the solid–liquid interface. The theoretical results verified the diameter-dependent slip length in carbon nanotubes and provided a plausible explanation for the difference in water slip on graphene and graphite surface. This provided a theoretical basis for understanding the interaction between microscopic particles at the solid–liquid interface in confined structure.

These reported mechanism of nanofluid energy conversion is still controversial, and various mechanisms require further experimental and theoretical backup. In the future, we expect more powerful technical means to reveal the mystery in confined nanofluidics. Understanding it will not only shorten our cognitive gap in nanofluidics, but also provide important ideas for us to push the nanofluid energy conversion to practical applications.

2.4 Water Transport

Water, as a ubiquitous solvent, plays an important role in many applications, ranging from energy harvesting to molecular sieving. Water under extreme confinement has been widely explored in the past few decades and various extraordinary phenomena have come into our sight including ultra-low dielectric constant, phase transitions, layering and ultra-fast water flow. Here we will introduce these unusual behaviors combined with the development of molecular dynamics simulation and experimental technology.

2.4.1 Ultra-Low Dielectric Constant

The dielectric response is an important element for energy conversion and molecular sieving. Water, a familiar polar solvent, shows dielectric constant reduction in the confined structure. Even under ambient conditions, water molecules near the solid–liquid interface could exhibit a prominent layered structure and form ordered phase like ice. The dipoles of water molecules are aligned regularly by solid surfaces, leading to small polarizability of highly ordered water molecules and reconfiguration resistance by an electric field.

When the height of channels is scaled down to nanoscale or angstrom scale, water molecules under extreme confinement start to show unusual chemical and physical properties, featuring various novel phenomena. Ahmadabadi et al. reported that the residence time and the diffusion constant of extreme nanoconfined water molecules are consistent with the structured water molecules near the solid–liquid surface (Fig. 2.10a) [176]. However, due to the lack of direct experimental technologies that probe interfacial water properties, most of the evidence were from molecular dynamics simulations over the past several years.

The dielectric constant of bulk water is $\varepsilon_{\perp} \approx 80$. Senapati et al. carried out the molecular dynamics simulation about confined water in the 12 Å nanocavity and reported about 50% decrease in dielectric constant [201]. In the model, they did not consider the electrostatic interaction between water molecules and the surface, and assumed that the reduction was only caused by strong confinement.

However, as the scale of confined water systems reaching the range of intermolecular interactions, we need to take not only water-water but also water-charge, water-substrate interactions into consideration. It has been reported that dipolar fluctuations of confined water were obviously suppressed by an order of magnitude in z direction (the direction perpendicular to graphene planes) but enhanced about 50% in the x–y plane (Fig. 2.10b) [177]. The symmetry breaking stems from the strong confinement rather than the interaction between water and graphene surfaces. The anisotropies have been further explored. Under strong confinement, parallel polarization correlations of adjacent water molecules are increased mildly but the perpendicular dipoles are primarily anticorrelated, leading to negative contribution [202]. Although several studies have reported the dielectric constant reduction, there is no clear exploration of contribution of electronic and dipolar dielectric constant. Shekarforoush et al. found that the proportions of dipolar contribution and electronic contribution were almost equal for dielectric constant in the strong confined water (Fig. 2.10c) [178].

All research of dielectric constant reduction mentioned above is based on asymmetric systems. What will happen in the nanotubes which is totally symmetric at cross profile? It has been found that the dielectric constant of confined water shows a radial distribution [203]. The confined water in the nanotubes can be divided into two parts based on location and behavior, the middle water (with very small density perturbations) and the hydration water (with two clear density peaks near the nanotube wall) on the basis of density perturbation. The hydrogen bond orientation of middle part water is isotropic but that of hydration layers are apt to parallel along the nanotube axis. Zhu et al. even divided the cylindrical nanoconfined water into n concentric radial shells but found that the average dielectric constant of strong confined water molecules has no connection with the number of concentric radial shells [204]. The water dipoles show strong orientation only in the vicinity of the surface which leads to dielectric constant reduction.

The ability to change the dielectric properties of confined water is vital and helpful for its practical applications, for example water desalination [205, 206]. By changing conditions such as channel height, oxidation concentration, surface charge density and so on, the dielectric constant of confined water can be tunable.

The first factor considered is the height of nanochannels. Itoh et al. found the nonlinear decrease in dielectric constant of water molecules as the height of nanochannels was narrowing [207] but the channel height was still larger than 10 Å. However, when the height of channels is 8 ~ 10 Å or smaller than 6.7 Å, the dielectric constant will be enhanced greatly by about 120 times (Fig. 2.10d) [179]. The dielectric constant of nanoconfined water molecules reaches the minimum value ($\varepsilon \approx 1.93$) at $h = 7.4$ Å. The extreme low value of dielectric constant can be attributed to the formation of proton-disordered ice.

The influence of oxidation concentration in graphene oxide nanochannel has been reported. With the oxidation concentration increasing, the dielectric constant can drop [208]. The existence of hydrogen bonds between water and graphene oxide should be responsible for the decrease, because it will lead to the reduction of water dipole fluctuation (Fig. 2.10e) [180]. But the influence is still smaller than the nanochannels' interval space. And the effects of surface charge density were also investigated in molecular dynamics simulations. Renou et al. carried out the simulation in silica nanopores and found that the raise of surface charge density would bring about a decrease of the dielectric constant of confined water into the axial direction (Fig. 2.10f) [181, 209]. They also detected an obvious slowdown in the rotational dynamics of water. These phenomena can be attributed to the strong orientation of water molecules caused by interaction with surface charge.

It is known that the dielectric constant of electrolyte solution is smaller than that of pure solvent, which is caused by excluded volume effects and the screening of solvent dipoles due to the existence of ions [210]. The dynamic contribution to the static dielectric constant has also been studied in molecular dynamics simulations where the contribution is finite and small [211]. However, in the two-dimensional slits of MXene, the confined water showed abnormal dielectric constant and polarization that has been confirmed by experiments and simulations [212, 213]. The hydration shell radius of intercalated cations played an important role in the system. At the critical hydration shell and ionic radius, the over-screened external electric field would give rise to the negative dielectric constant.

The surrounding media would also influence the dielectric response of confined water. Olivieri et al. have investigated that the permittivity decrease of confined water is not due to the structural alignment of interfacial water molecules [214]. They found that long-range anisotropic dipole correlations and excluded-volume effect of surrounding materials are the causes of permittivity reduction. Similarly, Bonthuis et al. have proved that the hydrophobic and hydrophilic surfaces could also influence the dielectric profile of confined water (Fig. 2.10g) [182]. Therefore, changing the confining material of different dielectric constants or hydrophilic and hydrophobic properties could control the dielectric response of confined water molecules.

The reduction of dielectric response is not unique and only exists in the confined water molecules. Motevaselian et al. have demonstrated universal reduction under extreme confinement in the organic liquid (Fig. 2.10h) [183]. This is of paramount importance for better understanding of biological system and electrochemical devices.

However, all theories of abnormal dielectric constant in confined water mentioned above are predicted by simulations [179–183, 207–214]. Experimental confirmation about dielectric response reduction requires further work.

The structure of oriented water molecules at the solid–liquid interface have been verified by X-ray scattering experiments [215]. In the experiment of neutral reverse micelles, the confined water molecules showed a polarity reduction with the size of micelles decreasing [216–218]. However, the results are reported based on absorption frequency shifts of dye molecules but not the dielectric constant variation straightly. Because of rapid development of nanofabrication and composite detection methods, a lot of experimental studies of artificial nanofluidic channels under strong confinement have come into our sight now. Dielectric constant measurement by atomic force microscope (AFM) has been applied to diverse materials such as viruses [219], single-layer biomembranes [220] and carbon nanotubes [221]. Fumagalli et al. used AFM to probe dielectric constant of water confined in nanochannels. With the height of channels decreasing to 1.4 nm, the dielectric constant of strong confined water is sharply reduced to $\varepsilon_{\perp} \approx 2.1$ (Fig. 2.10i) [105]. The result is even notably smaller than predicted value ($\varepsilon_{\perp} \approx 10$) in MD simulations and experimental result in low-temperature proton-ordered ices ($\varepsilon_{\perp} \approx 3to4$) [222]. The abnormal dielectric constant of strong confined water may originate from the complete suppression of dipole rotational contribution and hydrogen-bond contribution [223, 224].

There is another interesting dielectric property observed in the confined water molecules. Ferroelectricity in the nanoconfined water was detected in the beryl crystal lattices (Fig. 2.10j) [184]. The confined water molecules were located separately to prevent the formation of hydrogen bonding but still closely enough to keep the dipole–dipole interaction. And special structures finally lead to ferroelectricity which may play a key role in the functioning of biological systems. In the MD simulation, Ahmadabadi et al. further predicted the particular ice phase formation with stable ferroelectricity of confined water molecules in the channels whose height was only 7.4 Å. [176]

2.4.2 *Structural Phase Transition*

With the decrease of channel height down to the angstrom scale, the mean-field approach can no longer work in the nanofluidic systems and the microscopic details need to be included. The formation of amorphous ice which we have mentioned before could be anticipated. The formation of n-gonal (n = 4–6) ice nanotubes inside the confined carbon nanotubes was proposed by Koga et al. by MD simulation (Fig. 2.11a) [225]. They ranged the diameters of carbon nanotubes and the applied axial pressuresand found that the confined water molecules could present a first-order freezing transformation or continues phase transition to ice nanotubes. N-gonal ice nanotubes were consist of two piled ice rings [226]. The O–H bonds were connected clockwise in one ring and counterclockwise in the other ring. Zangi and Mark reported that under external electric filed, the confined water molecules would have the freezing transition at the ambient temperature (Fig. 2.11b) [227]. There were two different density ice phases which are very similar to two bilayer phases of liquid water. With the height of confining systems increasing, the reduction of ordering degree could be observed. The two different water molecule density layers in the ice nanotubes were also reported by Takaiwa et al. [228]. They called the ice polymorphs as filled ice nanotubes. After that, Takaiwa et al. further proposed the phase diagram of confined water in the single-walled carbon nanotubes (SWCNTs) in the MD simulation (Fig. 2.11c) [229]. They summarized the dependence of ice structure and melting points as a function of nanotube diameter and found that there were at least nine ice phases in the extreme confined space. Shiomi et al. have reported that the both water-water interaction and water-carbon interaction played an important role in the phase transition [230]. Therefore, we could find the dependence of freezing temperature on nanotube diameter visually. And there are slightly larger number of hydrogen bonds and great decrease in the diffusion coefficient with the phase transition from liquid confined water to amorphous ice [231]. Tanaka and Koga further found that in the composite ice nanotubes, the existence of guest molecules could stabilize the structure of ice nanotubes (Fig. 2.11d) [232]. This is because that the guest molecules such as neon, argon and methane had attractive interaction with confined water molecules.

The prediction of confined water phase transition in the MD simulations is usually made under high pressure or low temperature. However, the environment of water transport in the biological systems or artificial nanofluidic devices were mild. The phase transition in the physiological condition remain unclear.

The phase transition of confined water has been verified by various experiments via X-ray diffraction (XRD) [234, 235], nuclear magnetic resonance (NMR) [235], neutron diffraction [236] and vibrational spectroscopy [233] (Fig. 2.11e and 2.12h). Maniwa et al. have studied the formation of pentagonal to octagonal ice nanotubes in the SWCNTs by XRD [234, 238]. The applied pressure and diameter of carbon nanotubes could both influence the temperature of phase transition. They found a new peak ($Q = 22\ nm^{-1}$) below 235 K which could not belong to bulk ice (marked with '*' labels). Ghosh et al. have reported the complete freezing of confined water in the carbon nanotubes by proton NMR [239]. They found that the water molecules would withdraw from the nanotube center before complete freezing at 212 K. And the transition from liquid to ice could be divided into two parts: the axial phase freezing first and then the wall phase solidification. Integrating XRD with NMR, a wet-dry type transition of confined water molecules was observed with lowing the temperature in the SWCNTs with diameters ranging from 1.46 to 2.40 nm [235]. And the transition temperature is greatly dependent on the diameter of carbon nanotubes. In all samples exaimed, there are narrow NMR peak of water molecules above about 240 K, which means fast molecular motion in the NMR time scale. When temperature is below 240 K, the NMR indensity declined sharply which means the freezing of water molecules at low temperature. Neutron diffraction studies in parallel with MD simulation were carried out by Kolesnikov et al. [236]. They detected an ice-shell structure with water-chain. This may be due to the fierce change of hydrogen-bond connection in the extreme confinement. Using the detection of IR spectroscopy, Byl et al. verified the orientation of O–H bonds in the confined water molecules and found the stacked-ring structure of ice nanotubes. Algara-Siller et al. have observed the formation of square ice at the physiology conditions by high-resolution electron microscopy imaging (TEM) as shown in Fig. 2.11i [237]. The water molecules were locked between two monolayer graphene sheets and formed sub-micrometer size square ice. The lattice constant of square ice with high packing density could be got from the visualized method. And deeper analysis and demonstrations are of great importance for potential implications of water in the biological systems.

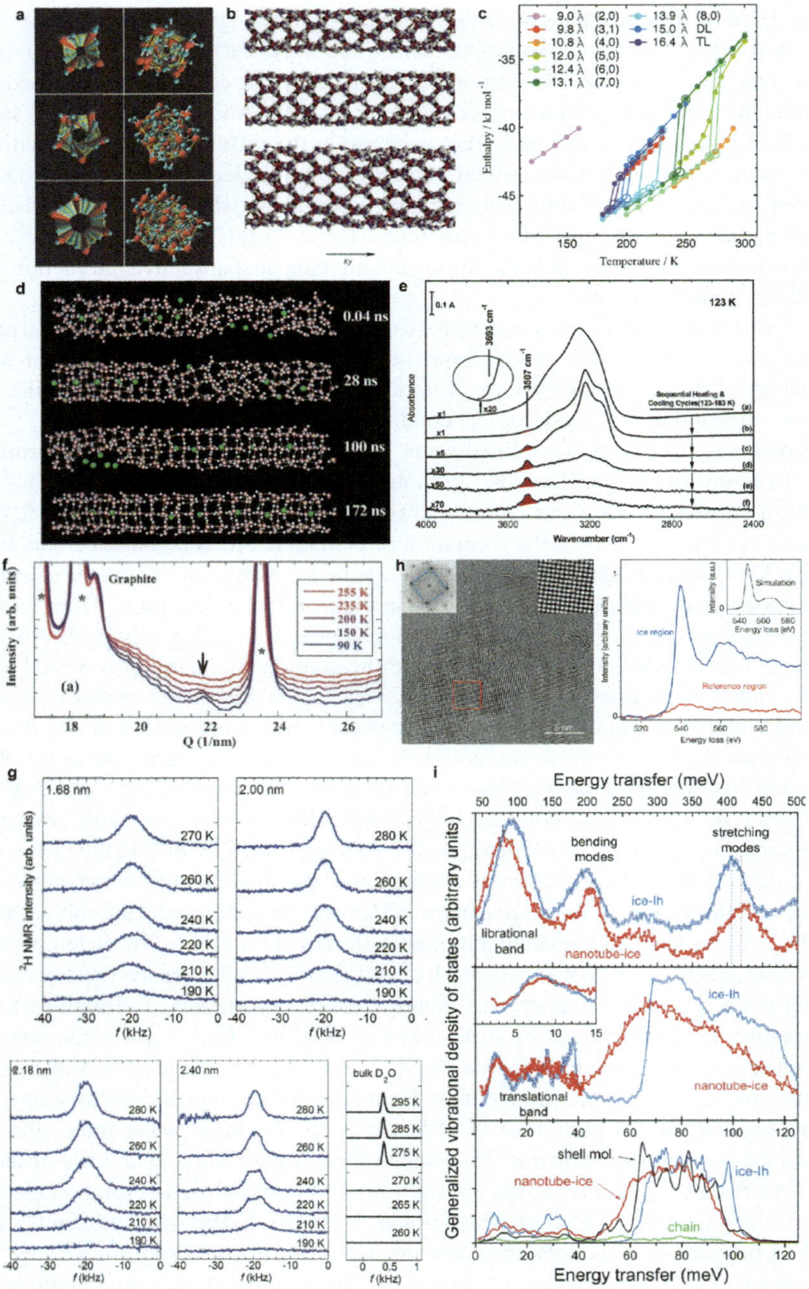
a
b
c
9.0 Å (2,0)
9.8 Å (3,1)
10.8 Å (4,0)
12.0 Å (5,0)
12.4 Å (6,0)
13.1 Å (7,0)
13.9 Å (8,0)
15.0 Å DL
16.4 Å TL
Enthalpy / kJ mol⁻¹
Temperature / K
d
0.04 ns
28 ns
100 ns
172 ns
e
0.1 A
123 K
3693 cm⁻¹
3507 cm⁻¹
Absorbance
Sequential Heating & Cooling Cycles(123-183 K)
Wavenumber (cm⁻¹)
f
Graphite
255 K
235 K
200 K
150 K
90 K
Intensity (arb. units)
(a)
Q (1/nm)
h
Ice region
Reference region
Simulation
Energy loss (eV)
Intensity (arbitrary units)
g
1.68 nm
2.00 nm
2.18 nm
2.40 nm
bulk D₂O
²H NMR intensity (arb. units)
f (kHz)
i
Energy transfer (meV)
bending modes
stretching modes
librational band
ice-Ih
nanotube-ice
translational band
shell mol.
chain
Generalized vibrational density of states (arbitrary units)

◂**Fig. 2.11** Structural phase transition of water under extreme confinement. **a** The structure of n-gonal ice nanotubes and corresponding liquid phases including square, pentagonal and hexagonal ice nanotubes [225]. Adapted with permission from Ref. [225] Copyright (2023) (The American Association for the Advancement of Science). **b** The ice nanotubes induced by electrofreezing at room temperature with three values of diameter: 1.14, 1.36, 1.86 nm [227]. Adapted with permission from Ref. [227] Copyright (2023) (AIP Publishing). **c** The influence of diameter and temperature on water enthalpy and ice nanotube formation [229]. Adapted with permission from Ref. [229] Copyright (2023) (the National Academy of Sciences of the USA). **d** The stability of ice nanotube enhanced by hydrophobic guests [232]. Adapted with permission from Ref. [232] Copyright (2023) (AIP Publishing). **e** The IR spectra changes of water molecules with temperature variation in a vacuum [233]. Adapted with permission from Ref. [233] Copyright (2023) (American Chemical Society). **f** The XRD profiles of confined water molecules with temperature decreasing [234]. Adapted with permission from Ref. [234] Copyright (2023) (the Physical Society of Japan). **g** The NMR spectra of carbon nanotubes (with mean diameter of 1.68, 2.00, 2.18, and 2.40 nm) filled with water molecules [235]. Adapted with permission from Ref. [235] Copyright (2023) (AIP Publishing). **h** The neutron diffraction of ice nanotube [236]. Adapted with permission from Ref. [236] Copyright (2023) (the American Physical Society). **i** The visible characterization by TEM of ice phase of water molecules under strong confinement [237]. Adapted with permission from Ref. [237] Copyright (2023) (Springer Nature)

2.4.3 Layering

The phenomenon of ultrafast water molecular transport in carbon nanotubes has been interpreted by large slip length. The slip length in the carbon nanotubes is affected by the friction between water molecules and surface and the viscosity of water flow. Therefore, the strong interaction between water molecules and capillary surfaces could have great impact on the viscosity and density of confined water. Under strong interaction with hydrophilic surfaces, the viscosity of nanoconfined water could be 7 orders of magnitude higher than that of bulk water. This phenomenon has implied the occurrence of layering and phase transition.

Neek-Amal et al. detected multilayered structures in nanoconfined water (Fig. 2.12a) [240]. They found that the uniform fluid density would break down under the nanoconfined water and distinct layers would form in the nano-scale confinement. By changing the channel heights down to the sub-angstrom scale, the variation of the shear viscosity could be observed in the confined water.

Although the phenomenon of layering in confined water has been widely studied, there are still some difference between simulation systems and realistic experiments for instance the inhomogeneous confinement in atomic scale existing in the experiments. Qiu et al. took this factor into account in the MD simulation and established the simulation system consisted of a convex plate and a flat plate (Fig. 2.12b) [241]. In the system, water molecules at low density presented three different coexisting liquid domains: monolayer, bilayer and trilayer. Moreover, with increasing density, the liquid water would transform into multiphase ice nanoribbons containing the structure of monolayer, bilayer and trilayer. The formation of bilayer square ice with AA-stacked structure has been reported by Algara-Siller et al. And the structure in the MD simulation is consistent with the experimental result [237].

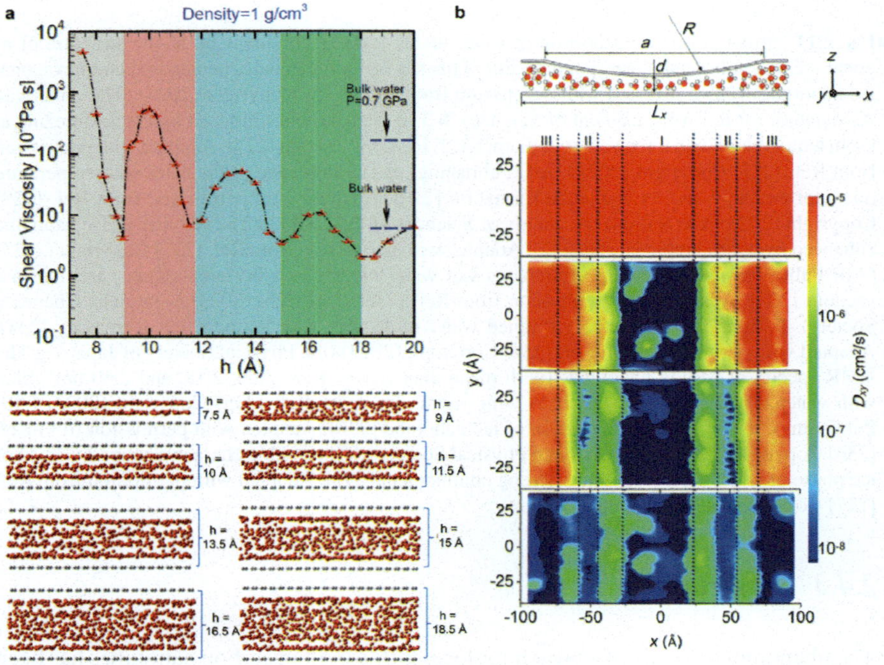

Fig. 2.12 Layering of strong confined water in nanochannels. **a** The shear viscosity changing and layering structure of confined water between two graphene sheets with different distance [240]. Adapted with permission from Ref. [240] Copyright (2023) (American Chemical Society). **b** The lateral diffusion coefficient image of water molecules under inhomogeneous nanoconfinement at 300 K [241]. Adapted with permission from Ref. [241] Copyright (2023) (American Chemical Society)

2.4.4 Ultrafast Water Flow

When the physical length scale of the confinement is comparable to the scale of water molecules, the continuum hypothesis, and a series of theories in the bulk phase are no longer sufficient to explain odd transport phenomena like ultrafast water flow [242]. The smooth surface, adjustable size, and the compatibility with biological membranes make carbon materials a key material for studying molecular transport and nanofluids. Ultrafast water transport was first discovered theoretically and experimentally in carbon nanotubes [243], then extended to, later in membrane form [244, 245], and more recently in two-dimensional planar channels [245] and nanoslits [81] (Fig. 2.13).

Because carbon nanotubes are hydrophobic, a low-concentration depletion layer is formed between the water molecules and the inner wall of the tube, which is the root of ultra-fast water flow in the CNTs. It has been found that the atomically smooth inner surface of carbon nanotubes is conducive to the rapid flow of water

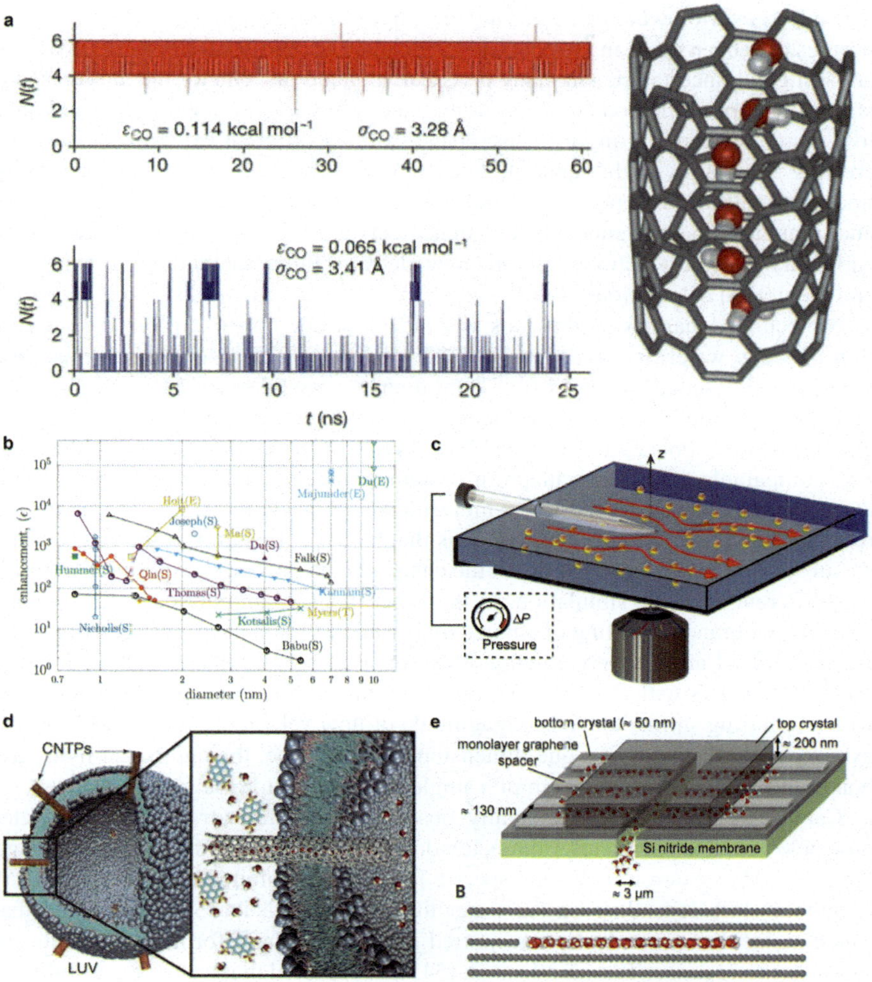

Fig. 2.13 Ultra-fast water flow under extreme confinement. **a** Molecular dynamics simulations reporting single-file arrangement and pulse-like transmission of water molecules inside CNT [246]. Adapted with permission from Ref. [246] Copyright (2023) (Springer Nature). **b** Current reports about flow rate enhancement of water in CNTs of diameter 0.81–44 nm [247]. Adapted with permission from Ref. [247] Copyright (2023) (Taylor & Francis). **c** Precise measurements of pressure-driven flow velocity in an individual nanotube which revealed radius-dependent surface slip [79]. Adapted with permission from Ref. [79] Copyright (2023) (Springer Nature). **d** Measuring the enhanced water flux in the carbon nanotube by the size change of the vesicles [54]. Adapted with permission from Ref. [54] Copyright (2023) (The American Association for the Advancement of Science). **e** 3.4 Å nano-slits allowing no ions but single layer water molecules to transport [44]. Adapted with permission from Ref. [44] Copyright (2023) (The American Association for the Advancement of Science)

through its interior, which attracted a wide range of research interest. Although great progress has been made in the physical mechanism of ultrafast water transport, there are still great uncertainties in many practical applications. One of the main reasons is that the current methods for measuring water transport are not uniform and have low accuracy, resulting in large discrepancy among different experimental results, creating uncertainty in the reported data. Here we reflect on some of the promising findings from various experimental and computational studies on water flow in CNTs and summarize the influencing factors that affect the water slip length and flow rate in carbon nanotubes, which will provide new ideas for future methodological research and theoretical calculations.

Molecular dynamics simulations were the first attempt to explore the interaction between water and hydrophobic CNT channels [225, 246]. It was predicted that water flow inside a CNT of 0.81 nm diameter would reach 51×10^{-14} cm^3/s, which is 2–3 orders of magnitude higher compared with the calculated result by the classical Hagen–Poiseuille equation [246] (Fig. 2.13a). The flow velocity calculated was comparable to that in aquaporin protein channels. And it was proposed that the one-dimensional ordered water molecule chain inside the CNT, which was the result of being extremely squeezed. This theoretical simulation result was verified by subsequent experiment [243], but the enhanced magnitude of experimental result was different from the simulation. In the experiment, Majumder et al. [243] fabricated a membrane consisting of aligned multiwalled carbon nanotubes and drove the water flow by 1 atm applied pressure. They reported a flow enhancement of $4 - 6 \times 10^4$ m^3/s in CNTs with diameter of 7 nm. After that, this result was shown in several works reporting similar enhanced magnitude of flow velocity [248, 249]. Under the extremely confinement of a one-dimensional carbon tube, the number of hydrogen bonds and the degree of freedom of a single water molecule are greatly reduced.

Cambre et al. first experimentally observed single-file arrangement of water molecules in CNT with tube diameter down to 5.5 nm by the radial breathing mode (RBM) vibration in Raman spectra [46]. They exploited the predicted RBM frequency shift depending on the water filling inside CNT [250] and proved that water molecules can only be transported in a single-chain form at this scale. In sub-nanometer range, Tunuguntla et al. [54] also reported that water permeability in 0.8-nm-diameter carbon nanotube porins (CNTPs), exceeds that of biological water transporters and of wider CNT pores by an order of magnitude (Fig. 2.13d). They embedded individual carbon nanotubes on the phospholipid bilayer of large vesicles and measured the flux of water molecules through the size change of vesicle scaling. Different from previous experiments where the average CNT diameter was less than 10 nm, Whitby et al. [251] tested whether flow rate enhancement can be achieved in CNTs with larger diameter (10–100 nm). The result shows that the magnitude of enhancement dramatically decreased in larger pores, which means the anomalously ultra-fast flow is strongly dependent on the diameter of CNTs. When the diameter of carbon nanotubes is in the range of 1–2.7 nm, water molecules form a "concentric-circle"-like layered structure in the tube [247]. Borg et al. [252] combined molecular dynamics (MD) simulations with a continuum flow resistance model and predicted the velocity of water flow in CNTs with different sizes. It was found that below 2 nm, the slip length, density, and viscosity all varied with diameter. In larger diameter carbon nanotubes—more than 10 times the size of single water molecule, the water inside the CNT assumes a bulk structure but still predicted higher flow rates than the classical continuum fluid model due to the small friction coefficient of water–solid surface [247].

These findings show that for the arrangement and transport of water in carbon nanotubes, diameter is one of the most important influencing factors, and there are quite a few studies on the dependence of the flow rate on the diameter (Fig. 2.14b). Secchi et al. [79] used the manipulation of SEM to insert single nanotube into the tips of glass nanocapillaries and detected the flux of water by indicator particles polystyrene, the sensitivity and precision of the detection are greatly improved compared with previous experimental methods (Fig. 2.13c). They verified the dependence of water permeability on the diameter of nanotubes—when the diameter of the nanotubes is less than 20 nm, the water flux increases dramatically and is accompanied by an increase of the slip length. It is worth noting that h-BN nanotubes, which are similar in structure to carbon nanotubes, did not show obvious diameter dependence, and this is supposed to the difference of atomic structures and their physiochemical properties.

Another major parameter used to investigate the confined water flow is the slip length. The slip length is the distance from the point where the liquid velocity distribution extends linearly to zero to the liquid–solid interface [247] (Fig. 2.14c). It is expressed by as

$$L_s = \frac{\eta}{\lambda}$$

where η is the fluid shear viscosity and λ is the interfacial friction coefficient at the fluid–solid interface. The relationship between fluid velocity enhancement and slip length is expressed as

$$\mu_s = \mu(R) = L_s \frac{\partial \mu(r)}{\partial r}\bigg|_{r=R}$$

$$\varepsilon = \frac{Q_{slip}}{Q_{no-slip}} = \left(1 + \frac{8L_s}{D}\right)$$

Q_{slip} is the observed flow rate, $Q_{no\text{-}slip}$ is the expected flow rate calculated from the Hagen–Poiseuille equation, and D is the diameter of the nanotube. The slip length of fluid in confined space has been studied for a long time. But because there is no unified measurement and evaluation standard, researchers still have different views on it. For example, only carbon nanotubes between 0.8 and 1 nm, slip lengths in the range of 1–50,000 nm have been reported [247].

In the first experimental verification of ultrafast water transport, Majumder et al. [243] reported the slip length of CNTs with a pore size of 3.5 nm as high as 3–70 um, a result obtained by linear extrapolation of the flow velocity at the tube wall.

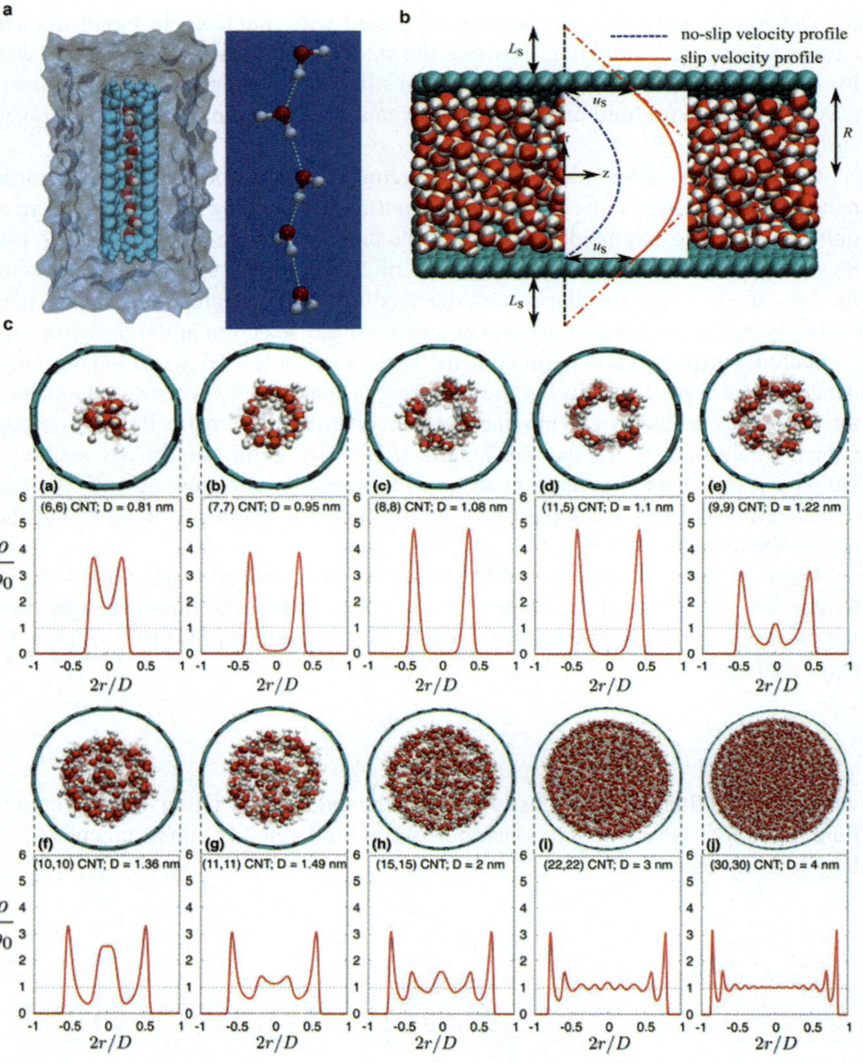

Fig. 2.14 Water structure arrangement in CNTs. **a** Single-file arrangement and hydrogen bond configuration of water molecules inside a single-walled CNT [253]. Adapted with permission from Ref. [253] Copyright (2023) (American Chemical Society). **b** Schematic depiction of velocity profiles of water flow through carbon nanotubes and definition of fluid–solid interfacial slip $L_{S.}$ **c** Cross-sectional view of water inside CNTs with different diameters [247]. Adapted with permission from Ref. [247] Copyright (2023) (Taylor & Francis)

Subsequently, Jason et al. [248] calculated slip lengths of 380–1400 nm in sub-2 nm carbon nanotube arrays, which is inconsistent with previous conclusion. A recent influential report proved that the length scale of slip length in carbon nanotubes is comparable to the diameter of the nanotubes (within 100 nm) and shows a strong tube diameter (wall curvature) dependence. This is because the friction coefficient of water transport in carbon nanotubes is positively correlated with the radius of curvature. That is, the smaller the carbon nanotube diameter, the smaller the friction coefficient, and the larger the corresponding flow enhancement [79, 254].

Ultrafast water transport is one of the most important and peculiar phenomena in the field of nanofluidics. Water acts as a medium for ions and other molecules and its behavior affects the behavior of other particles. Understanding the arrangement and transport of water molecules in strong confinement is of high significance for our understanding of sub-continuum fluid mechanics. If we can accurately understand the difference between the confined water and bulk water, ultrafast transport and some other anomalous water behaviors will be better applied to practical applications such as blue energy harvesting, desalination and interfacial catalysis.

2.5 Molecule Transport

In addition to ions and water, the behavior of molecules under strong confinement has gradually attracted widespread interest. From biomolecules involved in the popular nanopore sequencing, to the selective sieving of gas molucules, to the abnormal transport of uncharged neutral molecules, these phenomena provide a new perspective on our understanding of nanofluidics. Molecular transport involves complex many-body interactions, including electrostatic force between particles, van der Waals force, electric field force, and interfacial interaction of electric double layer (EDL) [53], which require us to consider the mutual coupling of ions, water, and molecules. Relevant content is discussed in previous sections and will not be repeated here. In this part we will briefly introduce the emerging phenomena of molecular transport in confined space and their applications, including the detection of biomolecules (Fig. 2.15a–c), selective gas sieving (Fig. 2.15d) and some other neutral molecules transport mechanism (Fig. 2.15e).

Sequencing and molecular sensing are of great significance in different fields such as molecular diagnosis and drug development, virus detection like SARS-CoV-2, forensic science, etc. Nanopore sequencing and molecular sensing have been developed for decades. It is based on the current blockage ($\Delta I/I0$) and the duration time of the event (t_{dwell}) when single molecule passes through the pore to distinguish different molecules or different units of a molecular sequence.

The transport of biomolecules was inspired by the Coulter Counter and channels embedded in the cell membrane, which undertake functions such as ion exchange and delivery of molecules such as proteins, DNA and RNA [256]. Kasianowicz et al. [98] firstly realized single-stranded DNA (ssDNA) and RNA (ssRNA) translocation through an α-hemolysin nanopore. They found that when the polymer molecule

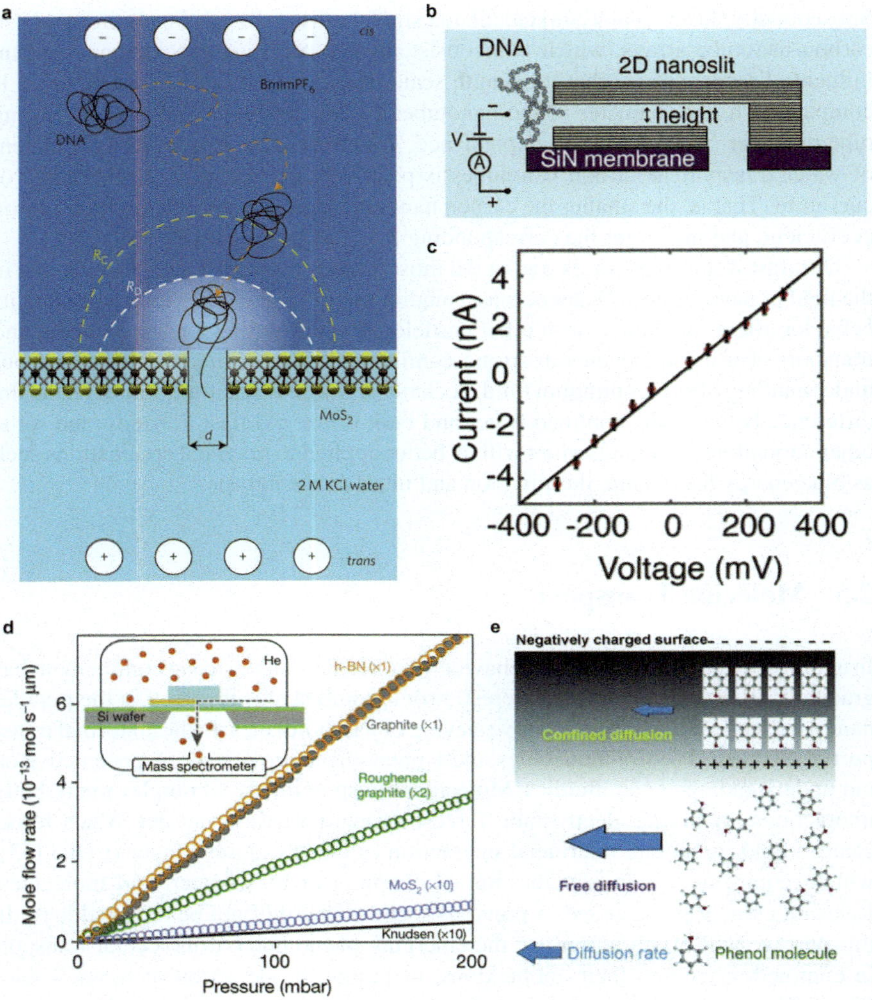

Fig. 2.15 Molecule transport under extreme confinement. **a** Identification of single nucleotides in MoS_2 nanopores [7]. Adapted with permission from Ref. [7] Copyright (2023) (Springer Nature). **b–c** The electrical signal used to study the process of DNA molecules passing through Nano slits [6]. Adapted with permission from Ref. [6] Copyright (2023) (John Wiley and Sons). **d** The specular reflection and ballistic transport of gas molecules on the inner wall observed in extremely narrow Nano slits [255]. Adapted with permission from Ref. [255] Copyright (2023) (Springer Nature). **e** Anomalous diffusion of neutral phenol molecules in charged nanochannels [14]. Adapted with permission from Ref. [14] Copyright (2023) (John Wiley and Sons)

was extended into a single chain and passed through the pore, the current across the pore was partially blocked, and the duration time of the current reduction was proportional to the length of the molecular chain. In 2008, Butler Tom et al. [257] used shorter MspA to distinguish DNA bases, achieving high sensitivity and fidelity. CsgG was also recently uncovered [258]. Concurrent with the development of biological nanopores, advances in micro- and nano-fabrication have allowed subsequent development of solid-state nanopores [259], atomically thin nanopores [7] and glass nanopores [260] to more or less compensate for the inadequacy of biological nanopore in molecule sensing (Fig. 2.15a). The development process of biomolecular sensing by nanopore has been elaborated by Deamer [261] and Varongchayakul et al. [262], therefore here we only introduce the recent reports of emerging one-dimensional and two-dimensional slits for biomolecule translocation in recent years. Liu et al. [5] fabricated a device of SWCNT connecting two reservoirs and PDMS used for the encapsulation of fluid pathway. The ionic conductivity through SWCNT is anomalously high and DNA translocation event caused large but transient increase in the ion current. Recently, Wang et al. [263] successfully differentiated four kinds of single nucleotide and proved translocation feature of dGTP in CNT is related to the tube diameter. Yang et al. [6] attempted to translocate DNA through nanoslits (Fig. 2.15b and c). They found that DNA loading, translation, and exit occurred in three stages during the current blocking process. Besides, during this process, the common clogging in graphene nanopores did not occur and the device was stable for several hours. Compared with DNA sequencing, the development of protein sequencing is seriously lagging, mainly due to the wide variety of amino acids that make up proteins and the complex hierarchical structure of proteins. Therefore, the biomolecule sensing in different confined structures is still in the early stage of exploration, and we look forward to better means and techniques to solve the existing problems in the future.

Before the strong confinement structure is applied for separation, porous membranes used for molecule sieving are usually thick (> 20 nm) [64], which is expected to maintain high selectivity by sacrificing partial flux. Since the atomically thin 2D materials such as graphene, molybdenum sulfide have thrived over the past few decades [35, 264], nanoscale confinement channels have made great strides in the field of selective sieving of small molecules. Varies of experimental and theoretical results have verified the gas selective transport as well as the enhanced permeability. Koenig et al. [265] introduced pores into graphene by ultraviolet-induced oxidative etching and realized selective molecule sieving based on the molecular weight measurement. Hyo et al. [266] exploited intrinsic defects in graphene for selective gas transport and controlled porosity by stacking different layers. For their prepared GO membranes, a highly interlocked layered structure was formed between the layers of the membrane, thus exhibiting extraordinary gas permeation behavior. Then in 2014 Celebi et al. [267] pushed the molecule permeability through atomically thin graphene to its limit. Recently, Keerthi et al. [255] reported frictionless molecule flow through ångström scale slit-like channels and they attributed the enhancing He gas flow by mirror-like reflection on the wall of the channel (Fig. 2.15d). This is an exciting discovery because the mechanism is beyond the traditional Knudsen theory. Besides, Sun et al. [268] showed that defect-free graphene was almost impermeability for most gas molecules except for H_2 and gave the explanation of H_2 dissociation at catalytically active graphene ripples. They fabricated high precise micrometer-size containers made from monocrystals of graphite or hexagonal boron nitride (hBN) to ensure no possibility of gas leakage.

Neutral molecules in solution are not electrically charged, and it is difficult to be driven by electric field forces, so research in this area is also relatively scarce. However, there are still sporadic reports focusing on the abnormal transport of neutral molecules. It has been reported that charged alumina nanochannels could also form induced dipoles in neutral molecules (phenol). Researchers found that the electric double layer of the inner wall interacted with phenol and induced phenol to form a dipole (Fig. 2.15e). Therefore, the diffusion rate of phenol would slow down, like a 'drag' effect. Recently, Cheng et al. [269] focused on the diffusion of organic solvent molecules across graphene in non-aqueous systems. They used dye molecules as the indicator and realized the separation of hexane isomers.

2.6 Conclusions and Perspective

Confined nanofluidics is the study of the properties or transport behavior of fluids confined in nanometer or even atomic scale channels and spaces. Due to the continuous development of micro/nano technology and measurement methodology, this field has led to countless applications for molecular separation and analysis, biological characterization, sensors, cell capture, energy harvesting, to name a few [10]. At the end of this whole chapter, we first summarize the enlightenment and related applications of some emerging fluid transport phenomena, and then briefly comment on the current research progress and future challenges. We hope that these humble few words will be helpful for guiding the way for future advances in the field of nanofluidics.

2.6.1 Molecular Sieving

The ion selectivity of nanopores, ultrafast water flow in confined channels such as carbon nanotubes, and the exclusion of ions by slit channels are emerging abnormal phenomena that have driven development of membrane separation science. Membrane separation research has important value in desalination, catalytic chemistry, petroleum industry, energy chemistry and other fields. In seawater desalination, graphene oxide (GO) may play an important role due to its porosity, hydrophilicity, and high chemical stability. The mechanism of seawater desalination is mainly based on size exclusion and electrostatic repulsion. The interlayer spacing of dry graphene oxide is about 0.76 nm, which is comparable to the size of hydrated cations, while the size of water molecules is about 0.28 nm, which can easily pass through the membrane channel [270]. However, graphene oxide films generally suffer from swelling problems [36]. To solve this problem, Chen et al. [13] used cations to precisely control the interlayer spacing of graphene oxide down to 1 Ångstrom, enabling specific ion repulsion. For gas molecule sieving, Li et al. [64] fabricated extremely thin GO film and achieved very high separation efficiency for H_2/CO_2 and H_2/N_2. Recently Keerthi et al. [255] reported that collisions of gas molecules with inner walls of Ångstrom-scale channels can be specular surface scattering, which also explained the ultrafast transport of He through nanochannels. In general, the main challenge in molecular sieveing is to improve selectivity while maintaining the high flux. At present, relying on the advancement of micro/nano processing technology, researchers have been able to achieve molecular sieving in thinner membrane materials, to a certain extent, to achieve a balance between the two. At the same time, in the process of scaling up and industrialization, the mechanical stability of the membrane and the problem of clogging need to be addressed. We look forward to a better solution in the future.

2.6.2 Energy Harvesting

Ion selectivity and solid–liquid interaction induced by the electrical double layers (EDLs) or carriers in solid and liquid have triggered many studies on nanofluidics for energy harvesting, which also involve different mechanisms, including reverse electrodialysis (RED), pressure-retarded osmosis (PRO), fluid–solid interaction, and asymmetric chemical modification etc. Nanopore-based osmotic power generation has shown remarkable advances at several time nodes [271]. Siria et al. [37] embedded single boron nitride nanotube (BNNT) into silicon nitride and harvested osmotic energy from the concentration difference. Later in 2016, Feng et al. [38] reported that a single MoS_2 nanopore of 15 nm can generate an energy of hundreds of pW. The scaling up osmotic energy conversion from single pores faces extra challenges [272], mainly because the energy dissipation caused by the concentration difference cannot be avoided in the process of scaling up. In addition, researchers have also reported different forms of fluid energy harvesting, including water vapor, droplets, waves, etc. The current problems with these energy harvesting studies are the inconsistent evaluation criteria and the unclear mechanism research. We look forward to developing more powerful theoretical and technical methods to study the mechanism of energy harvesting in the future, and to further optimize the way of energy harvesting on this basis. In addition, the integration of multiple energy harvesting technique may provide a better idea for practical applications.

2.6.3 Biomimic Channel for Information Storage and Transfer

Artificial nanochannels now exhibit some peculiar transport phenomena, including light/pressure/electricity-activated, ion selectivity, gating, rectification, and ion pumping based on Coulomb drag in CNTs. The physical parameters of the fluids in these phenomena including conductivity, mobility, ion selectivity etc., make the artificial channels featuring the functions of storing and transferring information like biological channels to a certain extent. Recently, Robin et al. [273] predicted the memristor effect of ion transport in 2D slits based on all-atom molecular dynamics (AAMD) simulations and show how it can be applied to artificial neural biosystems. This provided important insights for us to develop artificial neural module devices for information storage and transfer.

Confined fluids often exhibit phenomena different from those in the bulk phase, and these new emerging phenomena usually cannot be explained by traditional continuum fluid dynamics models [274, 275]. Attributable to the recent advances in micro/nano processing technology and the development of materials chemistry, various progresses have been made in the manipulation and visualization of confined nanofluidics. We have revealed peculiar phenomena that are different from bulk, including non-linear ion transport mechanisms, ultra-fast water transport, anomalous dielectric constant and low-temperature phase change of water, quantum effects etc. When in extremely confined structure, there are two factors needs to be paid attention. One is that when the length scale of the confinement is equal to the size of the molecule, the particle effect of the fluids including protons and electrons, needs to be considered; The other is the influence of water. When the confinement space is greatly narrowed, water polarization and ion hydration/dehydration are particularly important, and the coupling between water and ions will affect their transport behavior. However, the understanding of this problem currently is far from deep enough. There are some confusions in the research process of the transport behavior within confinement, which is mainly caused by various reasons. On the one hand, due to the confinement effect, quantum effect and small size effect, traditional mechanical theories cannot accurately describe the new emerging phenomenon of ion and water transport in nanochannels, and it is mainly because that the handling of boundary conditions weakens the decisive contribution of the surface and interface effects. On the other hand, more reasonable, essential, and universal theoretical models and systematic measurement methodology needs to be established to meet challenges in the field of confined nanofluidics. In addition. It can also be seen from all the previous discussions that the microscopic structure and transport of confined fluids are not solely the research category of a certain discipline. For nanofluidics to take a place in this field, it is necessary to combine quantum physics, chemistry, biological physics, material science and other disciplines to conduct cross-cooperative research, to build a deeper understanding of the nature and details of the surface and interface effects, which is crucial for the establishment of new understandings.

References

1. Hibara A, Fukuyama M, Chung M, Priest C, Proskurnin MA (2016) Interfacial phenomena and fluid control in micro/nanofluidics. Anal Sci 32:11–21
2. Han J, Yoon SG, Lee WH, Jin H, Cho YH, Kim YS (2022) Ionic diffusion-driven ionovoltaic transducer for probing ion-molecular interactions at solid-liquid interface. Advanced Science 9:2103038
3. Levin S, Fritzsche J, Nilsson S, Runemark A, Dhokale B, Ström H, Sundén H, Langhammer C, Westerlund F (2019) A nanofluidic device for parallel single nanoparticle catalysis in solution. Nat Commun 10:4426
4. Saha K, Deka J, Hens S, Saikia S, Raidongia K (2018) Chemical reactions under the nanofluidic confinement of reconstructed lamellar membranes. J Mater Chem A 6:22931–22939
5. Liu H, He J, Tang J, Liu H, Pang P, Cao D, Krstic P, Joseph S, Lindsay S, Nuckolls C (2010) Translocation of single-stranded DNA through single-walled carbon nanotubes. Science 327:64–67
6. Yang W, Radha B, Choudhary A, You Y, Mettela G, Geim AK, Aksimentiev A, Keerthi A, Dekker C (2021) Translocation of DNA through ultrathin nanoslits. Adv Mater 33:2007682
7. Feng J, Liu K, Bulushev RD, Khlybov S, Dumcenco D, Kis A, Radenovic A (2015) Identification of single nucleotides in MoS_2 nanopores. Nat Nanotechnol 10:1070–1076
8. van den Berg A, Craighead HG, Yang P (2010) From microfluidic applications to nanofluidic phenomena. Chem Soc Rev 39:899–900
9. Bocquet L (2020) Nanofluidics coming of age. Nat Mater 19:254–256
10. Yun TG, Bae J, Rothschild A, Kim I-D (2019) Transpiration driven electrokinetic power generator. ACS Nano 13:12703–12709
11. Delgado AV, González-Caballero F, Hunter RJ, Koopal LK, Lyklema J (2007) Measurement and interpretation of electrokinetic phenomena. J Colloid Interface Sci 309:194–224
12. Israelachvili JN (2011) Intermolecular and surface forces. Academic press
13. Chen L, Shi G, Shen J, Peng B, Zhang B, Wang Y, Bian F, Wang J, Li D, Qian Z, Xu G, Liu G, Zeng J, Zhang L, Yang Y, Zhou G, Wu M, Jin W, Li J, Fang H (2017) Ion sieving in graphene oxide membranes via cationic control of interlayer spacing. Nature 550:380–383
14. Chen W, Wu Z-Q, Xia X-H, Xu J-J, Chen H-Y (2010) Anomalous diffusion of electrically neutral molecules in charged nanochannels. Angew Chem Int Ed 49:7943–7947
15. Esfandiar A, Radha B, Wang FC, Yang Q, Hu S, Garaj S, Nair RR, Geim AK, Gopinadhan K (2017) Size effect in ion transport through angstrom-scale slits. Science 358:511–513
16. Levis RA, Rae JL (1993) The use of quartz patch pipettes for low noise single channel recording. Biophys J 65:1666–1677
17. Tabard-Cossa V, Trivedi D, Wiggin M, Jetha NN, Marziali A (2007) Noise analysis and reduction in solid-state nanopores. Nanotechnology 18:305505
18. Smeets RMM, Dekker NH, Dekker C (2009) Low-frequency noise in solid-state nanopores. Nanotechnology 20:095501
19. Bezrukov SM, Kasianowicz JJ (1993) Current noise reveals protonation kinetics and number of ionizable sites in an open protein ion channel. Phys Rev Lett 70:2352–2355
20. Balandin AA (2013) Low-frequency 1/f noise in graphene devices. Nat Nanotechnol 8:549–555
21. Hoogerheide DP, Garaj S, Golovchenko JA (2009) Probing surface charge fluctuations with solid-state nanopores. Phys Rev Lett 102:256804
22. Venta K, Shemer G, Puster M, Rodríguez-Manzo JA, Balan A, Rosenstein JK, Shepard K, Drndić M (2013) Differentiation of short, single-stranded DNA homopolymers in solid-state nanopores. ACS Nano 7:4629–4636
23. Jain T, Guerrero RJS, Aguilar CA, Karnik R (2013) Integration of Solid-state nanopores in microfluidic networks via transfer printing of suspended membranes. Anal Chem 85:3871–3878
24. Rosenstein JK, Wanunu M, Merchant CA, Drndic M, Shepard KL (2012) Integrated nanopore sensing platform with sub-microsecond temporal resolution. Nat Methods 9:487–492

25. Chen P, Gu J, Brandin E, Kim Y-R, Wang Q, Branton D (2004) Probing single DNA molecule transport using fabricated nanopores. Nano Lett 4:2293–2298
26. Steinbock LJ, Bulushev RD, Krishnan S, Raillon C, Radenovic A (2013) DNA translocation through low-noise glass nanopores. ACS Nano 7:11255–11262
27. Li W, Bell NAW, Hernández-Ainsa S, Thacker VV, Thackray AM, Bujdoso R, Keyser UF (2013) Single protein molecule detection by glass nanopores. ACS Nano 7:4129–4134
28. Green Y, Eshel R, Park S, Yossifon G (2016) Interplay between nanochannel and microchannel resistances. Nano Lett 16:2744–2748
29. Green Y, Abu-Rjal R, Eshel R (2020) Electrical resistance of nanochannel-microchannel systems: an exact solution. Phys Rev Appl 14:014075
30. Smeets RMM, Keyser UF, Krapf D, Wu M-Y, Dekker NH, Dekker C (2006) Salt dependence of ion transport and DNA translocation through solid-state nanopores. Nano Lett 6:89–95
31. Hille B (1968) Pharmacological modifications of the sodium channels of frog nerve. J Gen Physiol 51:199–219
32. Hall JE (1975) Access resistance of a small circular pore. J Gen Physiol 66:531–532
33. Kowalczyk SW, Grosberg AY, Rabin Y, Dekker C (2011) Modeling the conductance and DNA blockade of solid-state nanopores. Nanotechnology 22:315101
34. Geim AK, Novoselov KS (2007) The rise of graphene. Nat Mater 6:183–191
35. Geim AK (2009) Graphene: status and prospects. Science 324:1530
36. Sun P, Wang K, Zhu H (2016) Recent developments in graphene-based membranes: structure, mass-transport mechanism and potential applications. Adv Mater 28:2287–2310
37. Siria A, Poncharal P, Biance A-L, Fulcrand R, Blase X, Purcell ST, Bocquet L (2013) Giant osmotic energy conversion measured in a single transmembrane boron nitride nanotube. Nature 494:455–458
38. Feng J, Graf M, Liu K, Ovchinnikov D, Dumcenco D, Heiranian M, Nandigana V, Aluru NR, Kis A, Radenovic A (2016) Single-layer MoS_2 nanopores as nanopower generators. Nat 536:197–200
39. Gogotsi Y, Anasori B (2019) The rise of MXenes. ACS Nano 13:8491–8494
40. Li J, Stein D, McMullan C, Branton D, Aziz MJ, Golovchenko JA (2001) Ion-beam sculpting at nanometre length scales. Nature 412:166–169
41. Hsia KJ, Huang Y, Menard E, Park JU, Zhou W, Rogers J, Fulton JM (2005) Collapse of stamps for soft lithography due to interfacial adhesion. Appl Phys Lett 86:154106
42. Venugopal G, Jung M-H, Suemitsu M, Kim S-J (2011) Fabrication of nanoscale three-dimensional graphite stacked-junctions by focused-ion-beam and observation of anomalous transport characteristics. Carbon 49:2766–2772
43. Graf M, Lihter M, Thakur M, Georgiou V, Topolancik J, Ilic BR, Liu K, Feng J, Astier Y, Radenovic A (2019) Fabrication and practical applications of molybdenum disulfide nanopores. Nat Protoc 14:1130–1168
44. Gopinadhan K, Hu S, Esfandiar A, Lozada-Hidalgo M, Wang FC, Yang Q, Tyurnina AV, Keerthi A, Radha B, Geim AK (2019) Complete steric exclusion of ions and proton transport through confined monolayer water. Science 363:145–148
45. Jiang X, Zhao C, Noh Y, Xu Y, Chen Y, Chen F, Ma L, Ren W, Aluru NR, Feng J (2022) Nonlinear electrohydrodynamic ion transport in graphene nanopores. Sci Adv 8:eabj2510
46. Cambré S, Schoeters B, Luyckx S, Goovaerts E, Wenseleers W (2020) Experimental observation of single-file water filling of thin single-wall carbon nanotubes down to chiral index (5,3). Phys Rev Lett 104:207401
47. Marcotte A, Mouterde T, Niguès A, Siria A, Bocquet L (2020) Mechanically activated ionic transport across single-digit carbon nanotubes. Nat Mater 19:1057–1061
48. Comtet J, Grosjean B, Glushkov E, Avsar A, Watanabe K, Taniguchi T, Vuilleumier R, Bocquet M-L, Radenovic A (2020) Direct observation of water-mediated single-proton transport between hBN surface defects. Nat Nanotechnol 15:598–604
49. Thiruraman JP, Masih Das P, Drndić M (2020) Ions and water dancing through atom-scale holes: a perspective toward "Size Zero". ACS Nano 14:3736–3746

50. Feng J, Liu K, Graf M, Dumcenco D, Kis A, Di Ventra M, Radenovic A (2016) Observation of ionic Coulomb blockade in nanopores. Nat Mater 15:850–855
51. Macha M, Marion S, Nandigana VVR, Radenovic A (2019) 2D materials as an emerging platform for nanopore-based power generation. Nat Rev Mater 4:588–605
52. Kavokine N, Netz RR, Bocquet L (2021) Fluids at the nanoscale: from continuum to subcontinuum transport. Ann Rev Fluid Mech 53:377–410
53. Sahu S, Zwolak M (2019) Colloquium: ionic phenomena in nanoscale pores through 2D materials. Rev Mod Phys 91:021004
54. Tunuguntla RH, Henley RY, Yao YC, Pham TA, Wanunu M, Noy A (2017) Enhanced water permeability and tunable ion selectivity in subnanometer carbon nanotube porins. Science 357:792–796
55. Feng J, Liu K, Graf M, Lihter M, Bulushev RD, Dumcenco D, Alexander DT, Krasnozhon D, Vuletic T, Kis A, Radenovic A (2015) Electrochemical reaction in single layer MoS2: nanopores opened atom by atom. Nano Lett 15:3431–3438
56. Graf M, Lihter M, Unuchek D, Sarathy A, Leburton J-P, Kis A, Radenovic A (2019) Light-enhanced blue energy generation using MoS_2 nanopores. Joule 3:1549–1564
57. Guo S, Meshot ER, Kuykendall T, Cabrini S, Fornasiero F (2015) Nanofluidic transport through isolated carbon nanotube channels: advances, controversies, and challenges. Adv Mater 27:5726–5737
58. Mor GK, Shankar K, Paulose M, Varghese OK, Grimes CA (2005) Enhanced photocleavage of water using titania nanotube arrays. Nano Lett 5:191–195
59. Suetake J, Nosaka AY, Hodouchi K, Matsubara H, Nosaka Y (2008) Characteristics of titanate nanotube and the states of the confined sodium ions. J Phys Chem C 112:18474–18482
60. Daiguji H, Yang P, Majumdar A (2004) Ion transport in nanofluidic channels. Nano Lett 4:137–142
61. Hayden SC, Allam NK, El-Sayed MA (2010) TiO_2 nanotube/CdS hybrid electrodes: extraordinary enhancement in the inactivation of *Escherichia coli*. J Am Chem Soc 132:14406–14408
62. Ghosh S, Sood AK, Kumar N (2003) Carbon nanotube flow sensors. Science 299:1042–1044
63. Whitby M, Quirke N (2007) Fluid flow in carbon nanotubes and nanopipes. Nat Nanotechnol 2:87–94
64. Li H, Song Z, Zhang X, Huang Y, Li S, Mao Y, Ploehn Harry J, Bao Y, Yu M (2013) Ultrathin, molecular-sieving graphene oxide membranes for selective hydrogen separation. Science 342:95–98
65. Joshi RK, Carbone P, Wang FC, Kravets VG, Su Y, Grigorieva IV, Wu HA, Geim AK, Nair RR (2014) Precise and ultrafast molecular sieving through graphene oxide membranes. Science 343:752–754
66. Novoselov KS, Mishchenko A, Carvalho A, Castro Neto AH (2016) 2D materials and van der Waals heterostructures. Science 353:aac9439
67. Abraham J, Vasu KS, Williams CD, Gopinadhan K, Su Y, Cherian CT, Dix J, Prestat E, Haigh SJ, Grigorieva IV, Carbone P, Geim AK, Nair RR (2017) Tunable sieving of ions using graphene oxide membranes. Nat Nanotechnol 12:546–550
68. Xue Y, Xia Y, Yang S, Alsaid Y, Fong King Y, Wang Y, Zhang X (2021) Atomic-scale ion transistor with ultrahigh diffusivity. Science 372:501–503
69. Wang X, Xia F (2015) Van der Waals heterostructures: stacked 2D materials shed light. Nat Mater 14:264–265
70. Zhao F, Liang Y, Cheng H, Jiang L, Qu L (2016) Highly efficient moisture-enabled electricity generation from graphene oxide frameworks. Energy Environ Sci 9:912–916
71. Menni Y, Chamkha AJ, Azzi A (2019) Nanofluid transport in porous media: a review. Special Topics Rev Porous Media Int J 10
72. Cheng H, Lijie L, Wang B, Feng X, Mao Z, Vancso GJ, Sui X (2020) Multifaceted applications of cellulosic porous materials in environment, energy, and health. Prog Polym Sci 106:101253
73. Cheng C, Jiang G, Simon GP, Liu JZ, Li D (2018) Low-voltage electrostatic modulation of ion diffusion through layered graphene-based nanoporous membranes. Nat Nanotechnol 13:685–690

74. Yin K, Huang S, Chen X, Wang X, Kong J, Chen Y, Xue J (2018) Generating sub-nanometer pores in single-layer MoS_2 by heavy-ion bombardment for gas separation: a theoretical perspective. ACS Appl Mater Interfaces 10:28909–28917
75. Gierak J, Madouri A, Biance AL, Bourhis E, Patriarche G, Ulysse C, Lucot D, Lafosse X, Auvray L, Bruchhaus L, Jede R (2007) Sub-5 nm FIB direct patterning of nanodevices. Microelectron Eng 84:779–783
76. Fischbein MD, Drndić M (2008) Electron beam nanosculpting of suspended graphene sheets. Appl Phys Lett 93:113107
77. Whitney AV, Myers BD, Van Duyne RP (2004) Sub-100 nm triangular nanopores fabricated with the reactive ion etching variant of nanosphere lithography and angle-resolved nanosphere lithography. Nano Lett 4:1507–1511
78. Hu L, Hecht DS, Grüner G (2010) Carbon nanotube thin films: fabrication, properties, and applications. Chem Rev 110:5790–5844
79. Secchi E, Marbach S, Niguès A, Stein D, Siria A, Bocquet L (2016) Massive radius-dependent flow slippage in carbon nanotubes. Nature 537:210–213
80. Rabinowitz J, Cohen C, Shepard KL (2020) An electrically actuated, carbon-nanotube-based biomimetic ion pump. Nano Lett 20:1148–1153
81. Radha B, Esfandiar A, Wang FC, Rooney AP, Gopinadhan K, Keerthi A, Mishchenko A, Janardanan A, Blake P, Fumagalli L, Lozada-Hidalgo M, Garaj S, Haigh SJ, Grigorieva IV, Wu HA, Geim AK (2016) Molecular transport through capillaries made with atomic-scale precision. Nature 538:222–225
82. Cervera J, Ramirez P, Mafe S, Stroeve P (2011) Asymmetric nanopore rectification for ion pumping, electrical power generation, and information processing applications. Electrochim Acta 56:4504–4511
83. Hou X, Zhang H, Jiang L (2012) Building bio-inspired artificial functional nanochannels: from symmetric to asymmetric modification. Angew Chem Int Ed 51:5296–5307
84. Cervera J, Schiedt B, Neumann R, Mafé S, Ramírez P (2006) Ionic conduction, rectification, and selectivity in single conical nanopores. J Chem Phys 124:104706
85. Hou X, Dong H, Zhu D, Jiang L (2010) Fabrication of stable single nanochannels with controllable ionic rectification. Small 6:361–365
86. Kalman EB, Sudre O, Vlassiouk I, Siwy ZS (2009) Control of ionic transport through gated single conical nanopores. Anal Bioanal Chem 394:413–419
87. Hou X, Yang F, Li L, Song Y, Jiang L, Zhu D (2010) A biomimetic asymmetric responsive single nanochannel. J Am Chem Soc 132:11736–11742
88. Ali M, Yameen B, Cervera J, Ramírez P, Neumann R, Ensinger W, Knoll W, Azzaroni O (2010) Layer-by-layer assembly of polyelectrolytes into ionic current rectifying solid-state nanopores: insights from theory and experiment. J Am Chem Soc 132:8338–8348
89. Hummers WS, Offeman RE (1958) Preparation of graphitic oxide. J Am Chem Soc 80:1339–1339
90. Wang J, Fang R, Hou J, Zhang H, Tian Y, Wang H, Jiang L (2017) Oscillatory reaction induced periodic C-quadruplex DNA gating of artificial ion channels. ACS Nano 11:3022–3029
91. Geim AK (2021) Exploring two-dimensional empty space. Nano Lett 21:6356–6358
92. Menni Y, Chamkha AJ, Azzi A (2019) Nanofluid transport in porous media: a review 10:49–64
93. Zhang L, Jaroniec M (2020) Strategies for development of nanoporous materials with 2D building units. Chem Soc Rev 49:6039–6055
94. Andreeva DV, Trushin M, Nikitina A, Costa MCF, Cherepanov PV, Holwill M, Chen S, Yang K, Chee SW, Mirsaidov U, Castro Neto AH, Novoselov KS (2021) Two-dimensional adaptive membranes with programmable water and ionic channels. Nat Nanotechnol 16:174–180
95. Jia P, Wang L, Zhang Y, Yang Y, Jin X, Zhou M, Quan D, Jia M, Cao L, Long R, Jiang L, Guo W (2021) Harnessing ionic power from equilibrium electrolyte solution via photoinduced active ion transport through van-der-waals-like heterostructures. Adv Mater 33:2007529
96. Liu X, Gao H, Ward JE, Liu X, Yin B, Fu T, Chen J, Lovley DR, Yao J (2020) Power generation from ambient humidity using protein nanowires. Nature 578:550–554

97. Branton D, Deamer DW, Marziali A, Bayley H, Benner SA, Butler T, Di Ventra M, Garaj S, Hibbs A, Huang X, Jovanovich SB, Krstic PS, Lindsay S, Ling XS, Mastrangelo CH, Meller A, Oliver JS, Pershin YV, Ramsey JM, Riehn R, Soni GV, Tabard-Cossa V, Wanunu M, Wiggin M, Schloss JA (2008) The potential and challenges of nanopore sequencing. Nat Biotechnol 26:1146–1153
98. Kasianowicz JJ, Brandin E, Branton D, Deamer DW (1996) Characterization of individual polynucleotide molecules using a membrane channel. Proc Natl Acad Sci USA 93:13770–13773
99. Chen F, Gu Z, Zhao C, Chen Y, Jiang X, He Z, Lu Y, Zhou R, Feng J (2021) Ionic conductance oscillations in sub-nanometer pores probed by optoelectronic control. Matter 4:2378–2391
100. Keyser UF (2016) Enhancing nanopore sensing with DNA nanotechnology. Nat Nanotechnol 11:106–108
101. Xiao K, Chen L, Chen R, Heil T, Lemus SDC, Fan F, Wen L, Jiang L, Antonietti M (2019) Artificial light-driven ion pump for photoelectric energy conversion. Nat Commun 10:74
102. Yang J, Hu X, Kong X, Jia P, Ji D, Quan D, Wang L, Wen Q, Lu D, Wu J, Jiang L, Guo W (2019) Photo-induced ultrafast active ion transport through graphene oxide membranes. Nat Commun 10:1171
103. Yang J, Lu Y, Jin L, Zhao C, Chen Y, Xu Y, Chen F, Feng J (2022) Dynamic optical visualization of proton transport pathways at water-solid interfaces. Angew Chem Int Ed 61:e202112150
104. You Y, Ismail A, Nam G-H, Goutham S, Keerthi A, Radha B (2022) Angstrofluidics: walking to the limit. arXiv preprint arXiv:2203.13027
105. Fumagalli L, Esfandiar A, Fabregas R, Hu S, Ares P, Janardanan A, Yang Q, Radha B, Taniguchi T, Watanabe K, Gomila G, Novoselov KS, Geim AK (2018) Anomalously low dielectric constant of confined water. Science 360:1339–1342
106. Naguib N, Ye H, Gogotsi Y, Yazicioglu AG, Megaridis CM, Yoshimura M (2004) Observation of water confined in nanometer channels of closed carbon nanotubes. Nano Lett 4:2237–2243
107. Zhou KG, Vasu KS, Cherian CT, Neek-Amal M, Zhang JC, Ghorbanfekr-Kalashami H, Huang K, Marshall OP, Kravets VG, Abraham J, Su Y, Grigorenko AN, Pratt A, Geim AK, Peeters FM, Novoselov KS, Nair RR (2018) Electrically controlled water permeation through graphene oxide membranes. Nature 559:236–240
108. Thiruraman Jothi P, Dar Sidra A, Masih Das P, Hassani N, Neek-Amal M, Keerthi A, Drndić M, Radha B (2020) Gas flow through atomic-scale apertures. Sci Adv 6:eabc7927
109. Agrawal KV, Shimizu S, Drahushuk LW, Kilcoyne D, Strano MS (2017) Observation of extreme phase transition temperatures of water confined inside isolated carbon nanotubes. Nat Nanotechnol 12:267–273
110. Ferry D, Goodnick SM (1999) Transport in nanostructures. Cambridge University Press
111. Grabert H, Devoret MH (2013) Single charge tunneling: Coulomb blockade phenomena in nanostructures, vol 294. Springer Science & Business Media
112. Zwolak M, Lagerqvist J, Di Ventra M (2009) Quantized ionic conductance in nanopores. Phys Rev Lett 103:128102
113. Zwolak M, Wilson J, Di Ventra M (2010) Dehydration and ionic conductance quantization in nanopores. J Phys Condens Matter 22:454126
114. Krems M, Di Ventra M (2013) Ionic Coulomb blockade in nanopores. J Phys Condensed Matter 25:065101–065101
115. Kaufman IK, McClintock PVE, Eisenberg RS (2015) Coulomb blockade model of permeation and selectivity in biological ion channels. New J Phys 17:083021
116. Tanaka H, Iizuka H, Pershin YV, Ventra MD (2017) Surface effects on ionic Coulomb blockade in nanometer-size pores. Nanotechnology 29:025703
117. Sahu S, Elenewski J, Rohmann C, Zwolak M (2019) Optimal transport and colossal ionic mechano-conductance in graphene crown ethers. Sci Adv 5:eaaw5478
118. Kavokine N, Marbach S, Siria A, Bocquet L (2019) Ionic Coulomb blockade as a fractional Wien effect. Nat Nanotechnol 14:573–578
119. Kaufman IK, Fedorenko OA, Luchinsky DG, Gibby WAT, Roberts SK, McClintock PVE, Eisenberg RS (2017) Ionic Coulomb blockade and anomalous mole fraction effect in the

NaChBac bacterial ion channel and its charge-varied mutants. EPJ Nonlinear Biomed Phys 5:4
120. Ivanov AP, Instuli E, McGilvery CM, Baldwin G, McComb DW, Albrecht T, Edel JB (2011) DNA tunneling detector embedded in a nanopore. Nano Lett 11:279–285
121. Taniguchi M, Tsutsui M, Yokota K, Kawai T (2009) Fabrication of the gating nanopore device. Appl Phys Lett 95:123701
122. Neher E, Sakmann B (1976) Single-channel currents recorded from membrane of denervated frog muscle fibres. Nature 260:799–802
123. Chu LY, Li Y, Zhu JH, Chen WMJAC (2005) Negatively thermoresponsive membranes with functional gates driven by zipper-type hydrogen-bonding interactions 117:2162–2165 (2005)
124. Xia F, Guo W, Mao Y, Hou X, Xue J, Xia H, Wang L, Song Y, Ji H, Ouyang Q, Wang Y, Jiang L (2008) Gating of single synthetic nanopores by proton-driven DNA molecular motors. J Am Chem Soc 130:8345–8350
125. Zhai G, Toh SC, Tan WL, Kang ET, Neoh KG, Huang CC, Liaw DJ (2003) Poly(vinylidene fluoride) with grafted zwitterionic polymer side chains for electrolyte-responsive microfiltration membranes. Langmuir 19:7030–7037
126. Alexander S, Mathie A, Peters J (2011) Ligand-gated ion channels. Br J Pharmacol 164:S115–S135
127. Perozo E, Cortes DM, Sompornpisut P, Kloda A, Martinac B (2002) Open channel structure of MscL and the gating mechanism of mechanosensitive channels. Nature 418:942–948
128. Zhou Y, Pearson JE, Auerbach A (2005) Φ-value analysis of a linear, sequential reaction mechanism: theory and application to ion channel gating. Biophys J 89:3680–3685
129. Goychuk I, Hänggi P (2002) Ion channel gating: a first-passage time analysis of the Kramers type. Proc Natl Acad Sci 99:3552
130. Scherer POJ, Fischer SF (2017) In: Scherer POJ, Fischer SF (eds) Theoretical molecular biophysics. Springer Berlin Heidelberg, pp 113–118
131. Jain T, Rasera BC, Guerrero RJS, Boutilier MSH, O'Hern SC, Idrobo J-C, Karnik R (2015) Heterogeneous sub-continuum ionic transport in statistically isolated graphene nanopores. Nat Nanotechnol 10:1053–1057
132. Powell MR, Cleary L, Davenport M, Shea KJ, Siwy ZS (2011) Electric-field-induced wetting and dewetting in single hydrophobic nanopores. Nat Nanotechnol 6:798–802
133. Powell MR, Sullivan M, Vlassiouk I, Constantin D, Sudre O, Martens CC, Eisenberg RS, Siwy ZS (2008) Nanoprecipitation-assisted ion current oscillations. Nat Nanotechnol 3:51–57
134. Beckstein O, Biggin PC, Sansom MSP (2001) A hydrophobic gating mechanism for nanopores. J Phys Chem B 105:12902–12905
135. Beckstein O, Sansom MSP (2003) Liquid-vapor oscillations of water in hydrophobic nanopores. Proc Natl Acad Sci USA 100:7063–7068
136. Smirnov SN, Vlassiouk IV, Lavrik NV (2011) Voltage-gated hydrophobic nanopores. ACS Nano 5:7453–7461
137. Polster JW, Acar ET, Aydin F, Zhan C, Pham TA, Siwy ZS (2020) Gating of hydrophobic nanopores with large anions. ACS Nano 14:4306–4315
138. Zhao Q, Wu K, Geng J, Chi S, Wang Y, Zhi P, Zhang M, Xiao B (2016) Ion permeation and mechanotransduction mechanisms of mechanosensitive piezo channels. Neuron 89:1248–1263
139. Lan W-J, Holden DA, White HS (2011) Pressure-dependent ion current rectification in conical-shaped glass nanopores. J Am Chem Soc 133:13300–13303
140. Mouterde T, Keerthi A, Poggioli AR, Dar SA, Siria A, Geim AK, Bocquet L, Radha B (2019) Molecular streaming and its voltage control in ångström-scale channels. Nature 567:87–90
141. Wei C, Bard AJ, Feldberg SW (1997) Current Rectification at Quartz Nanopipet Electrodes. Anal Chem 69:4627–4633
142. Siwy Z, Fuliński A (2002) Fabrication of a synthetic nanopore ion pump. Phys Rev Lett 89:198103
143. Woermann D (2003) Electrochemical transport properties of a cone-shaped nanopore: high and low electrical conductivity states depending on the sign of an applied electrical potential difference. Phys Chem Chem Phys 5:1853–1858

144. Kovarik ML, Zhou K, Jacobson SC (2009) Effect of conical nanopore diameter on ion current rectification. J Phys Chem B 113:15960–15966
145. Duleba D, Dutta P, Denuga S, Johnson RP (2022) Effect of electrolyte concentration and pore size on ion current rectification inversion. ACS Meas Sci Au (2022)
146. Liu Q, Wang Y, Guo W, Ji H, Xue J, Ouyang Q (2007) Asymmetric properties of ion transport in a charged conical nanopore. Phys Rev E 75:051201
147. Ramírez P, Apel PY, Cervera J, Mafé S (2008) Pore structure and function of synthetic nanopores with fixed charges: tip shape and rectification properties. Nanotechnology 19:315707
148. Apel PY, Blonskaya IV, Orelovitch OL, Ramirez P, Sartowska BA (2011) Effect of nanopore geometry on ion current rectification. Nanotechnology 22:175302
149. Zhang H, Hou X, Yang Z, Yan D, Li L, Tian Y, Wang H, Jiang L (2015) Bio-inspired smart single asymmetric hourglass nanochannels for continuous shape and ion transport control. Small 11:786–791
150. Siwy Z, Heins E, Harrell CC, Kohli P, Martin CR (2004) Conical-nanotube ion-current rectifiers: the role of surface charge. J Am Chem Soc 126:10850–10851
151. Yameen B, Ali M, Neumann R, Ensinger W, Knoll W, Azzaroni O (2009) Single conical nanopores displaying pH-tunable rectifying characteristics. manipulating ionic transport with Zwitterionic polymer brushes. J Am Chem Soc 131:2070–2071
152. White HS, Bund A (2008) Ion current rectification at nanopores in glass membranes. Langmuir 24:2212–2218
153. Yusko EC, An R, Mayer M (2010) Electroosmotic flow can generate ion current rectification in nano- and micropores. ACS Nano 4:477–487
154. Perera RT, Johnson RP, Edwards MA, White HS (2015) Effect of the electric double layer on the activation energy of ion transport in conical nanopores. J Phys Chem C 119:24299–24306
155. Guerrette JP, Zhang B (2010) Scan-rate-dependent current rectification of cone-shaped silica nanopores in quartz nanopipettes. J Am Chem Soc 132:17088–17091
156. Jung J-Y, Joshi P, Petrossian L, Thornton TJ, Posner JD (2009) Electromigration current rectification in a cylindrical nanopore due to asymmetric concentration polarization. Anal Chem 81:3128–3133
157. Lin C-Y, Yeh L-H, Siwy ZS (2018) Voltage-induced modulation of ionic concentrations and ion current rectification in mesopores with highly charged pore walls. J Phys Chem Lett 9:393–398
158. Kalman EB, Vlassiouk I, Siwy ZS (2008) Nanofluidic bipolar transistors. Adv Mater 20:293–297
159. Vlassiouk I, Smirnov S, Siwy Z (2008) Ionic selectivity of single nanochannels. Nano Lett 8:1978–1985
160. Vlassiouk I, Siwy ZS (2007) Nanofluidic diode. Nano Lett 7:552–556
161. Karnik R, Duan C, Castelino K, Daiguji H, Majumdar A (2007) Rectification of ionic current in a nanofluidic diode. Nano Lett 7:547–551
162. Yan R, Liang W, Fan R, Yang P (2009) Nanofluidic diodes based on nanotube heterojunctions. Nano Lett 9:3820–3825
163. Nie J, Ren Z, Xu L, Lin S, Zhan F, Chen X, Wang ZL (2020) Probing contact-electrification-induced electron and ion transfers at a liquid–solid interface. Adv Mater 32:1905696
164. Snapp P, Kim JM, Cho C, Leem J, Haque MF, Nam S (2020) Interaction of 2D materials with liquids: wettability, electrochemical properties, friction, and emerging directions. NPG Asia Mater 12:22
165. Yin J, Zhou J, Fang S, Guo W (2020) Hydrovoltaic energy on the way. Joule 4:1852–1855
166. van der Heyden FH, Stein D, Dekker C (2005) Streaming currents in a single nanofluidic channel. Phys Rev Lett 95:116104
167. Yin J, Zhang Z, Li X, Yu J, Zhou J, Chen Y, Guo W (2014) Waving potential in graphene. Nat Commun 5:3582
168. Huang Y, Cheng H, Shi G, Qu L (2017) Highly efficient moisture-triggered nanogenerator based on graphene quantum dots. ACS Appl Mater Interfaces 9:38170–38175

169. Yin J, Li X, Yu J, Zhang Z, Zhou J, Guo W (2014) Generating electricity by moving a droplet of ionic liquid along graphene. Nat Nanotechnol 9:378–383
170. Zhang N, Gu H, Zheng H, Ye S, Kang L, Huang C, Lu K, Xu W, Miao Q, Wang Z, Zhang J, Zhou X (2020) Boosting the output performance of volume effect electricity generator (VEEG) with water column. Nano Energy 73:104748
171. Ho Lee S, Jung Y, Kim S, Han C-S (2013) Flow-induced voltage generation in non-ionic liquids over monolayer graphene. Appl Phys Lett 102:063116
172. Kral P, Shapiro M (2001) Nanotube electron drag in flowing liquids. Phys Rev Lett 86:131–134
173. Zhang Z, Li X, Yin J, Xu Y, Fei W, Xue M, Wang Q, Zhou J, Guo W (2018) Emerging hydrovoltaic technology. Nat Nanotechnol 13:1109–1119
174. Kim SH, Haines CS, Li N, Kim KJ, Mun TJ, Choi C, Di J, Oh YJ, Oviedo JP, Bykova J, Fang S, Jiang N, Liu Z, Wang R, Kumar P, Qiao R, Priya S, Cho K, Kim M, Lucas MS, Drummy LF, Maruyama B, Lee DY, Lepro X, Gao E, Albarq D, Ovalle-Robles R, Kim SJ, Baughman RH (2017) Harvesting electrical energy from carbon nanotube yarn twist. Science 357:773–778
175. Kavokine N, Bocquet M-L, Bocquet L (2022) Fluctuation-induced quantum friction in nanoscale water flows. Nature 602:84–90
176. Ahmadabadi I, Esfandiar A, Hassanali A, Ejtehadi MR (2021) Structural and dynamical fingerprints of the anomalous dielectric properties of water under confinement. Phys Rev Mater 5:024008
177. Zhang C, Gygi F, Galli G (2013) Strongly anisotropic dielectric relaxation of water at the nanoscale. J Phys Chem Lett 4:2477–2481
178. Shekarforoush S, Jalali H, Yagmurcukardes M, Milošević MV, Neek-Amal M (2020) Optoelectronic properties of confined water in angstrom-scale slits. Phys Rev B 102:235406
179. Hamid I, Jalali H, Peeters FM, Neek-Amal M (2021) Abnormal in-plane permittivity and ferroelectricity of confined water: from sub-nanometer channels to bulk. J Chem Phys 154:114503
180. Zhu H, Hu B, Hu H, Zhu Y, He W, Huang J, Li G (2020) Influence of the oxidation rate on the dielectric properties of water confined in graphene bilayers. J Mol Liq 317:113995
181. Renou R, Szymczyk A, Ghoufi A (2015) Tunable dielectric constant of water at the nanoscale. Phys Rev E 91:032411
182. Bonthuis DJ, Gekle S, Netz RR (2011) Dielectric profile of interfacial water and its effect on double-layer capacitance. Phys Rev Lett 107:166102
183. Motevaselian MH, Aluru NR (2020) Universal reduction in dielectric response of confined fluids. ACS Nano 14:12761–12770
184. Gorshunov BP, Torgashev VI, Zhukova ES, Thomas VG, Belyanchikov MA, Kadlec C, Kadlec F, Savinov M, Ostapchuk T, Petzelt J, Prokleška J, Tomas PV, Pestrjakov EV, Fursenko DA, Shakurov GS, Prokhorov AS, Gorelik VS, Kadyrov LS, Uskov VV, Kremer RK, Dressel M (2016) Incipient ferroelectricity of water molecules confined to nano-channels of beryl. Nat Commun 7:12842
185. Yao Y-C, Taqieddin A, Alibakhshi MA, Wanunu M, Aluru NR, Noy A (2019) Strong electroosmotic coupling dominates ion conductance of 1.5 nm diameter carbon nanotube porins. ACS Nano 13:12851–12859
186. Shen D, Xiao M, Zou G, Liu L, Duley WW, Zhou YN (2018) Self-powered wearable electronics based on moisture enabled electricity generation. Adv Mater 30:1705925
187. Narozhny BN, Levchenko A (2016) Coulomb drag. Rev Mod Phys 88:025003
188. Yamamoto M, Stopa M, Tokura Y, Hirayama Y, Tarucha S (2006) Negative Coulomb drag in a one-dimensional wire. Science 313:204–207
189. Li JIA, Taniguchi T, Watanabe K, Hone J, Levchenko A, Dean CR (2016) Negative Coulomb drag in double bilayer graphene. Phys Rev Lett 117:046802
190. Song JCW, Abanin DA, Levitov LS (2013) Coulomb drag mechanisms in graphene. Nano Lett 13:3631–3637
191. Schütt M, Ostrovsky PM, Titov M, Gornyi IV, Narozhny BN, Mirlin AD (2013) Coulomb drag in graphene near the dirac point. Phys Rev Lett 110:026601

192. Hwang EH, Sensarma R, Das Sarma S (2011) Coulomb drag in monolayer and bilayer graphene. Phys Rev B 84:245441
193. Zhong H, Li X, Wu Z, Zhang S, Xu Z, Chen H, Lin S (2015) Two dimensional graphene nanogenerator by Coulomb dragging: moving van der Waals heterostructure. Appl Phys Lett 106:243903
194. Král P, Shapiro M (2001) Nanotube electron drag in flowing liquids. Phys Rev Lett 86:131–134
195. Persson BNJ, Tartaglino U, Tosatti E, Ueba H (2004) Electronic friction and liquid-flow-induced voltage in nanotubes. Phys Rev B 69:235410
196. Zhao Y, Song L, Deng K, Liu Z, Zhang Z, Yang Y, Wang C, Yang H, Jin A, Luo Q, Gu C, Xie S, Sun L (2008) Individual water-filled single-walled carbon nanotubes as hydroelectric power converters. Adv Mater 20:1772–1776
197. Xu B, Chen X (2013) Liquid flow-induced energy harvesting in carbon nanotubes: a molecular dynamics study. Phys Chem Chem Phys 15:1164–1168
198. Dhiman P, Yavari F, Mi X, Gullapalli H, Shi Y, Ajayan PM, Koratkar N (2011) Harvesting energy from water flow over graphene. Nano Lett 11:3123–3127
199. Yin J, Zhang Z, Li X, Zhou J, Guo W (2012) Harvesting energy from water flow over graphene? Nano Lett 12:1736–1741
200. Boamah MD, Lozier EH, Kim J, Ohno PE, Walker CE, Miller TF 3rd, Geiger FM (2019) Energy conversion via metal nanolayers. Proc Natl Acad Sci USA 116:16210–16215
201. Senapati S, Chandra A (2001) Dielectric constant of water confined in a nanocavity. J Phys Chem B 105:5106–5109
202. Schlaich A, Knapp EW, Netz RR (2016) Water dielectric effects in planar confinement. Phys Rev Lett 117:048001
203. Qi W, Zhao H (2015) Hydrogen bond network in the hydration layer of the water confined in nanotubes increasing the dielectric constant parallel along the nanotube axis. J Chem Phys 143:114708
204. Zhu H, Yang F, Zhu Y, Li A, He W, Huang J, Li G (2020) Investigation of dielectric constants of water in a nano-confined pore. RSC Adv 10:8628–8635
205. Zhang L, Li X, Zhong Y, Leroy A, Xu Z, Zhao L, Wang EN (2022) Highly efficient and salt rejecting solar evaporation via a wick-free confined water layer. Nat Commun 13:849
206. Song W, Kumar M (2019) Artificial water channels: toward and beyond desalination. Curr Opin Chem Eng 25:9–17
207. Itoh H, Sakuma H (2015) Dielectric constant of water as a function of separation in a slab geometry: a molecular dynamics study. J Chem Phys 142:184703
208. Zhu H, Hu B, Hu H, He W, Huang J, Li G (2021) Tunable dielectric constant of water confined in graphene oxide nanochannels. J Mol Liq 324:115139
209. Renou R, Szymczyk A, Ghoufi A (2014) Water confinement in nanoporous silica materials. J Chem Phys 140:044704
210. Barthel JM, Krienke H, Kunz W (1998) Physical chemistry of electrolyte solutions: modern aspects, vol. 5. Springer Science & Business Media
211. Chandra A (2000) Static dielectric constant of aqueous electrolyte solutions: is there any dynamic contribution? J Chem Phys 113:903–905
212. Jalali H, Khoeini F, Peeters FM, Neek-Amal M (2021) Hydration effects and negative dielectric constant of nano-confined water between cation intercalated MXenes. Nanoscale 13:922–929
213. Sugahara A, Ando Y, Kajiyama S, Yazawa K, Gotoh K, Otani M, Okubo M, Yamada A (2019) Negative dielectric constant of water confined in nanosheets. Nat Commun 10:850
214. Olivieri J-F, Hynes JT, Laage D (2021) Confined water's dielectric constant reduction is due to the surrounding low dielectric media and not to interfacial molecular ordering. J Phys Chem Lett 12:4319–4326
215. Toney MF, Howard JN, Richer J, Borges GL, Gordon JG, Melroy OR, Wiesler DG, Yee D, Sorensen LB (1994) Voltage-dependent ordering of water molecules at an electrode–electrolyte interface. Nature 368:444–446

216. Zhu DM, Wu X, Schelly ZA (1992) Investigation of the micropolarities in reverse micelles of Triton X-100 in mixed solvents of benzene and n-hexane. J Phys Chem 96:7121–7126
217. Riter RE, Willard DM, Levinger NE (1998) Water immobilization at surfactant interfaces in reverse micelles. J Phys Chem B 102:2705–2714
218. Quist P-O, Halle B (1988) Water dynamics and aggregate structure in reversed micelles at sub-zero temperatures. A deuteron spin relaxation study. J Chem Soc Faraday Trans 84:1033–1046
219. Fumagalli L, Esteban-Ferrer D, Cuervo A, Carrascosa JL, Gomila G (2012) Label-free identification of single dielectric nanoparticles and viruses with ultraweak polarization forces. Nat Mater 11:808–816
220. Fumagalli L, Ferrari G, Sampietro M, Gomila G (2009) Quantitative nanoscale dielectric microscopy of single-layer supported biomembranes. Nano Lett 9:1604–1608
221. Zhao M, Gu X, Lowther SE, Park C, Jean YC, Nguyen T (2010) Subsurface characterization of carbon nanotubes in polymer composites via quantitative electric force microscopy. Nanotechnology 21:225702
222. Petrenko VF, Whitworth RW (1999) Physics of ice. OUP Oxford
223. Hill NE, Vaughan WE, Price AH, Davies M (1969) Dielectric properties and molecular behaviour. Van Nostrand Reinhold
224. Hill NE (1963) Interpretation of the dielectric properties of water. Trans Faraday Soc 59:344–346
225. Koga K, Gao GT, Tanaka H, Zeng XC (2001) Formation of ordered ice nanotubes inside carbon nanotubes. Nature 412:802–805
226. Koga K, Parra RD, Tanaka H, Zeng XC (2000) Ice nanotube: what does the unit cell look like? J Chem Phys 113:5037–5040
227. Zangi R, Mark AE (2004) Electrofreezing of confined water. J Chem Phys 120:7123–7130
228. Takaiwa D, Koga K, Tanaka H (2007) Structures of filled ice nanotubes inside carbon nanotubes. Mol Simul 33:127–132
229. Takaiwa D, Hatano I, Koga K, Tanaka H (2008) Phase diagram of water in carbon nanotubes. Proc Natl Acad Sci 105:39–43
230. Shiomi J, Kimura T, Maruyama S (2007) Molecular dynamics of ice-nanotube formation inside carbon nanotubes. J Phys Chem C 111:12188–12193
231. Jalali H, Ghorbanfekr H, Hamid I, Neek-Amal M, Rashidi R, Peeters FM (2020) Out-of-plane permittivity of confined water. Phys Rev E 102:022803
232. Tanaka H, Koga K (2005) Formation of ice nanotube with hydrophobic guests inside carbon nanotube. J Chem Phys 123:094706
233. Byl O, Liu J-C, Wang Y, Yim W-L, Johnson JK, Yates JT (2006) Unusual hydrogen bonding in water-filled carbon nanotubes. J Am Chem Soc 128:12090–12097
234. Maniwa Y, Kataura H, Abe M, Suzuki S, Achiba Y, Kira H, Matsuda K (2002) Phase transition in confined water inside carbon nanotubes. J Phys Soc Jpn 71:2863–2866
235. Kyakuno H, Matsuda K, Yahiro H, Inami Y, Fukuoka T, Miyata Y, Yanagi K, Maniwa Y, Kataura H, Saito T, Yumura M, Iijima S (2011) Confined water inside single-walled carbon nanotubes: global phase diagram and effect of finite length. J Chem Phys 134:244501
236. Kolesnikov AI, Zanotti J-M, Loong C-K, Thiyagarajan P, Moravsky AP, Loutfy RO, Burnham CJ (2004) Anomalously soft dynamics of water in a nanotube: a revelation of nanoscale confinement. Phys Rev Lett 93:035503
237. Algara-Siller G, Lehtinen O, Wang FC, Nair RR, Kaiser U, Wu HA, Geim AK, Grigorieva IV (2015) Square ice in graphene nanocapillaries. Nature 519:443–445
238. Maniwa Y, Kataura H, Abe M, Udaka A, Suzuki S, Achiba Y, Kira H, Matsuda K, Kadowaki H, Okabe Y (2005) Ordered water inside carbon nanotubes: formation of pentagonal to octagonal ice-nanotubes. Chem Phys Lett 401:534–538
239. Ghosh S, Ramanathan KV, Sood AK (2004) Water at nanoscale confined in single-walled carbon nanotubes studied by NMR. Europhysics Letters (EPL) 65:678–684
240. Neek-Amal M, Peeters FM, Grigorieva IV, Geim AK (2016) Commensurability effects in viscosity of nanoconfined water. ACS Nano 10:3685–3692

241. Qiu H, Zeng XC, Guo W (2015) Water in inhomogeneous nanoconfinement: coexistence of multilayered liquid and transition to ice nanoribbons. ACS Nano 9:9877–9884
242. Wang D, Tian Y, Jiang L (2021) Abnormal properties of low-dimensional confined water. Small 17:e2100788
243. Majumder M, Chopra N, Andrews R, Hinds BJ (2005) Enhanced flow in carbon nanotubes. Nature 438:44–44
244. Nair RR, Wu HA, Jayaram PN, Grigorieva IV, Geim AK (2012) Unimpeded permeation of water through helium-leak–tight graphene-based membranes. Science 335:442–444
245. Huang H, Song Z, Wei N, Shi L, Mao Y, Ying Y, Sun L, Xu Z, Peng X (2013) Ultrafast viscous water flow through nanostrand-channelled graphene oxide membranes. Nat Commun 4:2979
246. Hummer G, Rasaiah JC, Noworyta JP (2001) Water conduction through the hydrophobic channel of a carbon nanotube. Nature 414:188–190
247. Sam A, Hartkamp R, Kumar Kannam S, Babu JS, Sathian SP, Daivis PJ, Todd BD (2021) Fast transport of water in carbon nanotubes: a review of current accomplishments and challenges. Mol Simul 47:905–924
248. Holt Jason K, Park Hyung G, Wang Y, Stadermann M, Artyukhin Alexander B, Grigoropoulos Costas P, Noy A, Bakajin O (2006) Fast mass transport through sub-2-nanometer carbon nanotubes. Science 312:1034–1037
249. Du F, Qu L, Xia Z, Feng L, Dai L (2011) Membranes of vertically aligned superlong carbon nanotubes. Langmuir 27:8437–8443
250. Longhurst MJ, Quirke N (2006) The environmental effect on the radial breathing mode of carbon nanotubes. II. Shell model approximation for internally and externally adsorbed fluids. J Chem Phys 125:184705
251. Whitby M, Cagnon L, Thanou M, Quirke N (2008) Enhanced fluid flow through nanoscale carbon pipes. Nano Lett 8:2632–2637
252. Borg MK, Lockerby DA, Ritos K, Reese JM (2018) Multiscale simulation of water flow through laboratory-scale nanotube membranes. J Membr Sci 567:115–126
253. Chakraborty S, Kumar H, Dasgupta C, Maiti PK (2017) Confined water: structure, dynamics, and thermodynamics. Acc Chem Res 50:2139–2146
254. Falk K, Sedlmeier F, Joly L, Netz RR, Bocquet L (2010) Molecular origin of fast water transport in carbon nanotube membranes: superlubricity versus curvature dependent friction. Nano Lett 10:4067–4073
255. Keerthi A, Geim AK, Janardanan A, Rooney AP, Esfandiar A, Hu S, Dar SA, Grigorieva IV, Haigh SJ, Wang FC, Radha B (2018) Ballistic molecular transport through two-dimensional channels. Nature 558:420–424
256. Chung S-H, Anderson OS, Krishnamurthy VV (2007) Biological membrane ion channels: dynamics, structure, and applications. Springer Science & Business Media
257. Butler Tom Z, Pavlenok M, Derrington Ian M, Niederweis M, Gundlach Jens H (2008) Single-molecule DNA detection with an engineered MspA protein nanopore. Proc Natl Acad Sci 105:20647–20652
258. Brown CG, Clarke J (2016) Nanopore development at Oxford Nanopore. Nat Biotechnol 34:810–811
259. Gao P, Wang D, Che C, Ma Q, Wu X, Chen Y, Xu H, Li X, Lin Y, Ding D, Lou X, Xia F (2021) Regional and functional division of functional elements of solid-state nanochannels for enhanced sensitivity and specificity of biosensing in complex matrices. Nat Protoc 16:4201–4226
260. Ohshiro T, Tsutsui M, Yokota K, Furuhashi M, Taniguchi M, Kawai T (2014) Detection of post-translational modifications in single peptides using electron tunnelling currents. Nat Nanotechnol 9:835–840
261. Deamer D, Akeson M, Branton D (2016) Three decades of nanopore sequencing. Nat Biotechnol 34:518–524
262. Varongchayakul N, Song J, Meller A, Grinstaff MW (2018) Single-molecule protein sensing in a nanopore: a tutorial. Chem Soc Rev 47:8512–8524

263. Wang Y, Wang Z, Wang L, Tong T, Zhang X, Fang S, Xie W, Liang L, Yin B, Yuan J, Zhang J, Wang D (2022) Comparison study on single nucleotide transport phenomena in carbon nanotubes. Nano Lett 22:2147–2154
264. Li X, Zhu H (2015) Two-dimensional MoS_2: properties, preparation, and applications. J Materiomics 1:33–44
265. Koenig SP, Wang L, Pellegrino J, Bunch JS (2012) Selective molecular sieving through porous graphene. Nat Nanotechnol 7:728–732
266. Kim Hyo W, Yoon Hee W, Yoon S-M, Yoo Byung M, Ahn Byung K, Cho Young H, Shin Hye J, Yang H, Paik U, Kwon S, Choi J-Y, Park Ho B (2013) Selective gas transport through few-layered graphene and graphene oxide membranes. Science 342:91–95
267. Celebi K, Buchheim J, Wyss Roman M, Droudian A, Gasser P, Shorubalko I, Kye J-I, Lee C, Park Hyung G (2014) Ultimate permeation across atomically thin porous graphene. Science 344:289–292
268. Sun PZ, Yang Q, Kuang WJ, Stebunov YV, Xiong WQ, Yu J, Nair RR, Katsnelson MI, Yuan SJ, Grigorieva IV, Lozada-Hidalgo M, Wang FC, Geim AK (2020) Limits on gas impermeability of graphene. Nature 579:229–232
269. Cheng C, Iyengar SA, Karnik R (2021) Molecular size-dependent subcontinuum solvent permeation and ultrafast nanofiltration across nanoporous graphene membranes. Nat Nanotechnol 16:989–995
270. Li X, Zhu B, Zhu J (2019) Graphene oxide based materials for desalination. Carbon 146:320–328
271. Zhang Z, Wen L, Jiang L (2021) Nanofluidics for osmotic energy conversion. Nat Rev Mater 6:622–639
272. Wang L, Wang Z, Patel SK, Lin S, Elimelech M (2021) Nanopore-based power generation from salinity gradient: why it is not viable. ACS Nano 15:4093–4107
273. Robin P, Kavokine N, Bocquet L (2021) Modeling of emergent memory and voltage spiking in ionic transport through angstrom-scale slits. Science 373:687–691
274. Faucher S, Aluru N, Bazant MZ, Blankschtein D, Brozena AH, Cumings J, Pedro de Souza J, Elimelech M, Epsztein R, Fourkas JT, Rajan AG, Kulik HJ, Levy A, Majumdar A, Martin C, McEldrew M, Misra RP, Noy A, Pham TA, Reed M, Schwegler E, Siwy Z, Wang Y, Strano M (2019) Critical knowledge gaps in mass transport through single-digit nanopores: a review and perspective. J Phys Chem C 123:21309–21326
275. Nazari M, Davoodabadi A, Huang D, Luo T, Ghasemi H (2020) Transport phenomena in nano/molecular confinements. ACS Nano 14:16348–16391

Chapter 3
Detecting DNA-Binding Sites of Regulation Proteins with Ion Beam Sculpted Silicon Nitride Nanopores

Haopeng Li, Sachini P. W. Withanage, Harpreet Kaur, and Jiali Li

Abstract After briefly introducing silicon nitride nanopore fabrication, experimental setup, and sample preparation, we describe the details on data analysis especially how to extract the DNA-binding sites of regulation proteins. As an example, we will estimate the binding sites of RNA polymerase (RNAP) holoenzyme on λ DNA (48.5 kbp, ~ 16.5 μm). By analyzing the current drop amplitudes and dwell times of current blockage events and subevents recorded from sample molecules of RNAP, λ DNA, and RNAP + λ DNA, the binding locations, and their efficiencies of RNAPs on a λ DNA are estimated. Finally, new perspectives and complications to measure DNA-binding sites of regulation proteins will be discussed.

Keywords Nanopore · Protein · DNA · RNAP · Subevent

3.1 Introduction

3.1.1 Background of Low Energy Noble Gas Ion Beam Sculpted Silicon Nitride Nanopores and Their Applications

Motivated to replace protein channels/pores such as α-hemolysin inserted in lipid membrane structure [1, 2] with improved and superior property synthetic materials, Silicon Nitride Nanopores (SNN) were fabricated by low energy Ion Beam Sculpting (IBS) method in early 2000s [3, 4]. SNN and other Solid-state nanopores can be fabricated with desired geometry and dimensions. These nanopores are mechanically robust, have tunable dimensions, tolerate broad temperature, pH, and chemical

H. Li
Cell and Molecular Biology Program, University of Arkansas, Fayetteville, AR 72701, USA

S. P. W. Withanage · H. Kaur · J. Li (✉)
Department of Physics, University of Arkansas, Fayetteville, AR 72701, USA
e-mail: jialili@uark.edu

J.-P. Leburton (ed.), *Solid State Nanopores*, Nanostructure Science and Technology,
https://doi.org/10.1007/978-3-031-42336-9_3

variations, and are suitable for analysis of charged biopolymers [5–7]. In the past two decades, SNN have been developed as biosensors to detect and analyze charged biomolecules like DNA [4, 7–19], RNA [20, 21], proteins [22–36], and recently for protein oligomerization and aggregation [37–39] in close to their native aqueous salt solution conditions.

For DNA molecules, SNN are capable to scan and determine their primary and secondary structure features: their length, effective charge, persistence length, and folding state. For a protein molecule, SNN can detect a protein's shape, charge, dipole moment, and denaturing state. The applications of SNN are not limited to DNA, RNA, and protein detection and analysis, these nanopore devices are also suitable to study protein-DNA complexes [40–46].

3.1.2 The Importance of DNA–Protein Binding

DNA and Protein molecules are essential for life. Large scale as well as single molecule DNA sequencing has become a reality in recent years [47]. Knowing the sequences of DNAs enables humans to better understand the blueprints of life. However, life is not only about DNA, the blueprints also need to be read by proteins and synthesized into proteins to perform the function of life, just like a blueprint of a house needs workers to construct.

Protein-DNA interactions are of key importance in gene regulation [48]. The ability of DNA-binding proteins to recognize and bind to specific sites on a DNA molecule is fundamental to control the flow of genetic information. Regulated gene expression is vital for the proper growth and survival of an organism. Aberrations in the flow of genetic information that result from mutations that alter protein-DNA interactions can lead to disease progression. For example, several oncogene and tumor suppressor proteins bind DNA and when they are mutated result in malignant transformation: cancer. In addition, DNA-binding proteins bind to the function in other cellular events such as mitosis where they bind to the kinetochore and function in chromosome segregation during cell division. Defects in this process can lead to alterations in chromosome number leading to birth defects such as Down's syndrome. Understanding how these different proteins are able to find and bind selectively to only one, or just a few, specific sequences out of the millions present in a genome is a major goal of molecular biology [49–51].

3.1.3 Methods Used to Measure DNA–Protein Binding Sites Today

Great efforts have been made to develop methods to better understand and study the interaction of DNA and proteins. Typical methods used for analyzing specific protein-DNA interactions [52] are: DNase Footprinting, gel mobility shift assay, nitrocellulose filter-binding assay, genetic analysis, and X-ray crystallography. The gel mobility assay gives some information about the mass of the protein bound, and filter-binding assay can measure on and off rates of the binding proteins, but both methods do not localize the binding sites. X-ray crystallography can provide protein-DNA complex structural information at atomic resolution, but the process is expensive and time consuming, and the picture is static. Genetic analysis can identify important bases in the binding site, but it does not provide precise mapping of the binding site. The DNase Footprinting method can localize the binding site to within a few base pairs, but the process is tedious and technically demanding.

Recent development with microarray technologies have increased the speed of measuring protein-DNA interaction [53]. In addition, the advance in computer science and molecular modeling have revealed some of the most important detailed interactions between genome and protein [54]. Furthermore, recent advances in the detection and manipulation of single DNA and protein molecules also offer great promise for determining the protein-DNA binding sites [55, 56]. Some of the leading techniques on DNA–protein interaction in the field are atomic force microscopy (AFM) [57], Optical tweezers [58], Molecular combing [59], and nanofluidic devices together with DNA stretching and fluorescence single-molecule imaging [60]. AFM and STM allow imaging of single protein-DNA complex molecules on surfaces; high-resolution laser fluorescence microscopy allows real-time observations of molecular interactions. None of these methods can measure the in vivo dynamic movements and identify the locations of protein-DNA binding sites in real time at single molecule level.

3.1.4 Detecting the Binding Sites of Proteins on DNA Molecules with Nanopore Technology

In this chapter we review the current development of using nanopore technology to detect protein-DNA binding sites. We mainly focus on our recently published work by Kaur at el. on the detecting of binding locations of RNA Polymerase (RNAP) on long λ DNA (48.5 kb) molecules [46, 61].

There are many DNA-binding proteins include transcription factors, various polymerases, nucleases, and histones. In this chapter we use RNA polymerase (RNAP) and λ DNA binding as an example. RNAP is a DNA binding protein that binds to specific bases of a DNA to transcribe DNA to messenger RNA (mRNA) for future activities such as the translation of mRNA to protein. A single malfunction of these interactions will have ramifications on life, most of the time it manifests as a disease to humans [62]. Many studies have been conducted on the interaction between RNA polymerase (RNAP) and DNA due to the importance of this interaction in biology.

3.2 Principle of Nanopore Detecting DNA-Binding Sites of Regulation Proteins

The principle of using a SNN device to detect protein-DNA binding sites is illustrated in Fig. 3.1. A nanopore chip that contains a single nanopore fabricated in an insulating silicon nitride membrane supported by a silicon substrate is the main component of a nanopore sensing device. As shown in Fig. 3.1a, a nanopore containing membrane separates two ionic solution filled compartments. When a voltage across the membrane is applied by a pair of electrodes, the flow of ions generates a measurable steady open pore ionic current, I_o. When a biomolecule (BM): a DNA or protein bound DNA sample, or a mixture of both, is added to the *cis* side (grounded) of a nanopore chip, negatively charged BMs near the nanopore will be captured by the electric field and forced to pass through the nanopore to the positive side. When a BM is entering the pore, the BM partially blocks the ionic current flow in the nanopore and produces a drop in current amplitude ΔI_{b}. When the BM is leaving the pore, the ionic current goes back to I_o. The whole process is called a current blockage event that can be characterized by its amplitude ΔI_{b} and time duration t_{d} (Fig. 3.1b).

To a close approximation, the nanopore electrical behavior in an ionic solution obeys Ohm's Law. A BM inside a nanopore partially blocks the flow of ions which increases the pore resistance. The current drop amplitude ΔI_{b} is proportional to the applied voltage Ψ and can be described as below [6, 8].

$$\Delta I_b(t) \approx \frac{\boldsymbol{\sigma \Psi}}{\boldsymbol{H}_{eff}^2} \boldsymbol{\Lambda}_{BM}(\mathrm{t}) = I_0 \frac{\Lambda_{BM}(t)}{V_{po}} \tag{3.1}$$

$$\mathrm{I_o} = \Psi/\mathrm{R_0} = \left(\frac{\sigma A_{po}}{H_{eff}}\right)\Psi \tag{3.2}$$

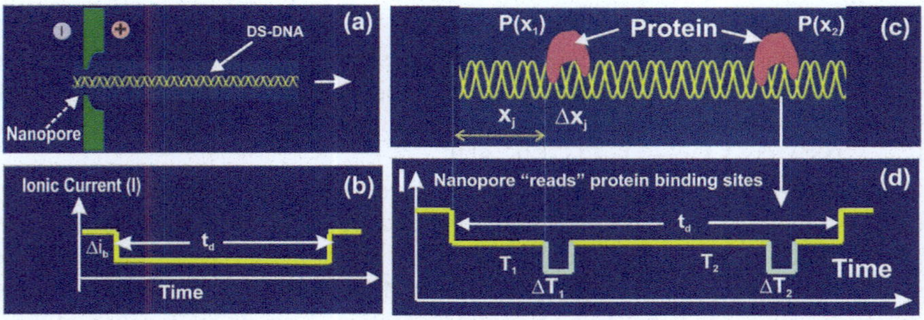

Fig. 3.1 The principle of nanopore detection for protein-DNA binding sites. **a** A dsDNA is translocation through a nanopore; **b** the induced ionic current blockage signal by the dsDNA; **c** an illustration of a DNA–protein complex with two protein molecules bound at X_1 and X_2; and **d** the expected current blockage signal induced by the protein-DNA complex

Here I_o is the open pore current, $R_0 = H_{eff}/\sigma A_{po}$ is the pore resistance, σ is the solution conductivity, H_{eff} is the effective thickness of a nanopore that has considered the pore access resistance, Λ_{BM} is the excluded volume of a BM in a nanopore, A_{po} is the average cross section area of the pore. As described in Eq. (3.1), when a segment of DNA with a bound protein passing a pore, its excluded volume Λ_{BM} (= Λ_{DNA} + Λ_{pro}) is larger than $\Lambda_{DNA} = A_{DNA} H_{eff}$ of a segment of DNA alone. This further increased current drop in ΔI_b is called a subevent as shown in Fig. 3.1d.

A sequence specific DNA binding protein (Fig. 3.1c) will bind at site $P(x_j) = i$ on a DNA, where i is the position number of a nucleotide in a DNA and it has a value of $i = 1$ to N, here N is the total number of base pairs (bp) for a double stranded DNA. For example, if a DNA has two protein molecules bound at x_1 and x_2 as shown in Fig. 3.1c, finding the subevent starting time T_1 and T_2 will lead to the determination of the locations x_1 and x_2 as illustrated in Fig. 3.1d. The subevent current drop amplitude $\Delta I_{sub}(t_j)$ and time duration ΔT_j contains the properties of the bound protein such as its size, shape, and charge.

The event time duration, t_d, is an important parameter for detecting protein binding sites on a DNA. Here we assume the DNA contour length is much longer than the dimension of the binding proteins, $L_{DNA} \gg L_{protein}$, therefore $t_d(BM) \sim t_d(DNA)$. If we use a simple model to describe a DNA molecule with linear charge density ζ passing through a nanopore of effective thickness H_{eff}, and assuming the electric driving force (F_E) balances the drag force in the pore so that the DNA molecule moves at an average terminal velocity v_d, then $F_E = F_{drag}$ [8, 63]. Here $v_d = L_{DNA}/t_d$, $F_{drag} = f v_d$, f is the frictional coefficient that is proportional to the solution viscosity η. The electric driving force is proportional to the local DNA segment total charge $Q = \zeta H_{eff}$ and the electric field strength in the nanopore (Ψ/H_{eff}), FE $= Q(\Psi/H_{eff}) = \zeta H_{eff}\left(\frac{\Psi}{H_{eff}}\right) = \zeta\Psi$. We obtain the following equation [35, 64] to describe the dwell time of a DNA molecule translocation through a solid-state nanopore,

$$t_d = \left(\frac{f}{\zeta}\right)\frac{L^{\alpha}{}_{DNA}}{\Psi} \tag{3.3}$$

Here α is a constant parameter measured empirically to indicate the unwinding of the DNA region just before its translocation through the pore, and its value is approximately α ~ 1.4 [12].

The time duration histograms of t_d can be fitted with a probability density function (PDF) of a 1D biased model reported earlier [64, 65] and is described below.

$$P(t) = \frac{CL * e^{-\frac{(L-vt)^2}{4tD}}}{t\sqrt{4\pi tD}} \tag{3.4}$$

where C is a normalized coefficient, $L = L_{DNA} + H_{eff}$ and $L_{DNA} \approx 16.5$ μm for a λ DNA molecule, v(orv_d) is the mean terminal velocity and D is the diffusion constant.

A parameter that represents the net excluded charges of an ionic current blockade event, A_{ecd} (*ecd*: event charge deficit, in the unit of ke: kilo electrons of charge), can also be used to characterize a DNA molecule with and without a bound protein. A_{ecd} can be calculated by the integrated area of a current drop event and is determined primarily by the identity of the molecule, its charge, geometry, and ionic solution parameters [11, 66].

$$A_{ecd} = \int_{event} \Delta I_b(t)dt \sim \langle \Delta I_b \rangle t_d \tag{3.5}$$

3.3 Experimental Setup and Sample Preparation

3.3.1 Silicon Nitride Nanopore Fabrication by Ion Beam Sculpting

The silicon nitride nanopores were fabricated by a combination of semiconductor device fabrication techniques followed by low energy ion beam sculpting [3, 4] as illustrated in Fig. 3.2. Briefly, the procedures include: (1) starting with a double-side polished silicon wafer with ~ 300 nm low stress silicon nitride deposited on each side (2a), free standing Si_3N_4 membrane windows in the center of each silicon substrate chip were created using photolithography, reactive ion etching, and wet chemical etching techniques. (2) a ~ 100-nm hole was milled in the center of each membrane window (2b) using a 50-keV Ga^+ focused Ion Beam (FIB) machine (Micrion 9500). (3) the large ~ 100-nm holes were controllably shrunk by feedback controlled low energy ion beam sculpting to desired size nanopores, 2–30 nm typically (3c–3f).

All noble gas ions, He^+, Ne^+, Ar^+, Kr^+, and Xe^+ have been used to shrink FIB holes [67]. The size of nanopores as well as FIB holes were imaged by transmission electron microscopy (TEM). Examples of TEM images of ~ 4 nm diameter He^+ and Kr^+ ion beam fabricated nanopores are shown in Fig. 3.2f. All nanopores after ion beam sculpting were thermally annealed at 750 °C in N_2 environment before the final TEM imaging. Our study showed that the best nanopores those were low noise and chemically stable were made by He^+ and Ne^+ ion beams [68].

3.3.2 DNA and Protein Sample Preparation

The RNAP holoenzyme used in this example has a 470 kDa molecular weight with an overall size of 12 nm × 15 nm × 18 nm (Fig. 3.3a) and a net charge of – 77e at pH 7 [43]. The RNAP has five subunits: two α, β, β´, and one ω. The RNAP enzyme [43, 69] requires a subunit σ_{70} (70 kDa) to bind a DNA specifically. The σ subunit has 6

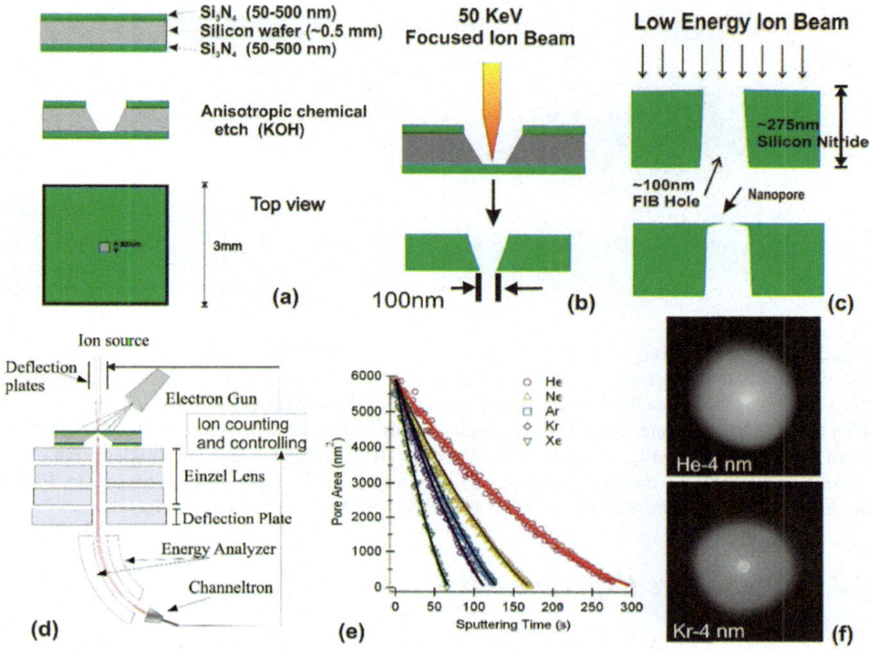

Fig. 3.2 Fabrication for solid-state nanopore chips (**a**). FIB milling (**b**). Ion beam sculpting nanopores by keV noble gas ions (**c**). Schematic drawing of the feedback-controlled IBS system (**d**). Shrinking prefabricated FIB holes with He^+, Ne^+, Ar^+, Kr^+, and Xe^+ noble gas ions at low flux 0.5–0.6 ions/nm^2 (**e**). TEM images of ~ 4 nm diameter He^+ and Kr^+ ion beam fabricated nanopores (**f**) [6, 67]

base pairs long specific promoter regions at − 10 (TATAAT) and − 35 (TTGACA) positions (The Promoter Sequence is 10 and 35 base pairs away from the transcript initial site). The holoenzyme can bind specifically to the promoter site and form a RNAP-DNA complex with a length of approximately 50 base pairs (Fig. 3.3a).

The DNA sample used here is a N = 48,502 bp (16.49 μm) long λ DNA (Fig. 3.3b, top panel). It has been reported earlier [60] that the RNAP has two strong binding sites at $P(x_{j1})$ = 38,003 bp (3.55 μm) and $P(x_{j2})$ = 35,602 bp (4.40 μm) close to the 3′ end, and three weaker binding sites between 7–9 μm at about $P(x_{j3})$ = 27,649 bp, $P(x_{j4})$ = 25,620 bp and $P(x_{j5})$ = 23,619 bp. A predicted current blockage event generated from a complex of λ DNA with bound RNAPs passing through a nanopore is drawn in Fig. 3.3b. Ideally, one should expect five RNAP subevents within one λ DNA current blockage level corresponding to the five RNAP binding positions, or at least one or two RNAP subevents within one λ DNA current blockage level corresponding to the two strong binding sites. To prevent separation of the bound RNAP from a λ DNA molecule in the strong electric field inside a nanopore, the RNAP proteins was cross-linked to λ DNA by formaldehyde [60, 70].

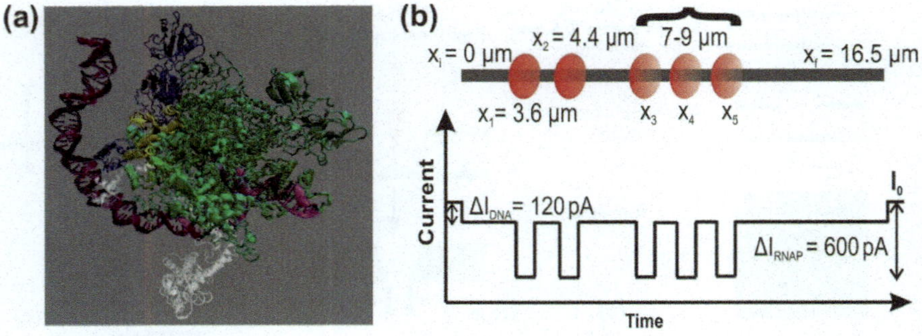

Fig. 3.3 **a** 3D X-ray structure of RNAP-DNA complex (3IYD pdb file). Structures shown in blue are the two α subunits, in dark green is β and light green is β' subunits, in yellow is ω, in white is σ_{70} factor, and in purple is DNA strands. **b** (top panel) Illustration of five possible RNAP binding positions on a 16.5 μm (48,502 bp) long λ DNA, and a predicted current blockage event (bottom panel) produced from the λDNA * $RNAP_{com}$ complex shown above. The current blockage amplitudes produced by λ DNA and RNAP were estimated by Eq. (3.1) based on the nanopore dimensions used. Figures are Reprinted with permission from [46]. Copyright [46] American Chemical Society

3.3.3 Nanopore Experimental Setup

The nanopore experiment setup for detecting protein-DNA interactions is illustrated in Fig. 3.4a and b. The preparation of nanopore chips were 15 min immersion in acetone, then 15 min in ethanol, and soaking in 20% ethanol for a few hours. After the preparation, a nanopore chip was aligned between two chambers made by Polydimethylsiloxane (PDMS). The chambers were filled with salt solution of 1 M KCl, 10 mM Tris, 1 mM EDTA, and 5% glycerol at pH = 8.0. The solution was filtered with a 20 nm membrane syringe filter (Whatman) and degassed at 40 °C. The solutions in both chambers could be changed with a fluidic system as shown in Fig. 3.4a. Each chamber was equipped with an AgCl electrode that a voltage bias was applied during experiments. The whole system was placed on a vibration isolated table and surrounded by a Faraday cage. DC voltage (Ψ) was applied by an integrated Axopatch 200B system which the current signal through a nanopore was also measured. The sampling rate for current signal recordings were set at 200 kHz and the low pass Bessel filter was set at 10 kHz. All the nanopores used in this work were characterized by IV curves and have an $I_{rms} \leq 10$ pA.

When an electric voltage ($\Psi = 120$ mV) is applied across a nanopore of ~ 22.1 nm in diameter (Fig. 3.4c), an open pore ionic current, $I_o = 11.7$ nA was measured. The open pore current amplitude can be estimated by a simplified ohm's law model as described in Eq. (3.2): $I_o = \sigma A_p \Psi / H_{eff}$. Where $\sigma = 10.2$ S/m is the conductivity of 1 M KCl salt solution with 5% glycerol, A_p is the area and H_{eff} is the effective thickness of a nanopore. Using $I_o = 11.7$nA, $H_{eff} \approx 39.8$ nm was calculated for the nanopore.

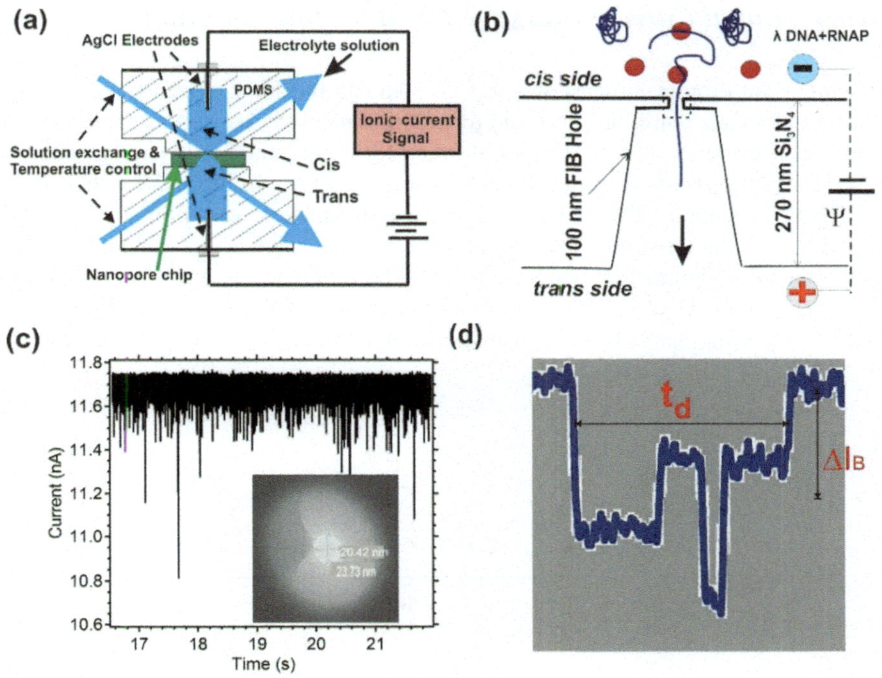

Fig. 3.4 **a** Schematic drawing for a nanopore experiment setup used to detect protein-DNA interactions. **b** Expanded view of a nanopore experiment with a mixture of λ DNA and RNAP added to the cis chamber. **c** A trace of current blockage events recorded from a λ DNA and RNAP mixture. The insert shows a TEM image of a 20.42 nm by 23.73 nm size nanopore used for the experiment. **d** An example of current blockage event recorded. Panels **c** and **d** Reprinted with permission from [46]. Copyright [46] American Chemical Society

Using the nanopore described in Fig. 3.4c, a series of measurements were performed for samples in the following order: (1) RNAP protein ($RNAP_{bare}$), (2) RNAP protein incubated (at 4 °C for 30 min) in formaldehyde ($RNAP_{fmd}$), and (3) a mixture of λ DNA plus RNAP incubated in formaldehyde ($\lambda DNA + RNAP_{mix}$). Approximately several thousands of current blockage events were recorded at each applied voltage for every sample measured. The $RNAP_{bare}$ sample was 1st added to the cis chamber. The chambers were rinsed with 1 M KCl (5% glycerol) solution until no more events were observed. The rinsing was performed after a sample was finished recording of events and then a new sample was added to the cis chamber. An example of recorded ionic current trace with many current blockage events for the $\lambda DNA + RNAP_{mix}$ is shown in Fig. 3.4c.

3.3.4 Initial Current Blockage Event Sorting and Analysis

Each current drop event as shown in Fig. 3.4d can be characterized by an average current blockage amplitude ΔI_b and time duration t_d (4d), as well as the integrated area of the evens A_{ecd} [38]. The plot of events, ΔI_b vs t_d, called scattered plot can be used for initial events sorting and classification as shown in Fig. 3.5 for the λDNA + $RNAP_{mix}$ sample. The histograms of these parameters were plotted, and curve fitted to find the most probable values with the respective errors. Table 3.1 below provides these parameters obtained for $RNAP_{bare}$, $RNAP_{fmd}$, λDNA + $RNAP_{mix}$, $λDNA_{bare}$, and $λDNA_{fmd}$. The control measurements of $RNAP_{bare}$ and $RNAP_{fmd}$ as well as $λDNA_{bare}$ and $λDNA_{fmd}$ were performed to help to identify event clusters.

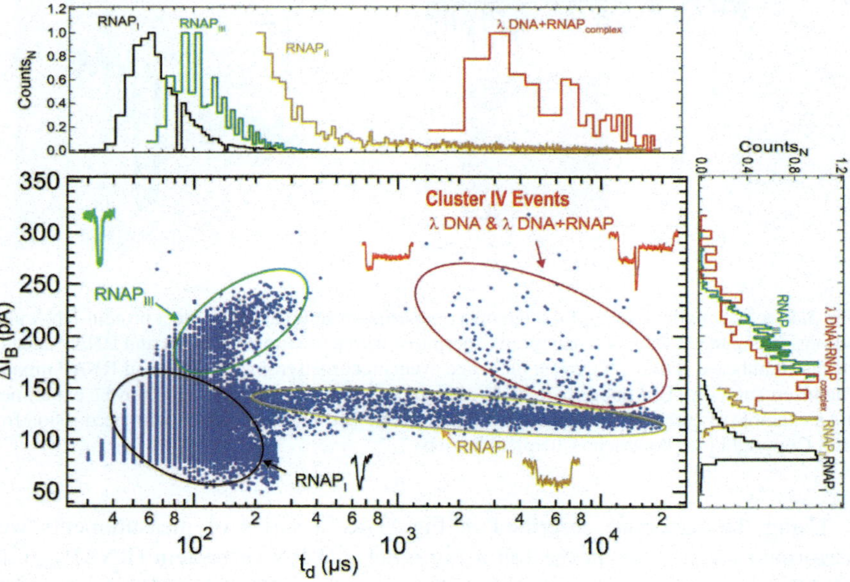

Fig. 3.5 Scatter plots of ΔI_b versus t_d measured with a 23.7 nm × 20.4 nm nanopore at 120 mV for the λDNA + $RNAP_{mix}$ sample, four different populations of events are observed and are circled and labelled $RNAP_I$, $RNAP_{II}$, $RNAP_{III}$, and λ DNA and λDNA + $RNAP_{comp}$. For comparison, current blockage histograms (right side) and dwell time histograms (above) are plotted for each population and are normalized to 1. An example of each population of events is shown next to the corresponding cluster. Reprinted with permission from [46]. Copyright [46] American Chemical Society

Table 3.1 Current amplitudes and dwell times for various samples at 120 mV

Sample	I_o (nA)	Populations observed	ΔI_B (pA)	t_{dwell} (μs)	A_{ecd} (pA μs) (Basic Charge: e)
$RNAP_{bare}$	11.0	$RNAP_I$	89.8 ± 11.3	57.3 ±1 3.8	
$RNAP_{fmd}$	11.0	$RNAP_I$ $RNAP_{II}$ $RNAP_{III}$	89.7 ± 7.9 121.1 ± 3.4 170.2 ± 27.8	64.7 4075.1 109.7	
λDNA + $RNAP_{mix}$	11.0	$RNAP_I$ $RNAP_{II}$ $RNAP_{III}$ λ DNA and λDNA * $RNAP_{comp}$	88.3 ± 7.4 124.1 ± 7.9 160.0 ± 51.8 142.6 ± 14.14 180.6 ± 35.26	65.83 5394.9 101.4 5795.7	5.12×10^5 (3.20×10^6 e) 1.00×10^6 (6.24×10^6 e)
λ DNA_{bare}	12.5		131.8 ± 6.46 148.6 ± 46.58	3000.00	5.55×10^5 (3.46×10^6 e)
λ DNA_{fmd}	12.5		118.7 ± 5.04 140.5 ± 38.30	4450.00	6.75×10^5 (4.21×10^6 e)

3.4 Results and Discussion: Classification and Selecting of λDNA * $RNAP_{comp}$ Events

To identify the events of λ DNA having RNAP bound, λDNA * $RNAP_{comp}$, from the scattered plot shown in Fig. 3.5 measured with the mixture λDNA + $RNAP_{mix}$ sample, we need to compare the scattered plots with control measurements $RNAP_{bare}$, $RNAP_{fmd}$, λDNA_{bare}, and λDNA_{fmd} measured at the same voltage $\Psi = 120$ mV [46].

The scatter plot in Fig. 3.5 displays four groups that are circled in different colors with labels. The top panel in Fig. 3.5 are the histograms of t_{dwell}. On the right panel, the current histograms of ΔI_b are presented. The populations of events labeled as $RNAP_I$, $RNAP_{II}$, and $RNAP_{III}$ were identified using the result of $RNAP_{fmd}$ sample. Events labeled under $RNAP_I$ were centered at $\Delta I_b = 88.3 \pm 7.4$ pA with $t_d = 65$ μs. $RNAP_{II}$ had a peak at 124.1 ± 7.9 pA and a mean t_d of 5.4 ms with a very broad time distribution. $RNAP_{III}$ had a main peak at $\Delta I_b = 160.0 \pm 51.8$ pA and $t_d = 101 \pm 38.4$ μs with wide ranges of current blockage amplitudes and dwells times. Comparing with the results of control samples, these values are very close to the three clusters of events observed for $RNAP_{fmd}$ sample. Therefore, the three populations ($RNAP_I$, $RNAP_{II}$, and $RNAP_{III}$) in Fig. 3.5 can be attributed to the events of RNAP. Our analysis suggested that these three populations are most likely produced by the subunits of RNAP protein which possibly denatured by the high electric force upon entering the nanopore in formaldehyde [46].

The cluster IV events have a maximum peak at $\Delta I_b = 142.6 \pm 14.8$ pA along with a small peak at about $\Delta I_b = 180.6 \pm 37.5$pA (Fig. 3.5 right axis), and a t_d distribution peak at $t_d \sim 2.8$ ms (Fig. 3.5 top), which are typical values for λ DNA events reported earlier. Therefore, the cluster IV events are most likely a mixture of events produced from λDNA_{fmd} and $\lambda DNA * RNAP_{comp}$.

Further analysis for the $\lambda DNA + RNAP_{mix}$ sample shows that both histograms of current drop amplitude ΔI_b (Fig. 3.6a) and the integrated area A_{ecd} (Fig. 3.6b) have an extra peak compared with λDNA_{bare} and $\lambda DNA_{fmd,}$ indicating the presence of RNAP bound on λ DNA for these events. Selecting these $\lambda DNA * RNAP_{comp}$ events and compare their ΔI_b (Fig. 3.6c) and t_d (Fig. 3.6d) histograms with the rest of λDNA_{fmd} events in the cluster IV, the histogram of $\lambda DNA * RNAP_{comp}$ displays two current drop peaks at 285 pA and 335 pA, and t_d histograms show two peak values at 2.7 ms and 6.0 ms with fitting parameters (Eq. 3.4) for the DNA traveling velocity approximately $v = 3.0 \pm 1$ nm/μs and $D \sim 10\ \mu m^2/ms$.

3.5 Subevent Analysis: Estimate the RNAP Binding Positions on a λ DNA

Next, we select events of $\lambda DNA * RNAP_{comp}$ in the cluster IV and extract the subevent starting times for the RNAP bound λ DNA molecules ($\lambda DNA * RNAP_{comp}$). The $\lambda DNA * RNAP_{comp}$ events were selected by their subevent features: $\Delta I_b > 400$ pA (indicating that the bound RNAPs may still be intact), these events have larger ΔI_b as shown in Fig. 3.6c and larger integrated area (Fig. 3.6b) which can be tagged in our data analysis program.

Three types of events were observed for $\lambda DNA * RNAP_{comp}$ molecules. Examples of typical events are shown in Fig. 3.7 with their possible configurations of $\lambda DNA * RNAP_{comp}$ and the nanopore. Each type of event has two possible configurations: either a $\lambda DNA * RNAP_{comp}$ enters with RNAP side first or with RNAP side at the end. The first type is when λ DNA with one RNAP bound enters the nanopore in a linear configuration as shown in Fig. 3.7a and stays unfolded during its passage. The second type of event (Fig. 3.7b) is when a λ DNA molecule was captured at a different location other than one of its ends, such that the first part of an event shows a folded level of DNA and the RNAP subevent follows after that. Figure 3.7c represents a third kind of event where a $\lambda DNA * RNAP_{comp}$ would be captured in such a way that an RNAP subevent level lies within the folded part of an event.

Out of 271 $\lambda DNA + RNAP_{mix}$ events, 137 events were from λDNA_{fmd} and 101 events were from $\lambda DNA * RNAP_{comp}$. This suggested that about 42.4% λ DNA molecules were bound with at least one RNAP molecule.

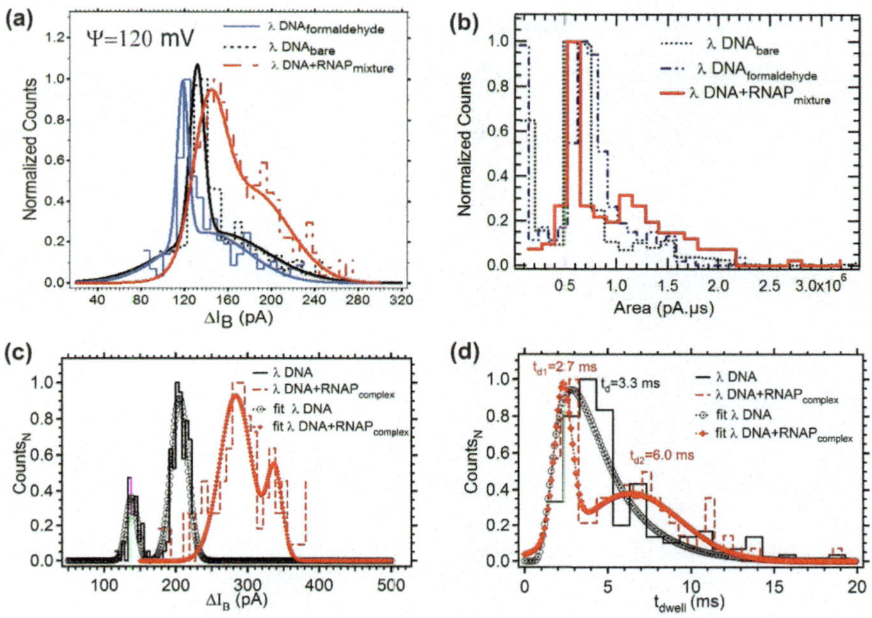

Fig. 3.6 Comparison of histograms of the 4th cluster of events with λ DNA control experiments at 120 mV. **a** ΔI_b histograms of λDNA_{bare} and λDNA_{fmd}, and the fourth population observed with $\lambda DNA + RNAP_{mix}$. Solid curves are fittings to double Gaussian functions. **b** Histograms for the integrated area of events for all three samples. For λDNA_{bare} and λDNA_{fmd}, there is a single peak at 5.55×10^5 pA μs and 6.75×10^5 pA μs, respectively. On the other hand, the fourth population of $\lambda DNA + RNAP_{mix}$ has two peaks at 5.13×10^5 pA μs and 1.15×10^6 pA μs. **c** The histograms of average current blockage ($\Delta \boldsymbol{I}_{\boldsymbol{B}}$) show two peaks at 137 pA and 204 pA for λ DNA only, corresponding to unfolded and partially folded translocations of λ DNA. Histogram of $\lambda DNA * RNAP_{comp}$ displays two current drop peaks at 285 pA and 335 pA. **d** t_d histograms without and with bound RNAPs are plotted together for comparison. Fitting the t_d histogram of λ DNA only with Eq. (3.4), a peak value centered at 3.3 ms was obtained. For the $\lambda DNA + RNAP_{mix}$, a greater diffusion coefficient or larger thermal motion $D = 18.9\ \mu m^2/ms$ and a moving speed of $v = 3.9\ \mu m/ms$ (nm/μs) were obtained. All counts are normalized for comparison. Solid curves ΔI_b histograms are fittings to double Gaussian functions. Reprinted (adapted) with permission from [46]. Copyright [46] American Chemical Society

The histograms of average current blockage (Fig. 3.7d) for these $\lambda DNA * RNAP_{comp}$ subevents show that $\Delta I_b = 400$–600 pA, which was consistent with the ΔI_b values estimated based on the excluded volume Λ of RNAP molecules. The dwell time histogram (Fig. 3.7e) for these RNAP subevents was centered at $t_d = 143$ μs with a smaller population around $t_d = 500$ μs.

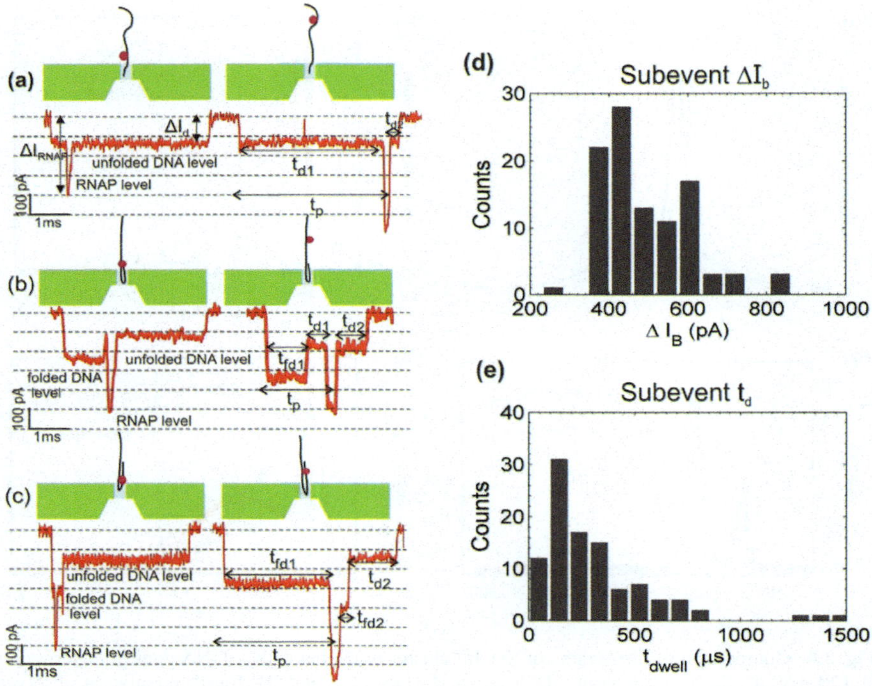

Fig. 3.7 Examples of typical λ DNA current blockage drops with subevents observed for λDNA * $RNAP_{comp}$ from the 4th population in Fig. 3.5. **a** Two examples of λDNA * $RNAP_{comp}$ events with one possible RNAP subevent in an unfolded DNA level. **b** Another type of event with a possible RNAP subevent (correct ΔI_b for RNAP level) in a partially folded DNA event. RNAP bound portion came after the folded λ DNA part. **c** Events with an RNAP subevent in the folded λ DNA portion. **d** $\Delta \boldsymbol{I}_B$histogram of RNAP subevents observed in λ DNA current drop events. **e** t_d histogram of RNAP subevents in λ DNA current drop events. Reprinted with permission from [46]. Copyright [46] American Chemical Society

3.5.1 Subevent Analysis for Estimating Protein Binding Position on DNA

To estimate the RNAP binding positions on a λ DNA, we need to find the subevent starting times, t_{sub} (from the beginning of an event to the time of a subevent across a set trigger line, Fig. 3.8a). The subevent trigger levels were set below the open pore current value at 220 pA for the λDNA + $RNAP_{mix}$ and 200 pA for the other two λ DNA samples. Figure 3.5 shows the histograms of the relative subevent starting times, $t_{rel} = t_{sub}/t_{total}$ (here t_{otal} was the total event time duration), for all the three types of samples measured: λDNA_{bare}, λDNA_{fmd}, and λDNA + $RNAP_{mix}$. For the λDNA_{bare} and λDNA_{fmd} samples (blue and green), their subevent starting times are random, almost evenly distributed within events. The subevent starting times of λDNA_{fmd} have small increased populations at some positions comparing with λDNA$_{bare}$, consistent with that formaldehyde could induce some structural changes like loops inside a λ DNA molecule that could produce more subevents. For the λDNA + RNAP$_{mix}$, the histogram of subevent relative start time shows one major broad peak centered at $t_{rel} = 0.3$. A small population between $t_{rel} = 0.6$ to 0.8 was also observed. Bases on these comparisons with the two control DNA samples, the two peaks produced by the λDNA + RNA$_{mix}$ sample were most likely from the binding of RNAP proteins on λ DNA molecules.

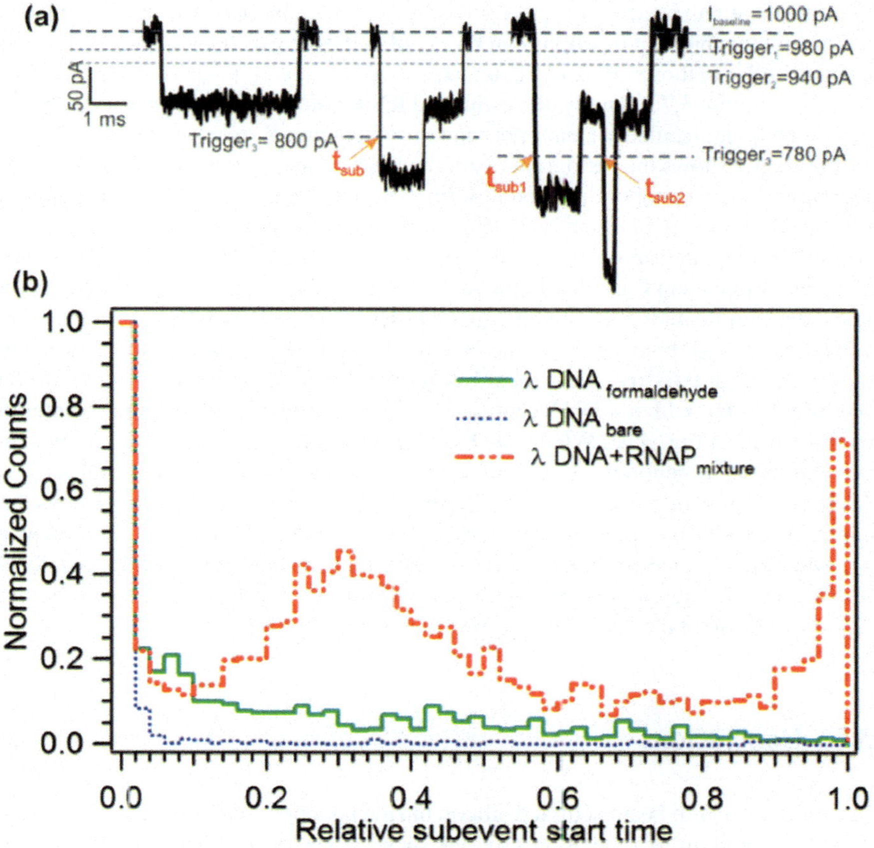

Fig. 3.8 **a** Examples for the selection of subevent start times. **b** Relative subevent starting times for λDNA + RNAP$_{mix}$, λDNA$_{fmd}$, and λDNA$_{bare}$. The starting time histogram for λDNA + RNAP$_{mix}$ shows a major peak at $t_{rel} = 0.3$, there are no major peaks for λDNA$_{fmd}$ and λDNA$_{bare}$. λDNA$_{fmd}$ has a little population at some t_{rel} values, but for λ DNA$_{bare}$ it is almost flat. Counts are normalized to 1 for comparison. Reprinted with permission from [46]. Copyright [46] American Chemical Society

3.5.2 Relative Binding Position of RNAP on λ DNA

The RNAP binding positions on a λ DNA molecule can be estimated for different subevent types. If a λ DNA with an RNAP bound to it passing through a nanopore linearly (Fig. 3.7a), the relative RNAP binding positions can be calculated as: $x_{rel} = t_p/t_d$. The time t_p is from the beginning of an event to the center of a RNAP subevent, and t_d is the dwell time of the whole event. For the events containing folded DNA configurations (with subevent time duration t_{fd}) shown in Fig. 3.7b and c, assuming the translocation speed of a λDNA * $RNAP_{comp}$ in a nanopore was a constant, The RNAP relative binding positions can be calculated as: $x_{rel} = (t_p + t_{fd1})/(t_{fd1} + t_d)$ for events similar to those in Fig. 3.7b, and $x_{rel} = (t_p + t_{fd1})/(t_d + t_{fd1} + 2_{tfd2})$ for events like those in Fig. 3.7c. Finally, the estimated RNAP binding positions on a 16.49 μm long λ DNA molecule by a nanopore measurement are: $x_j = x_{rel} \times 16.49\,\mu$m.

Figure 3.9 shows the histogram of x_j, the RNAP binding positions on the λ DNA measured by the nanopore method described in this chapter. The histogram shows a main peak at $x_j = 3.51 \pm 0.53$ μm (Fig. 3.9b), most likely corresponding to the two strong binding sites at $x_1 = 3.48$ and $x_2 = 4.43$ μm (38,003 and 35,602 bp) on λ DNA as displayed in Fig. 3.9a. In the region of $x_j = 7 - 9$ μm, the measurement did not show significant RNAP binding on the DNA, consistent with that $x_j = x_3, x_4, x_5$ were weak RNAP binding sites. The increased population appears at the exiting end at $x = 12.52 \pm 0.67$ μm were most likely corresponding to the two strong RNAP binding sites but with a λ DNA captured by the nanopore in the opposite orientation as illustrated in Fig. 3.9c. When a λDNA * $RNAP_{comp}$ was captured in an opposite orientation by a nanopore as several examples shown in Fig. 3.7, a peak at $x = 12.98$ μm (a mirror peak of $x_j = 3.51$ μm) would be expected. The data shows that capturing of λ DNA in the opposite orientation had larger errors in measurement of x_j, one possible explanation was that the data analysis here had assumed a λ DNA translocates through a nanopore with constant speed; however, as reported earlier, DNA speeds up at the end of its journey in a nanopore [71].

3.6 Conclusions

In summary, an ion beam sculpted silicon nitride nanopore based sensing device can detect the binding of regulation proteins such as RNAP or other proteins on long DNA molecules at single molecule level. In the case of RNAP and λ DNA, under the experimental conditions with the ratio of 6 to 1 for RNAP to λ DNA molecules, the probability of an RNAP molecule to bind a λ DNA is 42%; and the data analysis shows that a RNAP has a high tendency to bind around at 3.51 ± 0.53 μm, most likely corresponding to the two strong promoter regions at 3.48 and 4.43 μm (38,003 and 35,602 bp) of λ DNA, respectively.

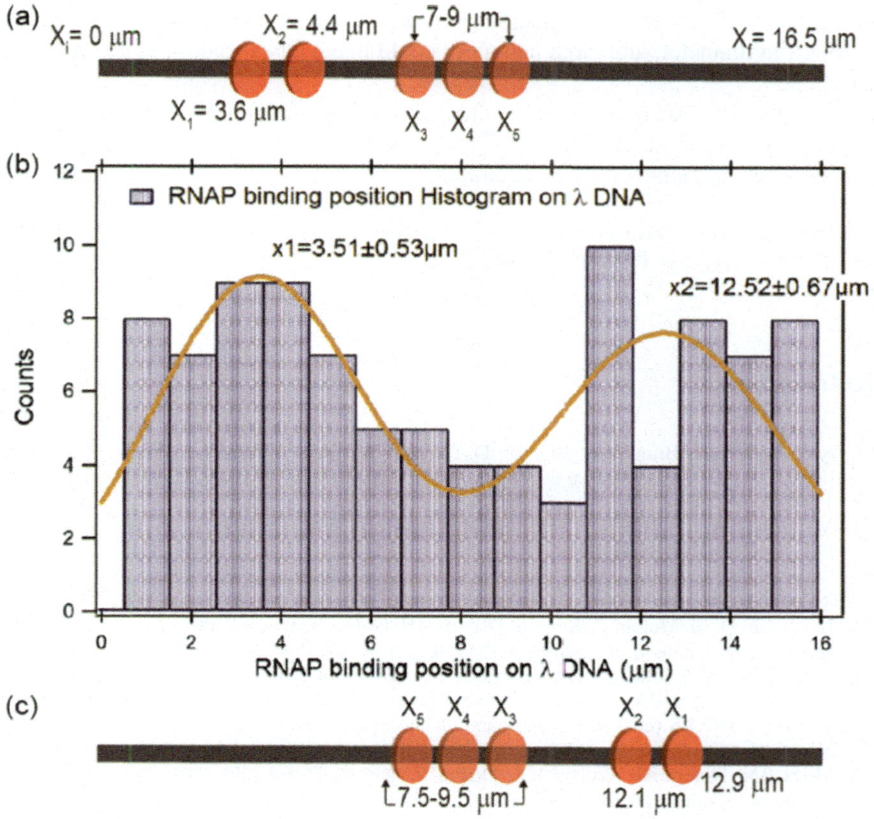

Fig. 3.9 **a** Idealized drawing of the expected RNAP binding locations. **b** Measured positions of RNAP bound on 48.5-kb-long λ DNA molecules. Two Gaussians fitting shows RNAP has a maximum tendency to bind at x = 3.51 ± 0.53 μm. The peaks at the exiting end around x = 12.52 ± 0.67 μm are possibly due to the capture of λ DNA molecules in opposite orientation by the nanopore. These are the counterparts of 3.5 μm. **c** Drawing to show if a λDNA * $RNAP_{complex}$ entered the nanopore from the opposite end as shown in **a**. Reprinted with permission from [46]. Copyright [46] American Chemical Society

However, measuring free translocation of protein bound DNA molecules with a silicon nitride sensing system could not distinguish between individual RNAP binding sites or could not resolve the five possible binding sites [60] accurately. Some possible reasons are: (1) DNA molecules could be captured and enter a nanopore from either ends or any location along the DNA that could make complicated DNA folding configuration and difficult data analysis, (2) low temporal resolution of the nanopore sensing system due to fast free DNA translocation speed, ~ 3000 bases per 100 μs [8], (3) Brownian motion of DNA molecules, (4) DNA passing speed in a nanopore not being constant, partially due to DNA rewinding and winding, and (5) the RNAP used in this example was a large protein, a 470 kDa with an overall size of 12 nm × 15 nm × 18 nm, that limited the choice of a nanopore had to be big, ~ 20 nm in diameter, and large nanopore could make DNA folding structures more favorable.

Possible future work and improvements: (1) using smaller DNA binding proteins; (2) using controlled translocation of DNA molecules that could solve most of the problems we encountered in this example.

Acknowledgements The authors would like to acknowledge financial support for this work from the Arkansas Bioscience Institute (ABI-0170).

References

1. Kasianowicz JJ, Brandin E, Branton D, Deamer DW (1996) Characterization of individual polynucleotide molecules using a membrane channel. Proc Natl Acad Sci USA 93:13770–13773
2. Meller A, Nivon L, Branton D (2001) Voltage-driven DNA translocations through a nanopore. Phys Rev Lett 86(15):3435–3438
3. Stein D, Li J, Golovchenko JA (2002) Ion-beam sculpting time scales. Phys Rev Lett 89(27):276106-1-4
4. Li J, Stein D, McMullan C, Branton D, Aziz MJ, Golovchenko JA (2001) Ion-beam sculpting at nanometre length scales. Nature 412(12 July):166–169
5. Dekker C (2007) Solid-state nanopores. Nat Nanotechnol 2:209–215
6. Li J, Golovchenko JA (2009) Solid-state nanopore for detecting individual biopolymers. In: Lee JW, Foote RS (eds) Micro and nano technologies in bioanalysis. Human Press, Totowa, NJ, pp 81–93
7. Miles BN, Ivanov AP, Wilson KA, Doğan F, Japrung D, Edel JB (2013) Single molecule sensing with solid-state nanopores: novel materials, methods, and applications. Chem Soc Rev 42(1):15–28
8. Li J, Gershow M, Stein D, Brandin E, Golovchenko JA (2003) DNA molecules and configurations in a solid-state nanopore microscope. Nat Mater 2:611–615
9. Storm AJ, Chen JH, Ling XS, Zandbergen HW, Dekker C (2003) Fabrication of solid-state nanopores with single-nnometre precision. Nat Mater 2:537–540
10. Fologea D, Uplinger J, Thomas B, McNabb DS, Li J (2005) Slowing DNA translocation in a solid state nanopore. Nano Lett 5(9):1734–1737
11. Fologea D, Gershow M, Ledden B, McNabb DS, Golovchenko JA, Li J (2005) Detecting single stranded DNA with a solid state nanopore. Nano Lett 5(10):1905–1909
12. Fologea D, Brandin E, Uplinger J, Branton D, Li J (2007) DNA conformation and base number simultaneously determined in a nanopore. Electrophoresis 28:3168–3192
13. Smeets R, Kowalczyk S, Hall A, Dekker N, Dekker C (2008) Translocation of RecA-coated double-stranded DNA through solid-state nanopores. Nano Lett 9(9):3089–3095
14. Hall AR, van Dorp S, Lemay SG, Dekker C (2009) Electrophoretic force on a protein-coated DNA molecule in a solid-state nanopore. Nano Lett 9(12):4441–4445
15. Dorvel B, Sigalov G, Zhao Q, Comer J, Dimitrov V, Mirsaidov U (2009) Analyzing the forces binding a restriction endonuclease to DNA using a synthetic nanopore. Nucleic Acids Res 37(12):4170–4179
16. Kowalczyk SW, Hall AR, Dekker C (2009) Detection of local protein structures along DNA using solid-state nanopores. Nano Lett 10(1):214–328
17. Sischka A, Spiering A, Khaksar M, Laxa M, König J, Dietz K-J (2010) Dynamic translocation of ligand-complexed DNA through solid-state nanopores with optical tweezers. J Phys Condens Matter 22(45):454121

18. Skinner GM, Hout M, Broekmans O, Dekker C, Dekker NH (2009) Distinguishing single- and double-stranded nucleic acid molecules using solid-state nanopores. Nano Lett 9(8):2953–2960
19. Uplinger J, Thomas B, Rollings R, Fologea D, McNabb D, Li J (2012) K+, Na+, and Mg2+on DNA translocation in silicon nitride nanopores. Electrophoresis 33(23):3448–3457
20. Wanunu M, Bhattacharya S, Xie Y, Tor Y, Aksimentiev A, Drndic M (2011) Nanopore analysis of individual RNA/antibiotic complexes. ACS Nano 5(12):9345–9353
21. Wanunu M, Dadosh T, Ray V, Jin J, McReynolds L, Drndic M (2010) Rapid electronic detection of probe-specific microRNAs using thin nanopore sensors. Nat Nano 5(11):807–814
22. Han A, Schurmann G, Mondin G, Bitterli RA, Hegelbach NG, de Rooij NF, Staufer U (2006) Sensing protein molecules using nanofabricated pores. Appl Phys Lett 88:093901–093903
23. Fologea D, Ledden B, McNabb DS, Li J (2007) Electrical characterization of protein molecules in a solid-state nanopore. Appl Phys Lett 91:053901
24. Oukhaled G, Mathe J, Biance A-L, Bacri L, Betton J-M, Lairez D, Pelta J, Auvray L (2007) Unfolding of proteins and long transient conformations detected by single nanopore recording. Phys Rev Lett 98:158101
25. Han A, Creus M, Schürmann G, Linder V, Ward TR, de Rooij NF (2008) Label-free detection of single protein molecules and protein–protein interactions using synthetic nanopores. Anal Chem 80(12):4651–4658
26. Talaga D, Li J (2009) Single-molecule protein unfolding in solid state nanopores. J Am Chem Soc 131:9287–9297
27. Firnkes M, Pedone D, Knezevic J, Doÿblinger M, Rant U (2010) Electrically facilitated translocations of proteins through silicon nitride nanopores: conjoint and competitive action of diffusion, electrophoresis, and electroosmosis. Nano Lett 10(6):2162–2167
28. Stefureac RI, Trivedi D, Marziali A, Lee JS (2010) Evidence that small proteins translocate through silicon nitride pores in a folded conformation. J Phys Condens Matter 17(22):454133
29. Niedzwiecki DJ, Grazul J, Movileanu L (2010) Single-molecule observation of protein adsorption onto an inorganic surface. J Am Chem Soc 132(31):10816–10822
30. Oukhaled A et al (2011) Dynamics of completely unfolded and native proteins through solid-state nanopores as a function of electric driving force. ACS Nano 5:3628–3638
31. Freedman KJ, Jürgens M, Prabhu A, Ahn CW, Jemth P, Edel JB, Kim MJ (2011) Chemical, thermal, and electric field induced unfolding of single protein molecules studied using nanopores. Anal Chem 83:5137–5144
32. Yusko EC, Johnson JM, Majd S, Prangkio P, Rollings RC, Li J, Yang J, Mayer M (2011) Controlling the translocation of proteins through nanopores with bioinspired fluid walls. Nat Nanotechnol 10:253–260
33. Rodriguez-Larrea D, Bayley H (2013) Multistep protein unfolding during nanopore translocation. Nat Nanotechnol 8(4):288–295
34. Plesa C, Kowalczyk SW, Zinsmeester R, Grosberg AY, Rabin Y, Dekker C (2013) Fast translocation of proteins through solid state nanopores. Nano Lett 13(2):658–663
35. Li J, Fologea D, Rollings R, Ledden B (2014) Characterization of protein unfolding with solid-state nanopores. Protein Peptide Lett 21(3):256–265
36. Yusko EC et al (2016) Real-time shape approximation and fingerprinting of single proteins using a nanopore. Nat Nanotechnol 12:360
37. Li X, Tong X, Lu W, Yu D, Diao J, Zhao Q (2019) Label-free detection of early oligomerization of α-synuclein and its mutants A30P/E46K through solid-state nanopores. Nanoscale 11(13):6480–6488
38. Hu R et al (2016) Intrinsic and membrane-facilitated α-synuclein oligomerization revealed by label-free detection through solid-state nanopores. Sci Rep 6:20776
39. Acharjee MC, Ledden B, Thomas B, He X, Giurleo J, Messina T, Talaga D, Li J (2022) Solid state nanopore characterization of ß-lactoglobulin aggregation and oligomerization (submitted)
40. Hornblower B, Coombs A, Whitaker RD, Kolomeisky A, Picone SJ, Meller A (2007) Single-molecule analysis of DNA-protein complexes using nanopores. Nat Methods 4(4):315–317
41. Zhao Q, Sigalov G, Dimitrov V, Dorvel B, Mirsaidov U, Sligar S (2007) Detecting SNPs using a synthetic nanopore. Nano Lett 7(6):1680–1685

42. Spiering A, Getfert S, Sischka A, Reimann P, Anselmetti D (2011) Nanopore translocation dynamics of a single DNA-bound protein. Nano Lett 11(7):2978–2982
43. Raillon C, Cousin P, Traversi F, Garcia-Cordero E, Hernandez N, Radenovic A (2012) Nanopore detection of single molecule RNAP–DNA transcription complex. Nano Lett 12(3):1157–1164
44. Raillon C, Granjon P, Graf M, Radenovic A (2013) Detection of RNAP-DNA complexes using solid state nanopores. Conf Proc IEEE Eng Med Biol Soc 2013:4106–4109
45. Ivankin A, Carson S, Kinney SR, Wanunu M (2013) Fast, label-free force spectroscopy of histone–dna interactions in individual nucleosomes using nanopores. J Am Chem Soc 135(41):15350–15352
46. Kaur H, Nandivada S, Acharjee MC, McNabb DS, Li J (2019) Estimating RNA polymerase protein binding sites on λ DNA using solid-state nanopores. ACS Sens 4(1):100–109
47. Wang Y, Zhao Y, Bollas A, Wang Y, Au KF (2021) Nanopore sequencing technology, bioinformatics and applications. Nat Biotechnol 39(11):1348–1365
48. Hippel PH, McGhee JD (1972) DNA-protein interactions. Annu Rev Biochem 41:231–300
49. Strauch MA (2001) Protein–DNA complexes: specific. In: Encyclopedia of life sciences
50. Garvie CW, Wolberger C (2001) Recognition of specific DNA sequences. Mol Cell 8:937–946
51. Schleif R (1988) DNA binding by proteins. Science 241(4780):1182–1187
52. Little JW (2004) Methods for studying protein-DNA interactions. http://www.biochem.arizona.edu/classes/bioc568/protein_dna_interactions.htm
53. Pugh BF, Gilmour DS (2001) Genome-wide analysis of protein-DNA interactions in living cells. Genome Biol **2**(4): reviews1013.1
54. Dey B, Thukral S, Krishnan S, Chakrobarty M, Gupta S, Manghani C, Rani V (2012) DNA–protein interactions: methods for detection and analysis. Mol Cell Biochem 365(1–2):279–299
55. Weiss S (1999) Fluorescence spectroscopy of single biomolecules. Science 283(5408):1676–1683
56. Elf J, Li G-W, Xie XS (2007) Probing transcription factor dynamics at the single-molecule level in a living cell. Science 306(5828):1191–1194
57. Ebenstein Y, Gassman N, Kim S, Weiss S (2009) Combining atomic force and fluorescence microscopy for analysis of quantum-dot labeled protein–DNA complexes. J Mol Recogn 22(5):397–402
58. Harada Y, Funatsu T, Murakami K, Nonoyama Y, Ishihama A, Yanagida T (1999) Single-molecule imaging of RNA polymerase-DNA interactions in real time. Biophys J 76(2):709–715
59. Gueroui Z, Place C, Freyssingeas E, Berge B (2002) Observation by fluorescence microscopy of transcription on single combed DNA. Proc Natl Acad Sci 99(9):6005–6010
60. Sriram K, Yeh J-W, Lin Y-L, Chang Y-R, Chou C-F (2014) Direct optical mapping of transcription factor binding sites on field-stretched λ-DNA in nanofluidic devices. Nucleic Acids Res 42(10):e85–e85
61. Kaur H (2017) Characterization of RNAP binding sites on a λ DNA using a solid state nanopore combined with a tuning fork force sensing probe tip. In: Physics. University of Arkansas: University of Arkansas Fayetteville, p 134
62. Bulyk ML (2006) DNA microarray technologies for measuring protein–DNA interactions. Curr Opin Biotechnol 17(4):422–430
63. Ghosal S (2007) Effect of salt concentration on the electrophoretic speed of a polyelectrolyte through a nanopore. Phys Rev Lett 98(23):238104
64. Li J, Talaga DS (2010) The distribution of DNA translocation times in solid-state nanopores. J Phys Condens Matter 22:454129 (8 pp)
65. Ling DY, Ling XS (2013) On the distribution of DNA translocation times in solid-state nanopores: an analysis using Schrödinger's first-passage-time theory. J Phys Condens Matter 25(37)
66. Storm AJ, Chen JH, Zandbergen HW, Dekker C (2005) Translocation of double-stranded DNA through a silicon oxide nanopore. Phys Rev E 71
67. Cai Q, Ledden B, Krueger E, Golovchenko JA, Li J (2006) Nanopore sculpting with noble gas ions. J Appl Phys 100

68. Rollings R, Graef E, Walsh N, Nandivada S, Benamara M, Li J (2015) The effects of geometry and stability of solid-state nanopores on detecting single DNA molecules. Nanotechnology 26(4):P044001
69. Geszvain K, Landick R (2005) The structure of bacterial RNA polymerase. In: Higgins NP (ed) The bacterial chromosome. ASM, Washington DC, pp 283–296
70. Brodolin K, Mustaev A, Severinov K, Nikiforov V (2000) Identification of RNA polymerase β′ subunit segment contacting the melted region of the lacUV5 promoter. J Biol Chem 275(February 4):3661–3666
71. Plesa C, van Loo N, Ketterer P, Dietz H, Dekker C (2014) Velocity of DNA during translocation through a solid-state nanopore. Nano Lett 15(1):732–737

Chapter 4
Solid-State Nanopore Sensing Enhanced by Designed DNA Nanostructures

Kaikai Chen and Ulrich F. Keyser

Abstract Solid-state nanopores are versatile single-molecule tools for label-free detection of biomolecules including DNA, RNA and protein by passing these molecules through nanopores using an applied electric potential. One of the main key challenges of solid-state nanopore in biosensing applications is their lack of specificity, as polymer chains with similar size and charge cause similar ionic current signals when they pass a nanopore. Thus, designed DNA nanostructures, named DNA carrier, which has specific binding sites that can selectively fish target molecules, were introduced to detect specific biomolecules such as protein and RNA. The DNA carrier attached with target protein generates a unique ionic current indicating the presence of the target and its location on the DNA carrier. Similarly, this system can be used to detect DNA sequence if functional protein molecules that can bind to specific DNA sequences are used to capture the DNA sequence. Moreover, the DNA carrier attached with DNA nanostructures has also find its way in the application as digital barcodes and for digital data storage. The use of DNA nanostructures significantly improves the specificity and versatility of solid-state nanopores in biosensing applications.

Keywords Nanopores · Solid-state nanopores · DNA nanotechnology · Biosensing · Single molecules · Data storage · DNA computing

Since the demonstration of DNA translocation through a natural protein pore [1], great effort has been invested in repurposing biological nanopores for DNA

K. Chen · U. F. Keyser (✉)
Cavendish Laboratory, University of Cambridge, Cambridge, UK
e-mail: ufk20@cam.ac.uk

K. Chen
e-mail: kc494@cam.ac.uk

K. Chen
School of Nanoscience and Nanotechnology, University of Chinese Academy of Sciences, Beijing, China

J.-P. Leburton (ed.), *Solid State Nanopores*, Nanostructure Science and Technology,
https://doi.org/10.1007/978-3-031-42336-9_4

sequencing applications. In particular, the funding available for an obvious billion-dollar application like DNA sequencing sped up the development of nanopore sensing. The main goal was to find a suitable nanopore that can distinguish the nucleotides and read out the sequence on a DNA chain. Nature has created many nanopores in biological systems, however, even if they have well-defined diameters in structural studies, many nanopores prove to be dynamic, displaying time-dependent current fluctuation that obscures the sensing signal. Thus, researchers familiar with semiconductor nanotechnology considered using use artificial pores with desired size and geometries made in silicon-based systems. With the advances in solid-state physics and nanofabrication, nanopores fabricated by sculpting a hole in a membrane were shown in 2001 by Li et al. [2]. The main aim was to mimic biological nanopores aiming for DNA sequencing, with diameter reaching the single nanometre range. Now, more than twenty years later, the nanopores used for sequencing remain biological. However, one may imagine that solid-state nanopores may one day surpass biological nanopores considering the controllable size and ever-widening range of material systems beyond silicon. Solid-state nanopores can be fabricated with a size smaller than 1 nm, equivalent to or even smaller than the diameter of single-stranded DNA (ssDNA), and they can be fabricated in atomic-thin 2D materials with a thickness below 1 nm for the monolayer membranes such as graphene and MoS_2 [3–6], comparable to the distance between neighbouring base pairs thus suitable for DNA sequencing. The undoubtedly large potential of solid-state nanopores for DNA sequencing lies in the long-term prospects of integrating the fabrication process into semiconductor fabs. DNA sequencing was firstly achieved by biological nanopores with a narrow reading head and a motor protein to unzip the double-stranded (dsDNA) and help control the translocation. This technology was then commercialised by Oxford Nanopore Technologies. To the best of our knowledge, solid-state nanopores have not achieved the function of DNA sequencing equivalent to biological nanopores. For the reasons stated above, the field is now moving towards other applications. Most of these applications are based on the translocation of DNA through nanopores which proves to be a complex problem due to the combination of forces acting on the polymer. DNA translocation through solid-state nanopores now allows for a plethora of biosensing experiments for many biomolecules. The field is moving beyond DNA sequencing [7, 8], from simply detecting DNA and protein molecules for biomarker detection to reading complex molecular structures for polymer-based digital data storage.

For biosensing with solid-state nanopores, one key challenge is that the nanopore signal can only provide information about the size and charge of the polymer, but not the sequence information, which is not enough to distinguish them. For example, for DNA with a different sequence but the same length or different protein molecules with a similar size, one cannot tell the difference from the nanopore signals. In other words, the lack of specificity of the ionic current signal is the main issue to solve. To this end, two popular methods could be used: (1) modification on the nanopore surface to create specific interactions [9]; (2) using a binding molecule that could bind to the target to generate unique signals [10]. The second approach provides a general detection of different molecules by designing a binding site on molecules getting rid of complex chemical modifications on the nanopore surface. For instance, a dsDNA molecule with a specific binding site was shown to be very useful, thus creating a long-strand chain bound with additional molecules, where the dsDNA was named the DNA carrier [10, 11].

In the DNA carrier system, a polymer chain with attached structures passes the nanopore. The molecule creates a signal with secondary spikes on the dsDNA level. The presence, as well as the positions of the additional structures on the linear polymer, could be decided from the ionic current signals. With this method, one can either detect the bound molecules if we design binding sites on a DNA carrier or detect the DNA sequences if we use molecules such as Cas9 that can specifically target a DNA sequence [12]. Moreover, by designing multiple binding sites or barcoding a DNA carrier with designed DNA nanostructures [13], multiplexed sensing of protein can be achieved. The design of the DNA barcode also allows for storing digital information on the polymer. By increasing the density of the bits, it can be used for digital data storage.

As we will discuss in this chapter, the introduction of the DNA carrier system provides an enhanced solid-state nanopore sensing platform by encoding information into the translocating molecule. We are discussing how the DNA carriers can be used to understand the translocation of DNA through solid-state nanopores, and how the resulting iconic signatures that reflect the 3D structure of DNA can be employed for sensing applications.

4.1 Conformation of DNA During Its Translocation Through Solid-State Nanopores

When the size of the nanopore is comparable to the diameter of the DNA strand, the DNA can only be dragged through the nanopore in a single-file format, i.e., the DNA is linearly threaded into the pore. When the nanopore diameter becomes larger, wider than double the diameter of the DNA double helix, the DNA can be folded to pass through the nanopore when the middle part is exposed to the electric field. Thus, for wide nanopores, unfolded and folded translocations are both observed, represented by distinct ionic current signatures shown in Fig. 4.1a and b, respectively. The plateau in the signal is caused by the unfolded translocation, while the two-level signal was caused by folded DNA. At the beginning of the folded DNA translocation, the current drop is about two times the DNA level, and then the current returns to the DNA level after the folded segment, Fig. 4.1b.

Apart from the double-folded DNA, knots in the DNA can also cause ionic signals deeper than the dsDNA level, Fig. 4.1c [14, 15]. Often the knots cannot be untied due to the force applied when the DNA passes through, so it causes an additional, characteristic spike. The knots appear more for longer DNA, for example with a length > 10 kbp, while it rarely appears for DNA with a short length.

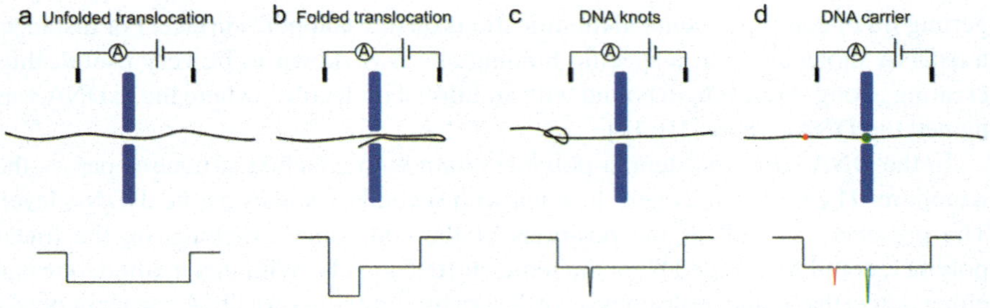

Fig. 4.1 A variety of DNA conformations as they translocate through nanopores and the resulting ionic current signals. **a** A single-file (unfolded) DNA translocation. **b** A folded DNA translocation. **c** Translocation of DNA with a knot. **d** Translocation of DNA with bound molecules

By attaching additional molecules, we can create more distinct current signals with a defined depth and position. In the example in Fig. 4.1d, two molecules represented by the blue and red dots are attached, at the ¼ and ½ positions on the DNA respectively. The molecules can be detected from the signal at the corresponding positions in the ionic current signal. Moreover, for the two molecules with different sizes, the signal strengths are different [13]. This is the key idea for the DNA carrier system. Using DNA nanotechnology, labelling enzymes or any targeted chemical attachment we can design multiple sites on the DNA and read them off from the nanopore ionic current signals.

One key question for the free-translocation nanopore-polymer system is whether we can accurately determine the position of the molecules on the DNA. The localisation of sites requires an understanding of how DNA passes through the nanopore: how the velocity changes, and how large the velocity fluctuations are. In the ideal constant speed assumption, the relative position of the attached molecule on the DNA carrier is equal to the position of the spike in the ionic current trace. However, this does not even work for biological nanopores and hence applications require a molecular machine that limits the step size to 1 nucleotide at a time. Already Lu et al. reported the fluctuations of the DNA through solid-state nanopores [16], where they found the wide distribution of the translocation times of 5 and 10 kb DNA, Fig. 4.2a, which were largely attributed to the conformation of the DNA outside of the nanopore. The fast and slow translocations were revealed by their model. The model suggested that the extension of the molecules if they were stretched or not, determined the translocation time. In the further experiments of Plesa et al. [17], the first experimental study of the intra-event velocity was demonstrated by attaching an overhang on the dsDNA strand, Fig. 4.2b. They showed the velocity fluctuation and the speed change, indicating a fast velocity at the end of the translocation. However, the evidence on how the speed changes during the whole translocation process was still missing. Only recently, the first direct experimental evidence was reported [18], showing that for nanopores with diameters of 5–10 nm the velocity slows down first and then speeds up quickly at the end, Fig. 4.2c. In this work, nanostructured DNA carriers were used to directly measure the velocity profile of the DNA translocation through the solid-state nanopores. The results reveal a two-stage DNA translocation, named the tension propagation stage (velocity decreasing) and the tail retraction stage (velocity increasing). In the initial slowing-down phase, the tension is only felt by a local section of the polymer and the tension front propagates out towards the end of the polymer as successive folds are straightened. In the second speeding-up phase, the tension has fully propagated through the polymer and the length of tensed polymer progressively decreases as it is sucked through the pore. Importantly, the experiments compare short and conical nanopores indicating the general applicability of the velocity profile for different nanopore shapes.

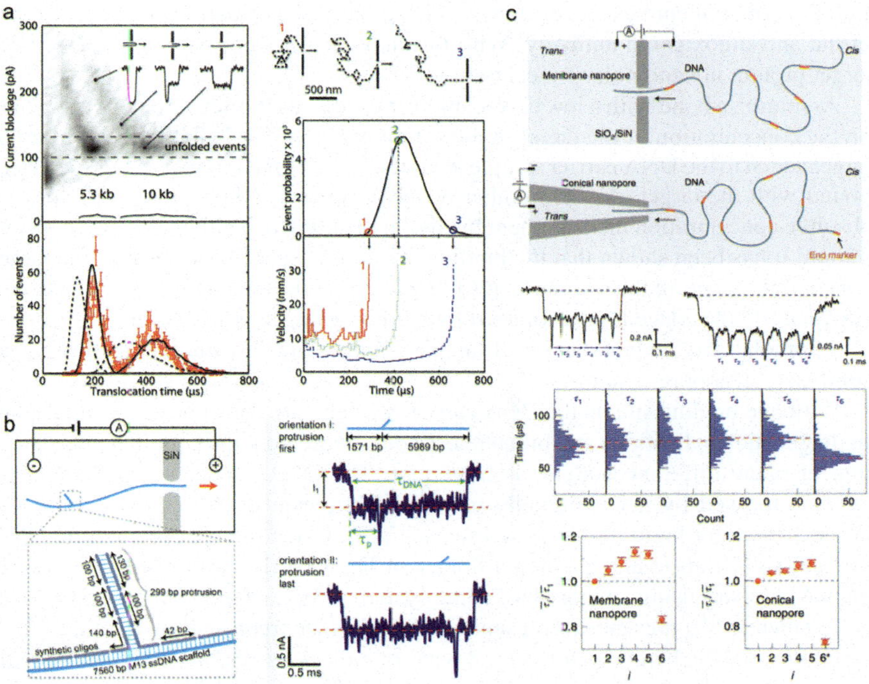

Fig. 4.2 The velocity of DNA translocation through solid-state nanopores. **a** Examples of the translocation time distributions of 5 and 10 kbp DNA and the model on the velocity distribution [16]. **b** Inconstant velocity reflected by measuring a dsDNA strand with an overhang [17]. **c** DNA velocity change measured by using the DNA ruler [18]. Panel **a** adapted from Ref. [16] with permission from Elsevier. Panel **b** adapted from Ref. [17] with permission from American Chemical Society. Panel **c** adapted from Ref. [18] from Springer Nature

4.2 DNA Carrier for Protein Sensing

As mentioned above, when a binding site is designed on a DNA carrier and the protein target has bound to DNA [19], Fig. 4.3a, the bound molecule together with the DNA could cause a well-defined signal, different from any other molecule in the solution. This signal is ideally suited to identify the protein in single-molecule detection. Bell et al. first demonstrated the DNA carrier system in the proof-of-concept work on its use for specific protein detection by designing a site with biotin molecules on the DNA carrier for the detection of streptavidin [11], Fig. 4.3b. This method can be widely expanded with antigen-antibody specific binding pairs. For example, digoxigenin can be designed as a tag on the DNA carrier for the detection of the anti-digoxigenin antibody. With this approach, it is possible to measure the target protein in nanomolar concentration.

Assuming a bond with a low dissociation constant, the binding efficiency depends on the concentration of the target in the solution. Hence, variation in the amount of target added to the DNA carrier sample leads to changing the number of DNA carriers bound with the target. As the number of DNA carriers is known this provides an absolute concentration measurement by digital counting using the reference concentration. It has been shown that the binding efficiency was almost linear to the target concentration before the saturation [20], Fig. 4.3c. Thus, the DNA carrier can detect and quantify the concentration of a strongly binding target, with concentration down to nanomolar and the inherent small sample volumes of a few microliters for typical solid-state nanopore systems.

With one binding site on the DNA carrier, we can only detect one type of protein. In diagnostic applications and bionanotechnology research, the simultaneous detection of many different analytes is important. The DNA carrier method could be easily improved by using digitally encoded DNA nanostructures to barcode the DNA carriers for highly multiplexed sensing [13]. Using the idea of building DNA origami [21], Bell et al. designed a library of DNA carries with nanostructures as unique barcodes using a binary barcoding system [13]. Each position in the barcode is signalled by the presence (representing 1) or absence (representing 0) of the DNA structures. A 3-bit barcode allows for 8 types of molecules, and the optimised signal-to-noise ratio allowed assignment with high accuracy using nanopore readout, as shown in Fig. 4.4a. The selected members of the library can be functionalized with a binding site for the detection of target molecules. The DNA carrier library enables the simultaneous detection of multiple targets, Fig. 4.4b. The advantage of the digital barcode carrier is the virtually unlimited number of molecules, where the maximum number of the DNA carries is 2^n, where n is the bit number. For example, with a 10-bit barcode, one can detect up to 1024 different targets assuming only one target per DNA carrier. This approach allows us to simultaneously detect different antibodies of the same isotype at nanomolar concentration levels.

With the antigen binding site designed on the carrier, the types of proteins that can be detected are determined by the available antigen-antibody reactions. To detect a

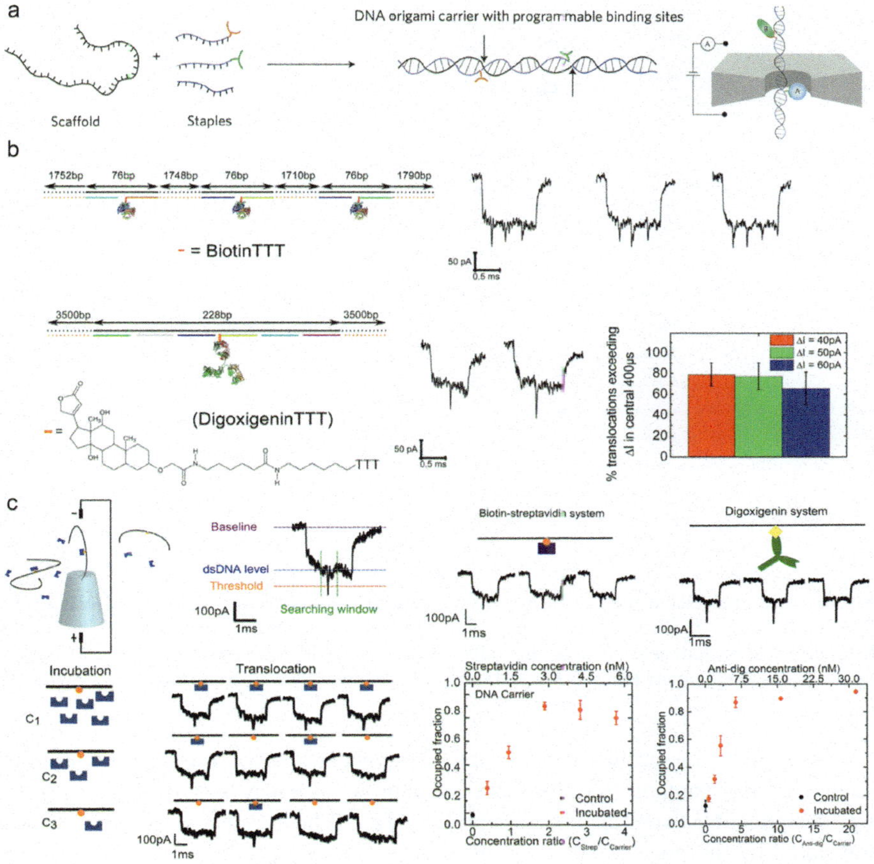

Fig. 4.3 DNA carrier system for detecting and quantifying nanomolar concentration proteins. **a** Sketch of the design of the DNA carrier system [19]. **b** Design of the DNA carrier with binding sites for the detection of specific proteins [11]. **c** Quantifying nanomolar concentration proteins using the DNA carrier and nanopore measurement [20]. Panel **a** adapted from Ref. [19] with permission from Springer Nature. Panel **b** adapted from Ref. [11] with CC-BY license. Panel **c** adapted from Ref. [20] with CC-BY license

wide range of proteins, the use of specifically designed DNA aptamers is especially attractive. A DNA aptamer is a short ssDNA strand that can bind to a specific protein, where the specificity lies in the DNA sequence, and it has been found that many proteins have unique DNA aptamer sequences that they specifically bind to. Following Bell's work on the DNA carrier [11], aptamers were used as baits for the detection proteins where the aptamer was attached at the end of a lambda DNA, Fig. 4.5a and b [22]. Several binding sites with aptamers can also be designed for multiplexed sensing. The DNA carrier system was used to perform single-molecule

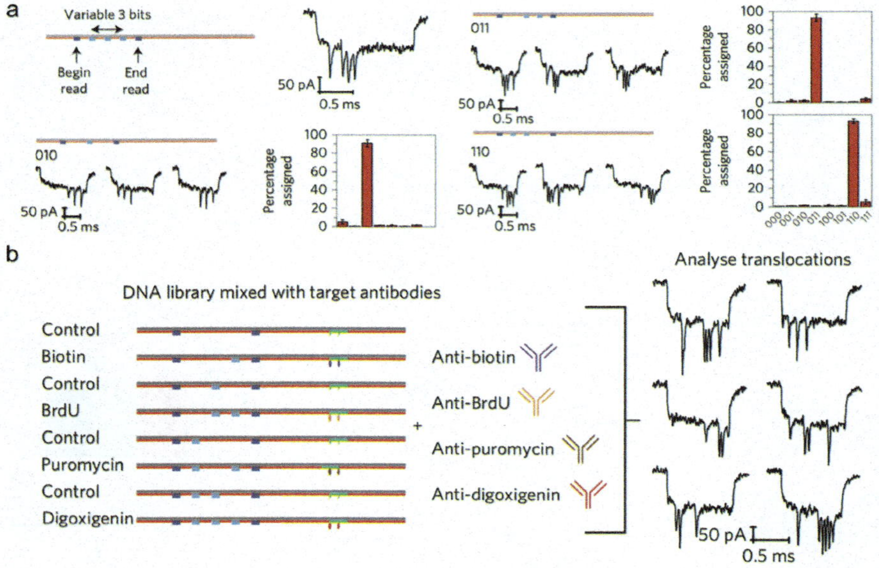

Fig. 4.4 Multiplexed DNA sensing using DNA nanostructures as the barcodes [13]. **a** Design of the DNA carriers with digitally encoded nanostructures and their signals from nanopore measurements. **b** Multiplexed protein sensing using digitally encoded DNA carriers functionalised with antigen binding sites. Panel **a** and Panel **b** adapted from Ref. [13] with permission from Springer Nature

screening in human serum at ultra-low protein concentrations. Kong et al. also demonstrated the use of aptamers for the detection of proteins, where the signal of the protein was designed in a two-step detection protocol [23]. The presence of the proteins was translated by the presence of the oligonucleotides using a DNA circuit. Thus, the environment of the target protein is not limited as we only need to detect the output of the circuits. In this work, it is also shown that the protein can be used as barcodes with different levels in the nanopore signal.

4.3 DNA Sequence Detection

The system with a linear DNA strand and a bound molecule can also be used to detect DNA sequences. If we have a molecule that can specifically target a DNA sequence, when we pass the bound molecule through a nanopore, a well-defined signal can also be observed. In this case, the system is used to detect the sequence with the DNA as the target and the protein as the bait. Several types of reactions are suitable for the detection of DNA sequences. In the pioneering work of Meller [24], Fig. 4.6a, the detection was achieved by hybridizing the peptide nucleic acid (PNA) probes with the dsDNA. This was further improved by using the probes to target multiple

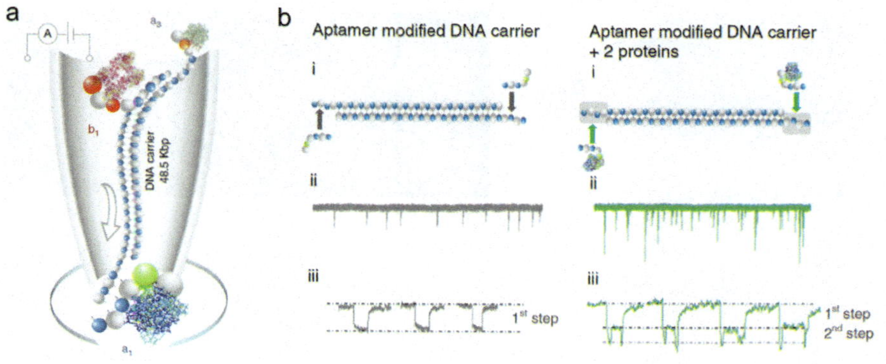

Fig. 4.5 Specific protein detection using nanopores and aptamers on DNA carriers [22]. **a** Schematic of a DNA carrier that contains aptamer sequences bound with three proteins passing through a nanopore. **b** Ionic signals caused by DNA carrier without and with proteins. Panel **a** and Panel **b** adapted from Ref. [22] with CC BY 4.0 license

sites on the DNA for electronic barcoding of a viral gene [25]. In recent work, they also demonstrated that the solid-state nanopores could be used for the detection of transcription factor (TF)-DNA duplex [26]. TF-binding plays an important role in the regulation of gene expression. Another option for targeting the specific sequence is to use the DNA-methyltransferase which specifically modifies a short sequence. Usually, the sequence is very short. For example, DNA-methyltransferase M.TaqI can specifically modify a short sequence 5′-TCGA-3′. Usually, this sequence appears several times in a unique pattern along dsDNA with the length of kilobase pairs or more. Thus sequence-specific labelling is a universal approach for barcoding DNA. Chen et al. showed that the binding site could be detected by adding streptavidin to bind to the biotin added by the enzyme. The streptavidin binds to the biotin and hence enlarges the current change in nanopore measurement [27]. This approach was improved by directly detecting the oligonucleotide (ODN) labels that were modified on the sequences with small nanopores down to the size of ~ 5 nm, Fig. 4.6b. In those experiments, the resolution was enhanced with a shorter sensing region of smaller nanopore diameter and position corrections to achieve higher accuracy with the known distance-time relation from the dynamics of the DNA translocation [28].

A powerful tool for editing and identifying DNA sequences which have attracted significant attention is the Clustered Regularly Interspaced Short Palindromic Repeats (CRISPR) system. The Cas9 molecules can specificity find a sequence usually with a length of 20 bp and edit the genome at these places. Deactivated Cas9 (dCAs9) does not cut or edit the sequence but just binds to the sequence, providing an option for detecting specific 20 bp sequences. The advantage of the methods is that the 20 bp sequence can be designed by changing the sequence of the guide RNA. A 20 bp sequence is highly unique, so one dCas9 with the specific

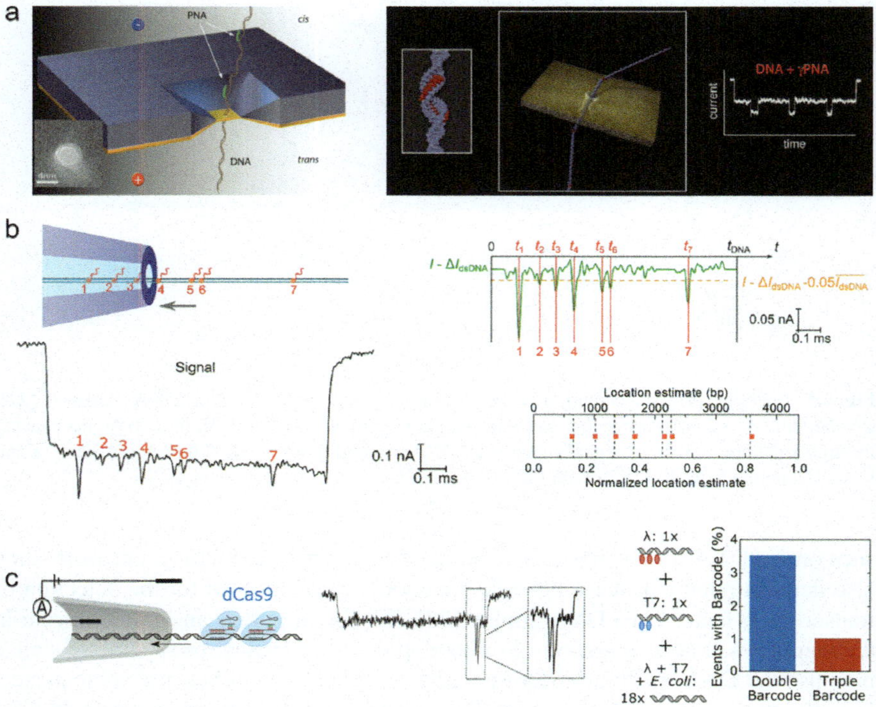

Fig. 4.6 DNA sequence detection using specifically bound molecules and nanopores. **a** Barcoding the DNA with PNA molecules [24, 25]. **b** Labelling the TCGA sequence with ODN using the DNA-methyltransferase M.TaqI measured ~ 5 nm nanopores [28]. The locations were corrected with the know distance-time relation. **c** Double and triple dCas9 barcoding system measured with nanopores [12]. Panel **a** adapted from Refs. [24, 25] with permission from American Chemical Society. Panel **b** adapted from Ref. [28] with CC-BY license. Panel **c** adapted from Ref. [12] with permission from American Chemical Society

sequence would be enough to identify a specific part of one genome. By combining the dCa9 and nanopores, the sequence could be easily detected. Yang et al. showed the translocation of theCas9 bound DNA through solid-state nanopores [29]. However, the signal of the DNA carrier bound with one Cas9 molecule could still be similar to that of a DNA fold or a knot. To increase the specificity and multiplex the detection, Weckman et al. used two or more Ca9 molecules to bind to the DNA sequence to form more unique patterns [12]. Double and triple barcodes were used to simultaneously identify two different DNA targets in a background mixture of bacterial DNA, Fig. 4.6c. The method forms the basis of a fast and versatile assay for multiplexed DNA sensing applications in complex samples.

4.4 Digital Data Storage on Polymers

Recently, DNA as an information carrier becomes popular as there is great demand for next-generation data storage platforms for the exponentially increasing digital data to be stored. DNA has shown great potential for next-generation data storage, with the advantages of ultra-high density, long lifetime, and low maintenance and hence energy cost. Inspired by nature, where genetic information is stored in the sequence of DNA, the idea of storing digital data in DNA is not new [30]. In recent years, the synthesis and sequencing of DNA become easier and more approachable, making DNA data storage promising for practical applications. The storage can be in two steps: the first step is to encode information in the polymer, and the second step is to decode the information using DNA sequencing [31]. More broadly, DNA is just one possible candidate for polymer data storage, as other polymers can also be repurposed for data storage. For instance, Cao et al. demonstrated the reading of the digital information encoded in tailored informational polymers with single-bit resolution using the engineered aerolysin nanopores [32].

Though the most straightforward method for polymer data storage appears to be to encode the information in the DNA sequence itself, the high cost and long-time consumption hinder the development of practical data storage so far. The data reading—DNA sequencing is becoming much cheaper and faster, while the data writing—DNA synthesis is still slow and expensive. Self-assembly may overcome this key bottleneck for applications of data storage. One option for simplifying the data writing is to borrow the idea of building DNA origami, where the structures are simply made by mixing the oligonucleotides and can be used to encode digital data. Based on Bell's work [13], the nanostructures could be more closely spaced to encode digital bits with higher density [33]. The distance of the structures was shortened to 76 bp to increase the data density, Fig. 4.7a [33]. Long (18 bp) and short (10 bp) hairpins represented 1 and 0 respectively and they were successfully read off by the nanopores. The data density could be further increased by using smaller nanopores. Even though the data density was much lower than that stored in DNA sequence, the ease of data writing and reading allows for the possibility of practical applications. Also, as the data density stored in the DNA sequence is ultrahigh, about six orders of magnitude of that in hard drives, so the density at 76 bp is still attractive [37]. Moreover, the reading is much faster because translocations through solid-state nanopores are much faster than that in biological nanopores for DNA sequencing. A typical read of a DNA carrier takes 1–2 ms while DNA sequencing requires 1 ms to read a single base. The data can be written by simply mixing the oligonucleotides, which could be conveniently achieved with microfluidics. The approach enables the encoding of a large amount of information with only a library with limited building blocks. For example, for a 100-bit DNA carrier, a maximum of 2^{10} bits of data could be stored. One route to improve the data density is to use different structures that create multi-level signals in nanopore measurements to encode ternary or quaternary data. Zhu et al. showed an example by using DNA junction structures of different sizes (4-way junction, 6-way junction, and 12-way junction) to generate a quaternary encoding system, Fig. 4.7b [34]. They also saved, encrypted and decoded a 2D grayscale image [34].

This data storage method was improved by using changeable bits in the data, mimicking the operation in hard drives, so it was named the DNA hard drive, Fig. 4.7c [35]. In this case, the data writing, erasing, and rewriting were achieved by attaching or removing molecules on the DNA carriers. It also has the potential for DNA computing based on the reactions on the DNA carrier. The possibility to build a computer based on biological molecules like DNA offers direct interfacing with biological materials like cells or even tissues. DNA carrier combined with nanopore readout of the structures would facilitate the interfacing with silicon-based information technology.

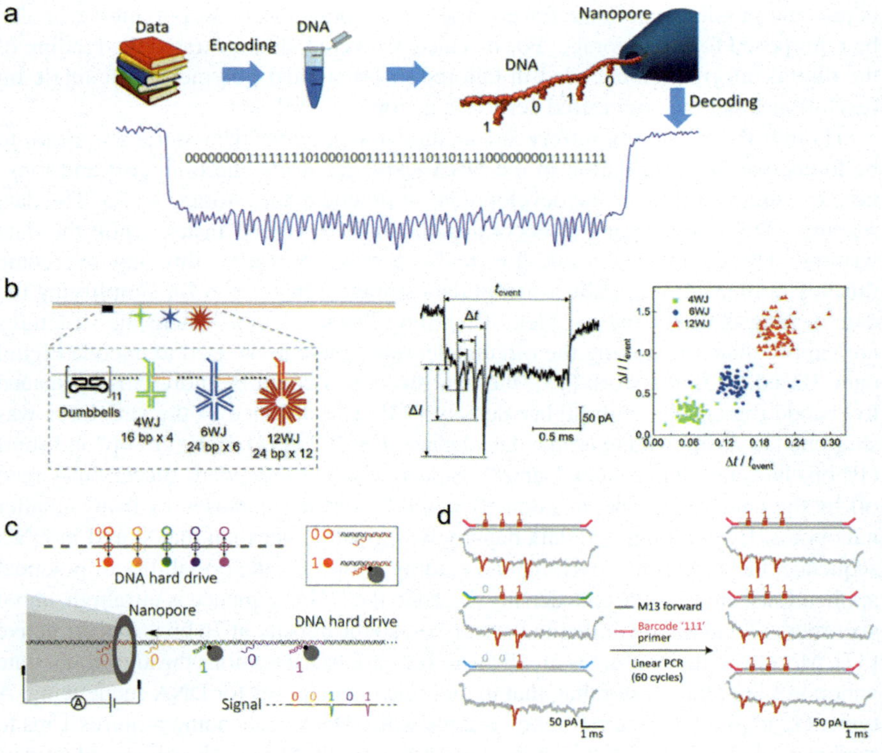

Fig. 4.7 Digital data storage using DNA and nanopores. **a** Digital data storage using DNA nanostructures and nanopores [33]. **b** Multi-level DNA barcodes for data storage [34]. **c** DNA hard drive achieved by attaching and removing molecules on DNA carriers [35]. **d** DNA structural barcode copying and random access [36]. Panel **a** adapted from Ref. [33] with permission from American Chemical Society. Panel **b** adapted from Ref. [34] with CC BY 4.0 license. Panel **c** adapted from Ref. [35] with permission from American Chemical Society. Panel **d** adapted from Ref. [36] with CC BY 4.0 license

Copying of the data is essential and is particularly important for molecular data storage because copying is the basis for random access to data in the system. The copying of data stored in DNA sequences can be easily performed using the polymerase chain reaction (PCR). For the DNA carrier build with an ssDNA scaffold and short oligonucleotides with structures, PCR cannot directly be achieved because of the nicks between the oligonucleotides. To solve this issue, the nicks between the oligonucleotides were linked by a DNA ligase so that the DNA carriers could be amplified using a modified PCR, Fig. 4.7d [36]. The study provides a roadmap for the convenient production of large quantities of self-assembled DNA nanostructures. In addition, this strategy offers access to specific targets by selectively amplifying the desired molecules.

In summary, the use of solid-state nanopores for biosensing enhanced by designed DNA carriers is discussed in the chapter. The combination of DNA nanotechnology and solid-state nanopores significantly improves the sensing specificity and expands the possible applications beyond DNA and RNA sequencing. First, specific protein detection was achieved by designing binding sites on the DNA carrier. DNA aptamer on the DNA carrier provides the possibility of detecting a variety of proteins. This approach shows the advantage of high specificity compared to protein detection by directly passing the protein molecules through nanopores. Second, the DNA–protein duplex also allows for the detection of DNA sequences when the protein molecules such as dCas9 can identify the DNA sequence. It can be used for diagnosis for rapid detection of DNA sequences of interest at a low concentration without sequencing. Thus, the electric detection of the DNA sequence is possible which could be measured with miniaturized devices for point-of-care testing. Finally, DNA nanostructures and solid-state nanopores show competitive advantages in polymer data storage and provide an alternative to data storage in DNA sequences with fast and easy data writing and reading. In future, solid-state nanopores could be further improved by combining them with more molecular tools to fully make full use advantages of solid-state materials and biomaterials. The applications in the detection of DNA, RNA and proteins will be improved and more applications will be shown beyond sequencing applications.

References

1. Kasianowicz JJ, Brandin E, Branton D, Deamer DW (1996) Characterization of individual polynucleotide molecules using a membrane channel. Proc Natl Acad Sci 93:13770–13773
2. Li J et al (2001) Ion-beam sculpting at nanometre length scales. Nature 412:166–169
3. Garaj S et al (2010) Graphene as a subnanometre trans-electrode membrane. Nature 467:190–193
4. Merchant CA et al (2010) DNA translocation through graphene nanopores. Nano Lett 10:2915–2921
5. Schneider GF et al (2010) DNA translocation through graphene nanopores. Nano Lett 10:3163–3167
6. Feng J et al (2015) Identification of single nucleotides in MoS2 nanopores. Nat Nanotechnol 10:1070–1076

7. Keyser UF (2012) Nanopores–mission accomplished and what next? Phys Life Rev 2:164–166
8. Ying Y-L et al (2022) Nanopore-based technologies beyond DNA sequencing. Nat Nanotechnol 17(11):1136–1146
9. Yusko EC et al (2017) Real-time shape approximation and fingerprinting of single proteins using a nanopore. Nat Nanotechnol 12:360–367
10. Xue L et al (2020) Solid-state nanopore sensors. Nat Rev Mater 5:931–951
11. Bell NA, Keyser UF (2015) Specific protein detection using designed DNA carriers and nanopores. J Am Chem Soc 137:2035–2041
12. Weckman NE et al (2019) Multiplexed DNA identification using site specific dCas9 barcodes and nanopore sensing. ACS Sens 4:2065–2072
13. Bell NA, Keyser UF (2016) Digitally encoded DNA nanostructures for multiplexed, single-molecule protein sensing with nanopores. Nat Nanotechnol 11:645–651
14. Plesa C et al (2016) Direct observation of DNA knots using a solid-state nanopore. Nat Nanotechnol 11:1093–1097
15. Kumar Sharma R, Agrawal I, Dai L, Doyle PS, Garaj S (2019) Complex DNA knots detected with a nanopore sensor. Nat Commun 10:1–9
16. Lu B, Albertorio F, Hoogerheide DP, Golovchenko JA (2011) Origins and consequences of velocity fluctuations during DNA passage through a nanopore. Biophys J 101:70–79
17. Plesa C, van Loo N, Ketterer P, Dietz H, Dekker C (2015) Velocity of DNA during translocation through a solid-state nanopore. Nano Lett 15:732–737
18. Chen K et al (2021) Dynamics of driven polymer transport through a nanopore. Nat Phys 17:1043–1049
19. Keyser UF (2016) Enhancing nanopore sensing with DNA nanotechnology. Nat Nanotechnol 11:106–108
20. Kong J, Bell NA, Keyser UF (2016) Quantifying nanomolar protein concentrations using designed DNA carriers and solid-state nanopores. Nano Lett 16:3557–3562
21. Rothemund PW (2006) Folding DNA to create nanoscale shapes and patterns. Nature 440:297–302
22. Sze JY, Ivanov AP, Cass AE, Edel JB (2017) Single molecule multiplexed nanopore protein screening in human serum using aptamer modified DNA carriers. Nat Commun 8:1–10
23. Kong J, Zhu J, Chen K, Keyser UF (2019) Specific biosensing using DNA aptamers and nanopores. Adv Funct Mater 29:1807555
24. Singer A et al (2010) Nanopore based sequence specific detection of duplex DNA for genomic profiling. Nano Lett 10:738–742
25. Singer A, Rapireddy S, Ly DH, Meller A (2012) Electronic barcoding of a viral gene at the single-molecule level. Nano Lett 12:1722–1728
26. Squires A, Atas E, Meller A (2015) Nanopore sensing of individual transcription factors bound to DNA. Sci Rep 5:11643
27. Chen K et al (2017) Ionic current-based mapping of short sequence motifs in single DNA molecules using solid-state nanopores. Nano Lett 17:5199–5205
28. Chen K, Gularek F, Liu B, Weinhold E, Keyser UF (2021) Electrical DNA sequence mapping using oligodeoxynucleotide labels and nanopores. ACS Nano 15:2679–2685
29. Yang W et al (2018) Detection of CRISPR-dCas9 on DNA with solid-state nanopores. Nano Lett 18:6469–6474
30. Davis J (1996) Microvenus. Art J 55:70–74
31. Ceze L, Nivala J, Strauss K (2019) Molecular digital data storage using DNA. Nat Rev Genet 20:456–466
32. Cao C et al (2020) Aerolysin nanopores decode digital information stored in tailored macromolecular analytes. Sci Adv 6:eabc2661
33. Chen K et al (2019) Digital data storage using DNA nanostructures and solid-state nanopores. Nano Lett 19:1210–1215
34. Zhu J, Ermann N, Chen K, Keyser UF (2021) Image encoding using multi-level DNA barcodes with nanopore readout. Small 17:e2100711

35. Chen K, Zhu J, Boskovic F, Keyser UF (2020) Nanopore-based DNA hard drives for rewritable and secure data storage. Nano Lett 20:3754–3760
36. Bošković F, Ohmann A, Keyser UF, Chen K (2021) DNA structural barcode copying and random access. Small Struct 2:2000144
37. Church GM, Gao Y, Kosuri S (2012) Next-generation digital information storage in DNA. Science 337:1628–1628

Chapter 5
Protein Profiling by a Confined Nanopore

Ru-Jia Yu, Shaochuang Liu, Yi-Lun Ying, and Yi-Tao Long

Abstract Nanopore analysis is an advanced recent development of single-molecule technology. In this chapter, we reviewed our recent studies on glass nanopipettes and SiN_x solid-state nanopores for protein analysis. We first focus on confined nanopipette sensing in biological samples and wireless nanopore electrode applications. Then we exhibit an example of profiling protein dynamics with SiN_x solid-state nanopore. Further, a nanopore combined measurement platform is provided to solve the comprehensive nanopore research for protein dynamics.

Keywords Nanopore · Nanopipette · Single molecule analysis · Protein profiling

5.1 Introduction

An emerging point of view in protein chemistry is that proteins perform their function as an ensemble of conformations, rather than a static structure determined by their specific amino acid sequences [1–4]. Protein dynamics are an intrinsic feature occurring throughout the whole life cycle of proteins whether the protein folding at the initial formation or in the process of carrying out their biological functions. The central problem of the protein dynamics is proper identification of the transition conformations and accurate description of the possible transitions or folding/unfolding pathways between these states to further understand the folding mechanism and the structure–function relationship of the protein [5–8]. Understanding the mechanism is essential for solving protein conformation-related diseases, such as Alzheimer's, Parkinson's, Huntington's disease, and type II diabetes [9–13]. Protein dynamics is a highly complex phenomenon contributed by different mechanisms of action, such as local flexibility, regional flexibility, or global flexibility caused by the bond vibrations, side-chain rotations, backbone fluctuations, and domain movement [14–19], that are happening with different timescales and amplitudes.

R.-J. Yu · S. Liu (✉) · Y.-L. Ying · Y.-T. Long
School of Chemistry and Chemical Engineering, Nanjing University, Nanjing, People's Republic of China
e-mail: shaochuangliu@nju.edu.cn

J.-P. Leburton (ed.), *Solid State Nanopores*, Nanostructure Science and Technology,
https://doi.org/10.1007/978-3-031-42336-9_5

At present, many studies, especially molecular dynamics (MD) simulations [20, 21], have been used to study the protein dynamics by mapping the transition conformations. However, experimentally characterizing the protein dynamics remains extremely difficult, owing to the short-lived transition and subtle conformation changes. Confined nanopore analysis has been emerging as a powerful tool that is capable of exploring the transient dynamics of single entities, specifically single molecules, single nanoparticles, and single cells [22–31]. The geometry of nanopore could be naturally formed by the pore-forming proteins, with typical lumens of 1–4 nm known from the exact crystal structures [32]. On the other hand, the nanopore geometrical structure could also be designed and fabricated in the laboratory. Generally formed on solid-state materials, such as SiN_x, SiO_2, PET, graphene, MoS_2 and other membranes and glass capillary substrate, these artificial nanopores shows promising features of tunable dimensions [33]. Those controllable geometries are more suitable to capture single protein molecules, with varying dimensions, inside the confined space. In recent years, SiN_x solid-state nanopores and the glass nanopipettes, covering the pore diameters of 1–100 nm, have been employed accommodate the critical issue of protein profiling.

In this chapter, we mostly emphasis on the confined electrochemical measurement through glass nanopipette and SiN_x solid-state nanopores in our laboratory, with the main focus on protein profiling. We first begin with the confined nanopipette sensing, including the biological sample delivery, wireless nanopore electrode applications and the electrochemical sensing of single molecules. The dynamic protein conformations in a solid-state nanopore are discussed in detail, revealing the transition pathways of individual protein molecules. Advanced measurement platform is further concluded, with future directions in the elaborate design of the intergrade instruments such as optical magneto and mass spectrometry.

5.2 Confined Nanopipette Sensing

Glass nanopipettes are generally prepared from capillaries with a central hollow geometry inside and a sharp tip at one terminal. In the last decades, with the rapid development on the field of microfabrication, nanopipettes could be easily obtained in a general chemistry laboratory. Nanopipettes with desirable aperture could be prepared with a laser pipette puller by elaborate control on the pulling parameters [34, 35]. Other chemical etching methods could also be used for nanopipette preparation but shows less control on the geometry and diameter of the tip pore. The nanopipette shows various application in the analytical and biological fields [36, 37]. The well-defined geometry of nanopipette with a hollow structure inside, a tunable sensing interface as the inner wall, and the typical nanopore structure at the very tip. Among the tunable surface, conductive metals could be introduced onto the pipette surface or exactly inside the nanopipette, that endow the nanopipette special properties of electrodes. Therefore, three aspects of applications are included in the section: biological sample delivery enabled by the hollow geometry of nanopipette, wireless nanopore electrode applications enabled by the conductive interface of nanopipette, and the nanopore-based single molecule sensing.

5.2.1 *Nanopipette for Biological Sample Delivery*

Single cell analysis has been attracted much attention for revealing the cell population heterogeneous, which are generally obscured in the average sampling. It promotes the understanding of unusual cellular processes that occur in the disease progression, such as that the intra-tumour cellular heterogeneity plays critical role in the therapeutic resistance. Therefore, single cell analysis would reveal these informative cellular behaviours, providing a unique landscape of the single cell behavior [38]. Nanopipette, with its typical geometries in needle-like tip and several centimetres capillary tail, serves as suitable tool for single cell monitoring [39, 40]. With the decreased tip diameter, reduced cellular destruction could be facilitated during the contact or piercing into living cells. It has been reported that quartz nanopipettes have been employed for precise delivery of fluorescent probes into single living cells with high robustness. These pipettes were prepared from quartz capillaries by a laser-mediated pulling method, exhibiting with high uniform of diameter of ~ 100 nm. An microoperation platform was integrated on a nanopipette backfilled with rhodamine B solution and moved slowly toward the cell surface. A pair of Ag/AgCl electrodes were respectively immersed in the inner and outer solutions to drive the liquid transfer. With a biased voltage, the fluorescent molecules inside the pipette were injected into a target cell and the whole injection process could be well controlled by the applied voltage (Fig. 5.1a) [41]. This nanopipette-based method offers an effective and direct way for single cell imaging. Specifically, further imaging of subcellular domain within one single cell could be enabled with the high spatial resolution of nanopipette. By filling the pipette with multiple fluorescent probes, a one-step simultaneous delivery was enabled into single living cell within 20 s. Target living cells were imaged in both the green and red channels by using Cy5 and FITC mixture, which shows a comparable imaging efficiency with traditional incubation method for 20 min. Thus, this method offers a highly effective and rapid method for multichannel single cell imaging. Additionally, an indicator dye for mitochondria, JC-10, was subsequently employed for specific imaging of subcellular organelles. It showed clear difference in the fluorescent signals between the MCF-7 control cells and drug treated ones, which exhibited as high red and green fluorescence intensity due to the active and inactive state of mitochondria [42].

Besides the direct liquid delivery, the confined space in the hollow cone tip of nanopipette provides a great restriction to the manipulated liquid samples. Ultra-small volume of liquid could repeatedly move back and forth inside the pore

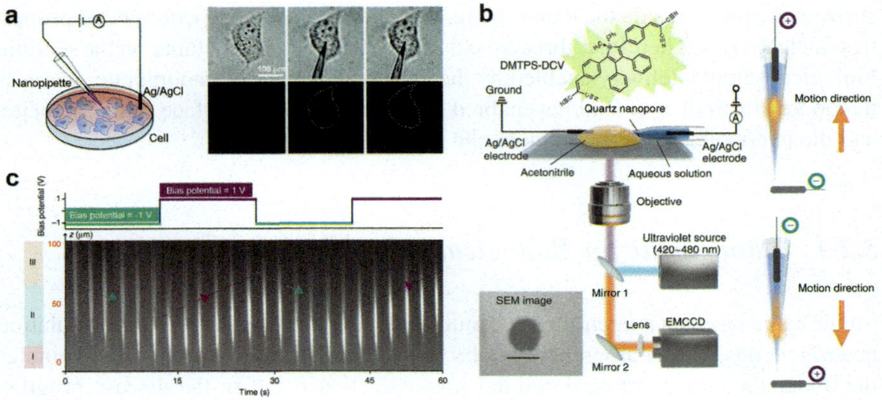

Fig. 5.1 **a** Schematic representation of the nanopipette based method for single cell injection and imaging (left). Bright field and fluorescence images showing the RhB injecting into a single living cell (right). "Reprinted with permission from [41]. Copyright [41] Royal Society of Chemistry." **b** The experimental setup for visualization of DMTPS-DCV emission inside a glass nanopipette. **c** The reversible manipulation of the AIEgens to exhibit the dynamic "on-to-off" and "off-to-on" emission process. "Reprinted with permission from [43]. Copyright [43] Springer Nature"

cavity under electrochemical control. Surprisingly, the aggregation-induced emission (AIE) luminogens exhibits reversible conversion of "on-to-off" and "off-to-on" emission inside the nanopipette confinement, which could hardly recover into initial dispersed state once the aggregates formed in bulk system. As shown in Fig. 5.1b, a 100 nm quartz nanopipette filled with aqueous KCl solution was horizontally placed on a cover glass. An acetonitrile solution of 1,1-dimethyl-2,5-bis(4-benzylidenemalononitrile)-3,4-diphenylsilole (DMTPS-DCV) was dropped outside the pipette tip as surrounding bulk solution. An applied voltage of 1 V drove the directional motion of the DMTPS-DCV acetonitrile solution from the exterior to the nanopipette inside aqueous solution. An intensive emission was thus observed with the fluorescent light signals collected by a $\times$ 40 objective on the bottom glass. The AIEgen drove into the wider cavity with this voltage continues, leading the emission "on-to-off" switch. Whereas, $-$ 1 V drove the opposite directional motion and led to a "off-to-on" emission (Fig. 5.1c) [43]. Thus, this nanopipette confinement offers promising approach to modulate and visualize the dynamic emission from AIEgens.

5.2.2 *Application of Wireless Nanopore Electrode*

Nanopipettes with specific modification and rational surface design enable various applications for molecular interactions and beyond. Amongst, conductive nanopipette with metal film coated could not only serves as a nanopore, but also a nanoelectrode for electrochemical sensing [44]. In recent years, wireless nanopore electrode (WNE) has been widely employed for the electrochemical analysis of single entities [45]. It is generally obtained by the nanoscale metal deposition on the pipette tip, where the tip metals are out of touch with neither working electrode nor the reference electrode. The WNE functions under a bipolar electrochemical mechanism, with a pair of Ag/AgCl electrodes immersed in the inside and outside solutions, respectively. And due to the asymmetric geometry of nanopipette, most of the potential drop cross the pipette could be located on the coated metal. As show in Fig. 5.2a, a silver coated nanopipette was employed as bipolar WNE for the detection of single molecules. Briefly, under biased voltage, a pair of bipolar electrochemical reactions occurred at the two terminals of conductive metal: oxidation of Ag into Ag^+ at the polarized cathode and the reduction of H^+ to H_2 at polarized anode. Continuous generation of H_2 at the pore orifice formed nanobubbles in the aqueous solution and resulted in enhanced spike signals in the nanopipette ionic current. Finite element simulation was employed to confirm this process, demonstrating that about 14 H_2 molecules could form a nanobubble of 10 nm. With the addition of Hg^{2+} in this system, the generation of nanobubbles was greatly suppressed by the reduction of Hg^{2+}, leading to the decreased frequency of the ion current signals. This bipolar WNE enabled the direct and label-free detection mercury ions at pM level, accessing possible detection of small molecules or ions at the single molecule level [46]. Furthermore, the generation of silver ions could be devolved as a template for the electrochemically controlled self-assembly of microcyclic structures. It has been reported that alkynyl-functionalized gold nanoparticles would aggregate in the presence of silver ions [47], but such aggregations could hardly be controlled well in bulk solution. Actually, the self-assembly of nanoparticles especially in the micrometres scale is challenging due to the critical needs of high stability and reproducibility. To address this issue, silver coated WNE was employed as a template for controllable release of silver ions at a diffusion distance of ~ 15 μm. Alkynyl-functionalized gold nanoparticles in the bulk solution spontaneously self-assembled into microcyclic structures. The consistent diameter of ~ 15 μm were exhibited to be independent of neither the pipette diameter nor the functionalized terminal alkyne molecules. These microcyclic structures show high uniformity and reproductivity both in the morphology and diameter and this nanopipette-templated method exhibits as promising way for controllable nanoparticles assemblies [48].

The bipolar electrochemical reactions across the metal deposit could further be engineered for specific electrochemical measurement in single cells. Taking advantage of the syringe structures of the WNE, specific design on the WNE surface could enable the imaging of electron transfer in a single living cell. As shown in Fig. 5.2b, gold coated WNE was modified with 4-thiol catechol, which could be oxidized into o-benzoquinone at biased potential of 400 mV. With an applied voltage higher than 400 mV across WNE, the reduction of H^+ into H_2 occurs at the polarized cathode generating enhanced ionic current signals. Nicotinamide adenine nucleotide (NADH) served as a substrate for the electro-catalysis of oxidized o-benzoquinone, yielding reduced catechol. This catalytic process at the polarized anode greatly promoted

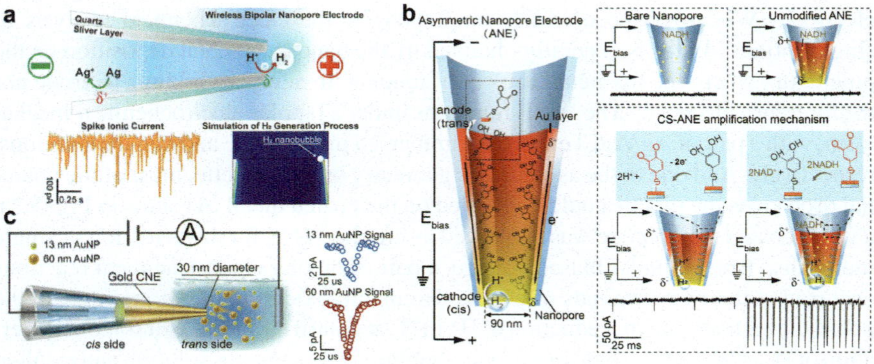

Fig. 5.2 **a** Illustration of the bipolar electrochemistry based on the silver coated nanopipette. The generation of hydrogen nanobubble occurred at the pore orifice, resulted in typical spike signals in the ionic current trace. "Reprinted with permission from [46]. Copyright [46] American Chemical Society." **b** Asymmetric nanopore electrode for the electrochemical measurement of electron transfer dynamics in living cells. The nanobubble-induced current amplification was recorded as the bipolar electrochemical reactions occurred at the two terminals of ANE under a proper potential drop. "Reprinted with permission from [49]. Copyright [49] American Chemical Society." **c** A 30 nm confined nanopore electrode for capacitance-based detection of single gold nanoparticles. "Reprinted with permission from [52]. Copyright [52] John Wiley & Sons"

the H_2 generation at the cathodic pole, leading to an elevated signal frequency. The current signals were amplified near 3 orders of magnitude in the presence of NADH, achieving high sensitivity and selectivity for the NADH detection at concentration of 1 pM [49].

By shrinking the conductive metals on the nanopipette orifice into nanometre scale, single nanoparticle electrode could be further developed for electrochemical sensing. When the metal nanoparticle translocated through the pipette orifice, it served as a conductor in electrical field and the gather a high potential drop. A single silver nanoparticle polarized into two poles to mediate a pair of bipolar electrochemical reactions, the oxidation of AgNPs and the MoS_2 catalysed hydrogen evolution reaction. The generation of hydrogen nanobubbles on the QD surface and the dissolution of silver nanoparticle could be monitored by the ionic current signals in real time [50].

The nanopipette could not only include one single nanoparticle at the confined space, but also a length of metal to fully block the pore orifice. As shown in Fig. 5.2c, the gold filled nanopipette was prepared by the chemical-electrochemical reduction of chloroauric acid by $NaBH_4$. The length of gold metals could be several micrometres that ensure the non-contact of working electrode to the tip metals. Under a biased voltage across the pipette, the generated gold could be polarized into two poles. This kind of pipette was regarded as closed type WNE since the ionic current was fully blocked. Though single nanoparticle electrochemistry has been developed with nanopipettes showing advantages on the reduced sample consumption and controllable nanoparticles' delivery [51]. In this case, the WNE shows exceptional high current resolution with average RMS of 0.6 pA, and small capacitance variation could be identified through the current traces showing as small spike signals. Single nanoparticle collision was performed via this closed WNE and the nanoparticles in diameter of 13 and 60 nm were clearly discriminated via the current signals. This nanometre size dimension of the electrode surface exhibit great confinement on the electrochemical detection, enhancing the collision frequency to nearly two orders of magnitude compared with traditional ultramicroelectrode (UME) [52]. With further employing silver nanoparticles in this platform, the electron transfer between single nanoparticle and the electrode surface could be revealed under a bipolar electrochemical mechanism. The faradaic and capacitive currents of single nanoparticle provide informative characterizations on single nanoparticle, which shows higher temporal and current resolutions. Multistep oxidation of small nanoparticles in 30 and 5 nm were observed due to the reduced adsorption of nanoparticles on the WNE surface. The nanoconfinement of WNE shows more possibilities in recapturing the partial oxidized nanoparticles until a fully dissolution [53]. This method serves as a single entity electrochemistry approach, showing superiorities in better understanding the heterogeneity of single particles.

5.2.3 Sensing of Single Nanoparticles and Single Molecules by Nanopipette

Compared with biological and 2D nanopores, the mechanical stable nanopipettes show great advantages in the robust pore modification. It owns a channel-like structure with the inner wall as the controllable electrochemical sensing interface. This conical asymmetric geometry and the existing surface charge of nanopipette would induce an asymmetric I-V response especially at a low electrolyte concentration. This phenomenon is known as the ionic current rectification (ICR), which mainly due to the ion distribution when the thickness of diffuse double layer is comparable with the pore diameter. Therefore, the nanopipette surface is crucial in this ICR behaviour, and its modifications enables a variety of biosensors for application [54, 55]. As we previously demonstrated, a silver layer could be coated on the nanopipette surface through magnetron sputtering, which facilitated subsequent chemical modifications. Via the silver-thiolate interaction, macrocyclic dioxotetraamine derivative (C5), was immobilized on the inner wall of nanopipette. The surface charge of the C5-modified nanopipette changed from negative to neutral after the coordination of macrocyclic dioxotetraamine with Hg^{2+}, resulting in a distinct decrease in the ICR ratio. Hg^{2+} with concentrations from 10 pM to 1 mM could be detected with this ICR response, showing high selectivity (Fig. 5.3a) [56]. This ICR based nanopipette sensing methods offers plenty of opportunities in analytical applications and promising in molecular conformations.

The rational design of the nanopipette interface could further enable the profiling of molecule interactions at the nanometre scale [57]. Nanoparticles could serve as a

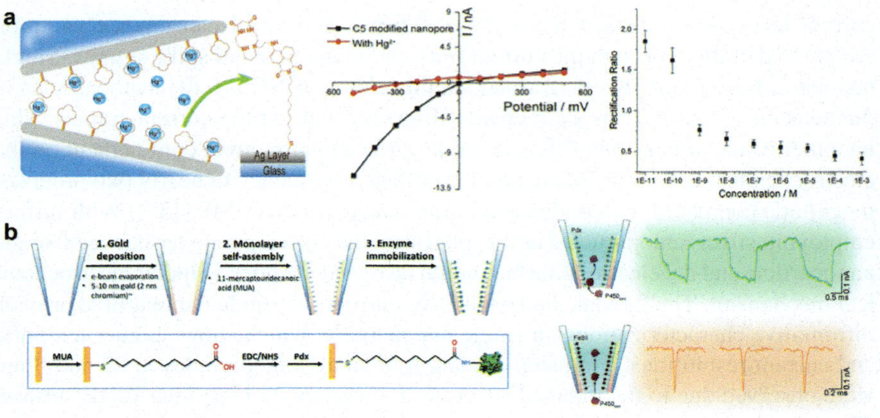

Fig. 5.3 **a** Representation of the macrocyclic dioxotetraamine functionalized nanopipette for the ICR sensing of Hg^{2+} (left). Typical I–V curves of the C5-modified nanopipette in 10 mM KCl with the presence an absence of 1 mM Hg^{2+} (middle). The rectification ratio of the C5-modified nanopipette with the addition of various concentrations of Hg^{2+} from 10 pM to 1 mM. "Reprinted with permission from [56]. Copyright [56] Springer Nature." **b** Step-by-step modification of the nanopipette with specific enzyme molecules (left). Schematic of the single molecule P450cam translocation through the Pdx-modified nanopipette and FeSII-modified nanopipette. "Reprinted with permission from [44]. Copyright [44] Royal Society of Chemistry"

nanocarrier to enable the measurement of molecular interactions at the pipette interface. The chemical reactions at the confined nanointerface could be largely amplified through the resistive-pulse current signals of single gold nanoparticles. Azide-functionalized thiol was immobilized on the gold coated nanopipette, which was prepared by electron-beam evaporation; alkyne-functionalized thiol was firstly modified on the surface of 13 nm gold nanoparticles. Click reactions at the nanoconfined interface was monitored in real time under the nanopipette electrochemical measurement. The Cu^{+} catalysed azide-alkyne 1,3-dipolar cycloaddition greatly prolonged the translocation of single nanoparticles under a bias voltage. This strategy could play a significant role in revealing the chemical reactions at a controllable nanoscale interface, especially the coupling reactions in organic chemistry [58].

Further exploiting the volume effects of biomacromolecule, non-covalent interactions could be directly identified from the single molecule events in nanopore sensing. Protein–protein interactions serve as an integral role in protein functions and biological processes. Understanding such interactions at single molecule level would help to explore the hidden heterogeneities of protein molecules in ensemble average. A step-by-step modification could be implemented on the gold coated nanopipette interface, which followed by the self-assemble of thiol monolayers with carboxylic acid functionality, and subsequent protein conjugation. Targeted antigen–antibody interactions were unravelled via the transient ionic current signals. The duration time of single antigen translocation across the specific antibody modified nanopipette was ten times larger than that through a bare nanopipette. The equilibrium dissociation constant or the binding affinity for these antigen–antibody interactions was obtained, offering single molecular characteristics [59]. It is widely known that the interactions between P450cam and putidaredoxin (Pdx) is important in the electron transport, which generally includes the protein recognition, binding and disassociation. Pdx was firstly immobilized onto the nanopipette surface with the following three steps: gold deposition, monolayer self-assembly and enzyme immobilization. Nanopipette modified with FeSII, which is the same size as Pdx but could hardly interact with P450cam, was employed as control. It showed clearly that the current blockage of P450cam in these two kinds of pipette interface are comparable, whereas, the statistic durations differ largely. Specific interactions between P450cam and Pdx induced a distinct long blockage duration (Fig. 5.3b) [44]. This strategy could be further applied for a wide range of molecular interactions and probable applications in single molecule biosensors.

Nanopipettes with controllable diameters from several to hundreds of nanometres could be elaborately prepared in a common chemical laboratory. Its orifice offers a nanometre scale aperture, that could be used as a typical single entity method for electrochemical measurement. Single molecules or even entities could be identified through the fluctuation of ionic current generated by analyte translocation across the nanopore [54]. Typically, combined mechanisms of electrophoresis, electroosmosis and diffusion that account for most of the analyte translocation, that resulted in a pore conductance change, either a blockage or enhancement. This analyte-induced conductance change is mainly based on the dimensions and physiochemical properties of the analyte, as well as the bulk electrolyte solution and the intrinsic nanopore properties [60]. As an example, the generation of hydrogen nanobubbles from $NaBH_4$ hydrolysis was firstly investigated within a confined nanopipette. $NaBH_4$ in ethanol solution with electrolyte was filled in the inside solution of nanopipette, meanwhile the aqueous PBS solution served as the outside solution. Spontaneous reaction between $NaBH_4$ and H_2O occurred at the nanopipette confinement, that include the oscillation in ionic current with the nanobubbles formation. It is generally considered that the dynamic formation of single nanobubbles experiences a bubble nucleation and subsequent bubble growth processes. And both changes on the surface charges and dimensions of nanobubble were included in such dynamic process. Absorption of BO_2^- on the formed nanobubbles would result in a distinct ion accumulation at the nanopore tip, while the simultaneously increased volume would trigger the exclusion of the nanopore current. As expected, a biphasic pulse-like signal was observed in the nanopore ionic current, demonstrating as a significant enhancement followed by a blockage on the pore conductivity (Fig. 5.4a) [61]. Therefore, the confined nanopipette offers superiority in the measurement of single entities, especially in charactering their dynamic behaviours.

During the electrochemical sensing of single entities, the characteristic ionic current signals triggered by their translocation through the confined nanopipette dominates the precise measurement. Besides the electrochemical sensing of nanobubbles and nanoparticles, single biomolecules could also be identified through the resistive-pulse sensing mechanism. Protein sensing, at the single molecule level,

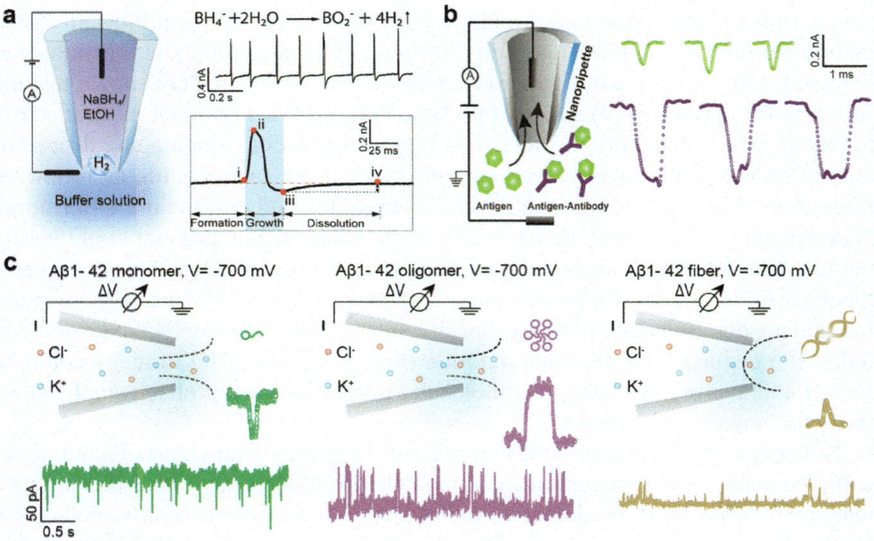

Fig. 5.4 **a** Schematic representation of the nanopipette method for charactering the dynamic growth of nanobubbles. Hydrogen bubbles were generation by the reaction between $NaBH_4$ and H_2O. "Reprinted with permission from [61]. Copyright [61] American Chemical Society." **b** Label-free monitoring of the immunoreactions at single molecule level by the glass nanopore method. Characteristic ionic current blockages were respectively from antigen and antigen–antibody complex. "Reprinted with permission from [62]. Copyright [62] American Chemical Society." **c** The nanopore electrochemical sensing of the dynamic aggregation of Aβ1-42 peptide. Distinct characteristic current fluctuations were observed from the Aβ1-42 monomers, oligomers and fibres. "Reprinted with permission from [63]. Copyright [63] Royal Society of Chemistry"

is of fundamental significance in understanding the heterogenous single molecular behaviours, which is critical in understanding the protein-mediated biological functions. α-fetal protein, an important biomarker for hepatocellular carcinoma was employed as a model protein for single molecule sensing. It could be discriminated from the ionic current trace with a 30 nmnanopipette. Characteristic current blockages with duration time of 0.24 ms and blockade amplitude of 0.049 were observed in the current traces with the addition of AFP in the outside solution of nanopipette. As comparation, AFP firstly incubated with its specific antibody was further employed for single molecule characterization. An enhanced blockade amplitude with ~ 2.7 folds of that AFP was identified as the translocation of single AFP-antibody complex, which is about twice volume larger than AFP. Besides, the durations times for single AFP-antibody complex was greatly longer than the AFP due to the enlarged volume of target molecules (Fig. 5.4b) [62]. Thus, single protein molecules could be discriminated directly under a typical volume exclusion model, presenting as a label-free method to monitor the single molecule immunoreaction in aqueous solution.

It is widely recognized that some proteins experience dynamic conformational change or disorders in the human bodies, that leads to diverse conformational diseases. For instance, the Alzheimer's disease is known to be closely related with the dynamic aggregation of β-amyloid 42 peptide, especially the structural conformation the oligomers. During the Aβ 42 aggregation process, the dimension of single molecules ranges from several nanometres to sub-micrometres, regarded as monomers, oligomers and fibres. A combined sensing mechanism was proposed for precisely discriminate these three states of peptides, and the electroosmotic forces was modulated to mainly drive the peptide traversing across the pore. Different current fluctuations were observed in the ionic current recording due to the differences both in size and surface charge density of the Aβ 42 aggregates. Aβ 42 monomers, in smaller size and higher surface charge generated ionic current enhancement, while the oligomers and fibres which were in larger volume, generated current blockage. Further oligomers and fibres could be discriminated via a distinct translocation or collision process at the pore confinement. This dynamic monitoring of the protein conformations plays significance role in understanding the protein neurotoxicity and the related disease mechanism [63]. Other functional proteins, such as enzymes could also be detected with glass nanopipettes at single molecule level. Nanopipettes with varying tip orifice were employed for the detection of single glucose oxidase in a near physiological solution. It showed that the duration time of single glucose oxidase could be effectively extended by controlling the interactions between the analyte and the inner surface of pipette (Fig. 5.4c) [64]. This shows potential possibilities in capturing transient conformational change during enzymatic processes.

5.3 Snapshotting the Transient Conformation and Dynamics of Single Peptides

Capturing the transient and stable conformations of a protein is essential for the study of protein dynamics. In traditional nanopore studies, the proteins experience a fast translocation under the electric field force that is usually beyond the bandwidth of the current amplifier. This makes the signal only could reflect an instantaneous stage of a protein, and the transition pathways are neglected. So, it is difficult to use the nanopore to observe the dynamic changes of proteins for a long time. Therefore, a new method should be developed to prolong the observation time of protein in nanopores and study protein dynamics. In this section, we will provide an example for the study of protein folding and dynamics with an extended duration of time [65].

5.3.1 Confining a Single Protein into a Nanopore with High Throughput

We developed a new method to capture the transient conformations and resolve the multiple transition pathways of a single peptide inside the nanopore. A β-hairpin peptide, consisting of 16 amino acids with sequence GEWTYDDATKTFTVTE from the C-terminal fragment (41–56) of protein G at the B1 domain [66], was taken as the model peptide. This peptide had been widely used to study the folding mechanism by molecule dynamics (MD) [66–68]. In this method, the protein is modified with a biotin tag at the C terminus to bind to a monovalent streptavidin (mSA) [69] protein and form a single peptide-mSA complex (Fig. 5.5a). The solid-state nanopore size we used is about 4 nm, which is smaller than mSA with a hydrodynamic diameter of ~ 5.6 nm. Therefore, this complex will be captured into the nanopore under a constant voltage and reside in a nanopore other than translocating through it (Fig. 5.5a). Then the signal with a long duration time will be acquired (Fig. 5.5b). After a while, we inverted the voltage and this complex would be released for the next peptide capture. By automatically repeating this process with a periodic voltage, high-throughput protein dynamics analysis had been achieved (Fig. 5.5c).

5.3.2 Investigation of the Dynamic Conformation Changes

As shown in Fig. 5.6, we could find that the signals of peptides residing in nanopores have unique current distributions with several distinct steps. Although all peptide molecules have the same amino acid sequence, each captured single peptide exhibits different characteristics in the current signal, indicating that the peptide molecules have different conformational dynamics, which suggested the heterogeneity of the peptide in a solid-state nanopore.

Due to the spatial size, different electric charges, and hydrophilicity on the surfaces, solid-state nanopores could form a nanoscale space that has strong weak interactions with single molecules into the nanopores. In this space, the ionic current formed by the action of the electric field could respond sensitively to the changes in ionic conductance or capacitance generated by the interactions. Therefore, this nanoscale space could be called nanoconfined space, and this interaction could be named to spatially confined effect. Due to the volume exclusion effect of peptides and the interaction between peptides and the inner surface of the nanopore, we observed characteristic blocking current signals that were different from those of the perforation experiment. The conformation of a peptide is dynamic changed, the different conformations usually have different exclusion volumes, surface charge, and interactions in the confined space of the nanopore. These effects will produce different

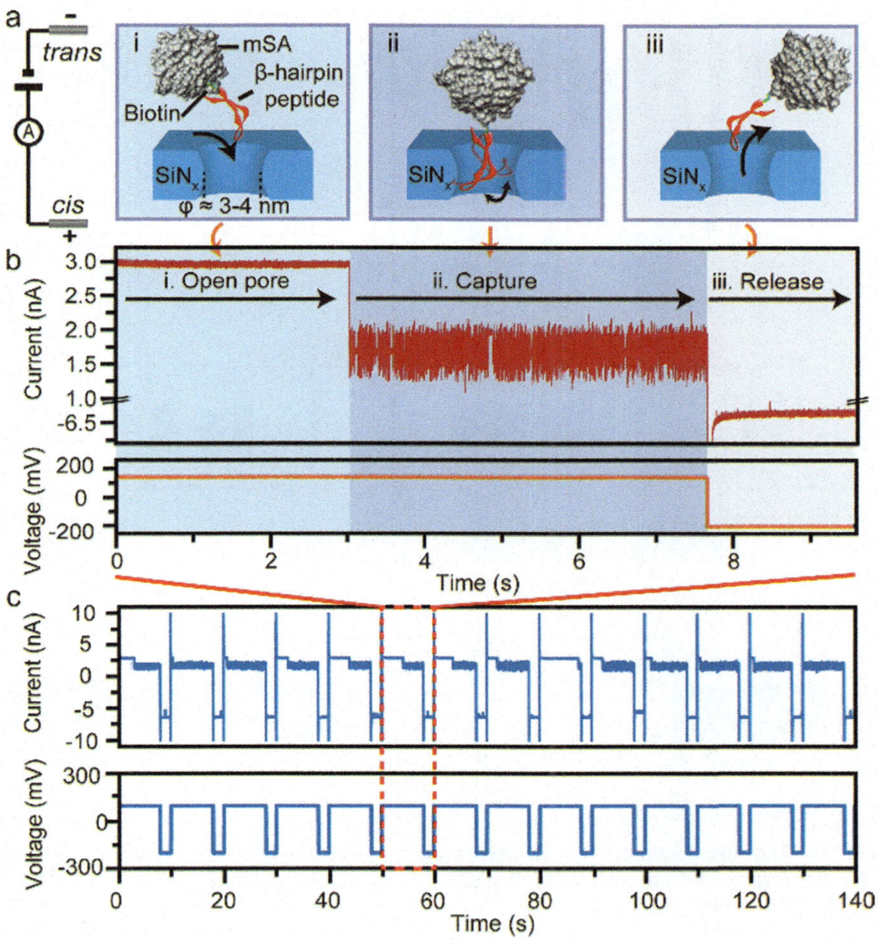

Fig. 5.5 **a** The three-step process of electrochemically confining a peptide into a nanopore that consists of an open-pore state (i), capturing state (ii), and releasing state (iii). **b** Representative continuous ionic-current trace and corresponding voltage protocol of a single peptide in nanopore with three steps. **c** High-throughput capturing of the single peptide by periodic voltage with 7 s positive voltage for capturing and 3 s negative voltage for releasing. "Reprinted with permission from [65]. Copyright [65] Royal Society of Chemistry"

types of current signals. Besides, each conformation has a different lifetime and could transfer to other conformations quickly. Therefore, we could deduce that the characteristic current signal contains the dynamic conformation information of the peptide.

To investigate the characteristics of the current signal generated by individual peptides, the current distribution of each blocking signal was calculated. We find that every peptide-mSA molecule could produce a unique current distribution. As

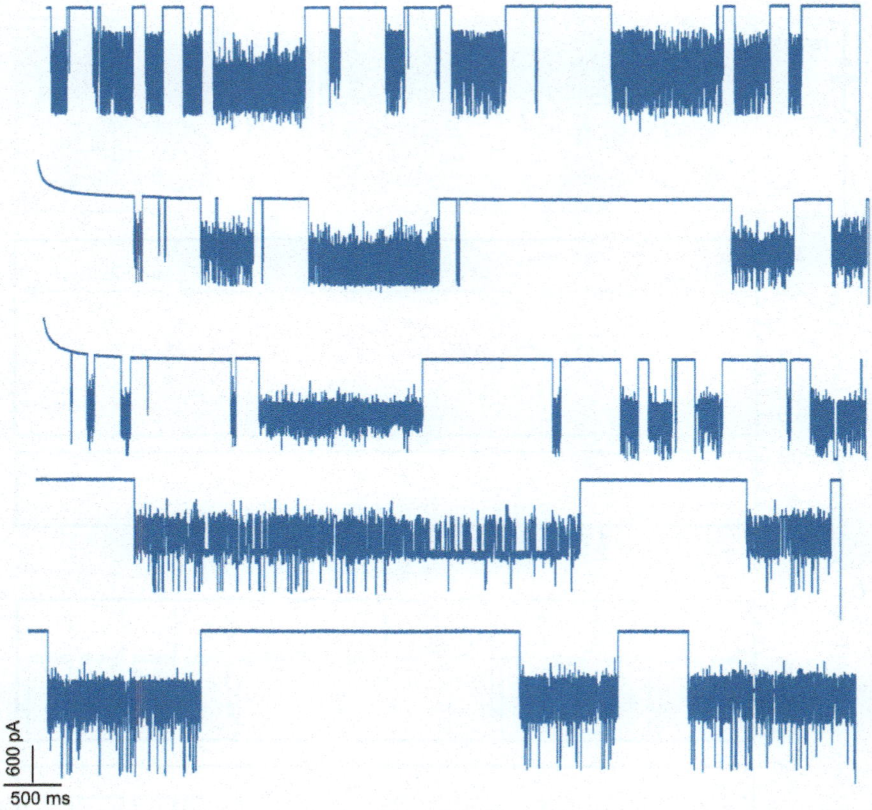

Fig. 5.6 The ionic current trace of the mSA-peptide complex captured by a solid-state nanopore with a diameter of 4 nm at voltage + 150 mV. The data is acquired with a sample rate of 100 k and with a low-pass filter of 5 k

shown in Fig. 5.7a–d, the current of every signal could be divided into four stages as stage 1 of $\Delta I/I_0 = 0.1–0.24$, stage 2 of $\Delta I/I_0 = 0.24–0.46$, stage 3 of $\Delta I/I_0 = 0.46–0.58$ and stage 4 of $\Delta I/I_0 = 0.58–0.80$ by evaluating the current blockage histogram from all the events. But for a peptide individual, only one dominated stage occupies the main superiority. According to these features, all the signals could be classified into four patterns with different dominated stages. Statistic results showed that the peptide with stage 2 as the main peak occupied the highest proportion in all the collected single molecule signals, up to 78%. Therefore, stage 2 is the most likely current step generated by the stable β-hairpin conformation of the peptide, while the other three stages could be attributed to the metastable conformation of the peptide. This metastable conformation has a low probability of occurrence in ensemble analysis, but can still exist stably in a certain peptide in single molecule analysis.

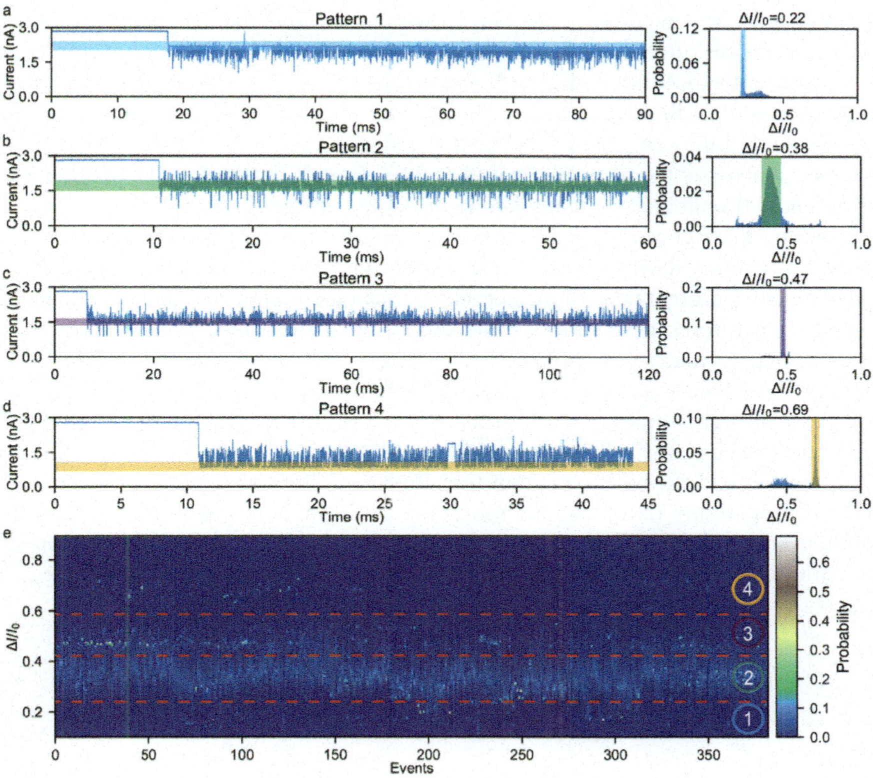

Fig. 5.7 **a–d** The ionic current trace and current histogram of four special patterns. **e** The heatmap of the current distribution includes 300 peptides. "Reprinted with permission from [65]. Copyright [65] Royal Society of Chemistry"

To verify this perspective, the molecular dynamics (MD) simulation was used to investigate the folding/unfolding kinetics of the β-hairpin peptide and to predict the possible conformation of the peptide. The peptide we used is from the B1 domain of protein G, which had been widely used to study the formation mechanism of β structure (Fig. 5.8a). This peptide has a hydrophobic cluster formed by W43, V54, Y45, and F52 to stable the β-hairpin structure [70]. As shown in Fig. 5.8b, the root mean square deviation (RMSD), solvent accessible surface area (SASA), and van der Waals volume of the peptide throughout the whole 60 ns simulation were analyzed. The RMSD results show the difference between the current conformation and the initial structure. The SASA characterizes the solvent-impregnable surface area of the peptide conformation, which mainly reflects the hydrophobicity of the peptide. According to these results, we find that there are four main representative conformations of the β-hairpin peptide, shown as compressed, stable β-hairpin, partially unfolded and unfolded conformation, respectively (Fig. 5.8c). In solid-state nanopores, the generation of current blockage is mainly due to the exclusion volume of molecules, which could change the conductance within the nanopore. In addition to the exclusion volume, the surface charge and hydrophobicity of a single molecule also could influence the conductance within the nanopore. As shown in Fig. 5.8d, the conformations of this peptide have almost the same van der Waals volume, yet the SASA are not the same, which is because different conformations lead to different surface charges of the peptide exposed to the solution. This makes the peptide have different hydrophobicity and ultimately different current blockage.

Among these conformations, conformation 2 is the stable hairpin structure with a 60% possibility during 60 ns of MD simulations. And correspondingly, the dominating stage 2 current peak of the peptide has the maximum percentage in all current signals. Therefore, these results could confirm that the folded β-hairpin results in the current fluctuation of stage 2, while stages 1, 3, and 4 may be related to the short-lived metastable conformations. A previous study had demonstrated that the blockage degree would increase proportionally as the protein unfolded gradually from the folded state [71]. Therefore, we could deduce that the four current stages 1, 2, 3, and 4 could attribute to the four transition conformations as compressed, stable β-hairpin, partially unfolded, and unfolding conformation, respectively. By evaluating every current spike from one blockage event, each folding/unfolding pathway and transition conformations of a single peptide could be acquired in real-time.

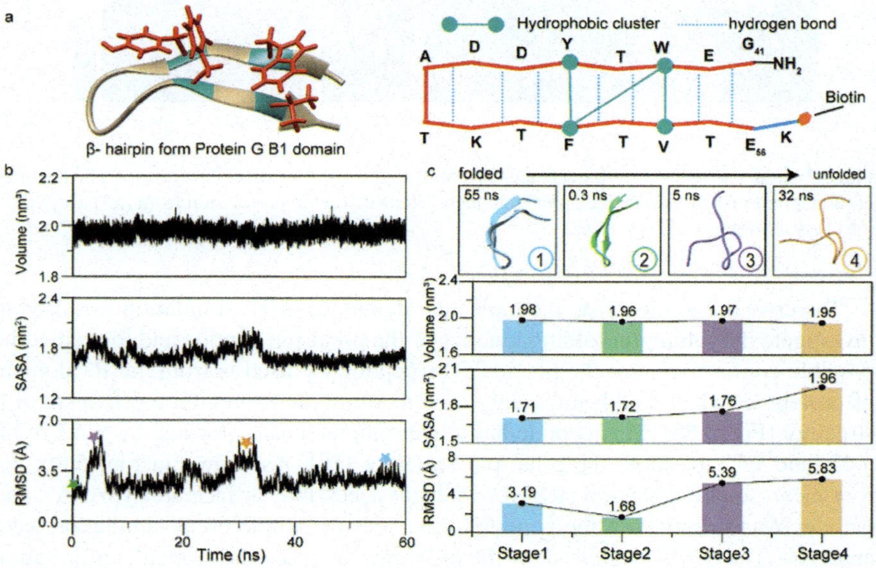

Fig. 5.8 Molecule dynamics simulation of the beta-hairpin peptide. **a** The stable wild-type conformation of the beta-hairpin peptide from 41 to 56 of the B1 domain of protein G. **b** The RMSD, SASA, and volume of the peptide as a function of simulation time in total 60 ns simulation. **d** The representative four metastable and stable conformations of the peptide. And the corresponding RMSA, SASA, and volume for each conformation. "Reprinted with permission from [65]. Copyright [65] Royal Society of Chemistry"

5.3.3 Folding Pathway Model for a Single β-Hairpin Peptide

Revealing the pathways of a protein transition in different conformations is a crucial question in protein folding. Generally, the transition between the four states of the peptide could be described by a four-stage Markov chain model [72]. The Markov chain model is a statistical tool to describe state transfer in stochastic processes. The process of peptide transition between different conformations can be regarded as a stochastic process conforming to a discrete Markov chain, as the next conformation of the peptide is only related to the current state other than the previous one. For example, if the current blockage of the peptide is stage 2, at the next moment it may transfer to any of the blockage states 1, 3, or 4, while it may also continue to transfer to blockage state 2.

As shown in Fig. 5.9a, we first calculated the stage sequence of a peptide. Each peak segment of the blockage current signal (each horizontal segment in the red line) can be attributed to one of the four blockage stages. Thus, the entire sequence of the signal transformed between the four blockage stages can be obtained. Then, according to the definition of the Markov chain, we counted the number of current steps transferred from one state to another according to the time sequence of the blockage stage and calculated the probability of each transfer path to obtain the transfer probability matrix of the peptide conformation transfer. Finally, the pathways of peptides transferred between different conformations are acquired based on the obtained transfer probability matrix, which leads to the most probable folding pathways for each peptide.

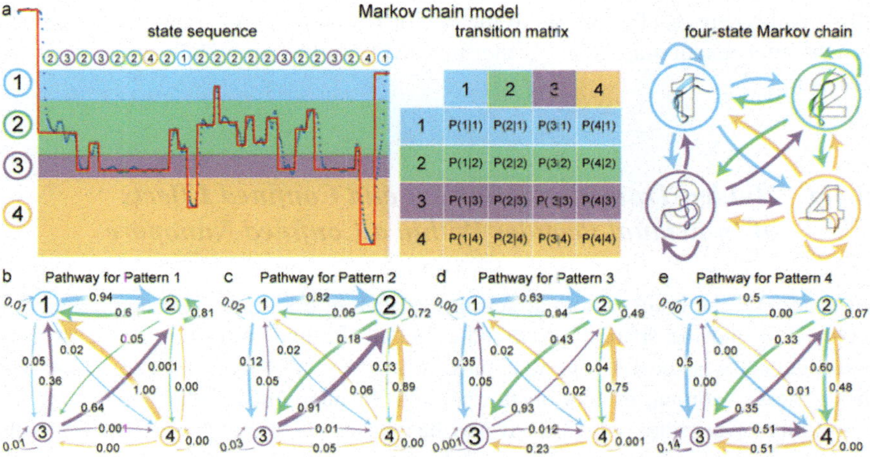

Fig. 5.9 The Markov Chain model for peptide folding. **a** The schematic of the Markov Chain model. **b–e** The folding pathways of the four special current patterns. "Reprinted with permission from [65]. Copyright [65] Royal Society of Chemistry"

By this method, the four pathway patterns were analyzed as shown in Fig. 5.10b–e. We found that the compact conformation in pathway 1 has the highest percentage, due to the longer average duration time of current stage 1. Structurally, conformation 1 is the state of contraction of the hairpin structure, which requires high energy to form and only could keep stable for a short time. Pattern 2 is the most possible pathway for all signals, and more than 70% of the conformation is the stable β-hairpin structure. In this pathway, the other three conformations occur with only a relatively low probability. This is because conformation 2 is a structure stabilized by a hydrophobic cluster, and a higher force is required to disrupt the hydrophobic core in the nanopore and under the perturbation of the electric field. In the Pathway of pattern 3, the partially unfolded conformation is the dominant structure. The higher probability transitions of stage 2 or 2 to stage 2 indicate the fast transformation between the partially unfolded conformation and the stable β-hairpin conformation. In contrast, the pathway of pattern 4 is taking the unfolded conformation indicated by state 4 as the dominated intermediate. Further, we have carefully examined more than 300 events and obtained the transition pathways. As shown in Fig. 5.11, each peptide displays characteristic folding/unfolding pathways, which indicated the heterogeneity of the protein at single molecule level.

Except for the pathways, we also could study the conformation transition dynamics by this method. According to previous studies [72], the process of single-molecule conformational conversion within nanopores can be regarded as a chemical reaction process. As shown in Fig. 5.11a, we first obtain a scatter plot of the current steps of a single peptide blockage signal and divide it into four parts according to Gaussian distribution. Then the time distribution also could be acquired from this result, and we found that the time distribution conforms to a first-order exponential distribution (Fig. 5.11b and c). So the conformation transition could be considered a first-order reaction and then further be used to futher study the state transfer rate constant.

5.3.4 Hidden Transition Pathways and Confined Effects of Individual Peptides Within a Confined Nanopore

In nanopore, the main factors affecting the peptide folding are the applied electric field force and the spatially confined effect of the nanopore. For single peptides with a specific size, the confined space formed by nanopores with different sizes has different confine effects on the peptide. In nanopores with a small size, the unfolding of peptides is restricted and has strong interactions with the nanopore surface, thus limiting the generation of some free conformations. For nanopores with a larger size, the peptides have more space to transfer between different conformations. To further understand the folding process of β-hairpin structured peptides, voltage-dependent experiments under different nanopore sizes were conducted to investigate the dynamic conformational transition of peptides.

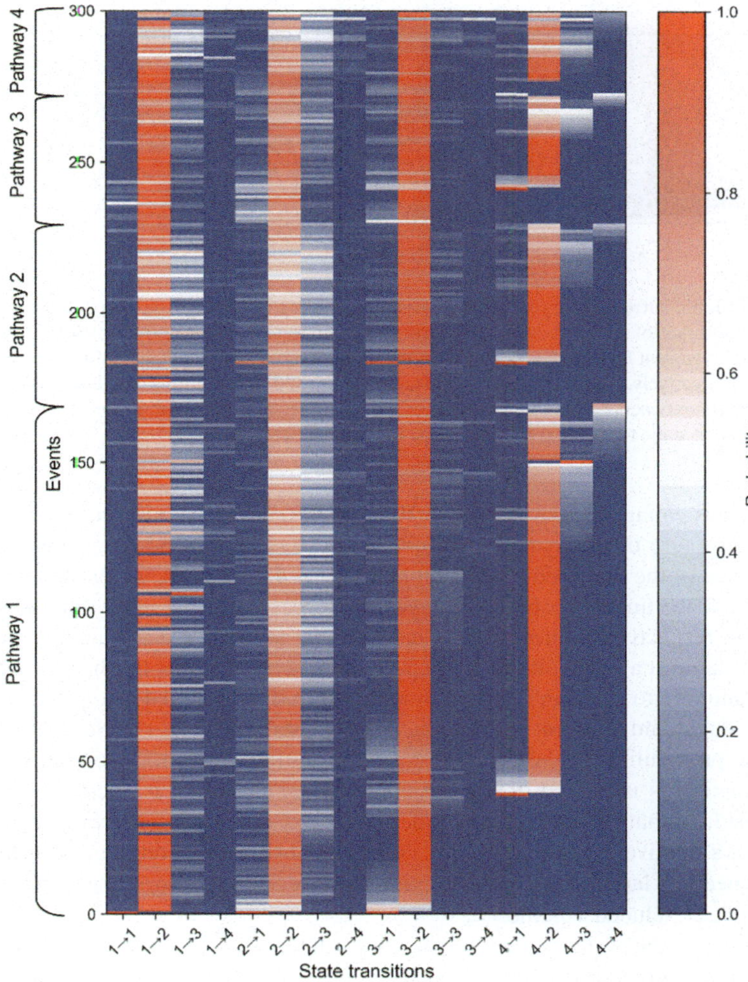

Fig. 5.10 The fingerprint map of folding/unfolding pathways for each β-hairpin peptide. The heatmap showing the transition probability of State n (n = 1, 2, 3 and 4) → n (n = 1, 2, 3 and 4) for each β-hairpin peptide. Color represents transition probability related to each transition. The total 300 events could be classified into Pathway from I to VI. "Reprinted with permission from [65]. Copyright [65] Royal Society of Chemistry"

To study the influence of the applied voltage, we also conducted voltage-dependent experiments from + 100 to + 200 mV with the ~ 4 nm nanopore. As shown in Figs. 5.12a–d and 5.13a, b, we calculated the probability of the entire duration spent on each stage in one single event to evaluate the stabilization of each conformation. We found that the time probability of stage 1 decreases, while that of stage 4

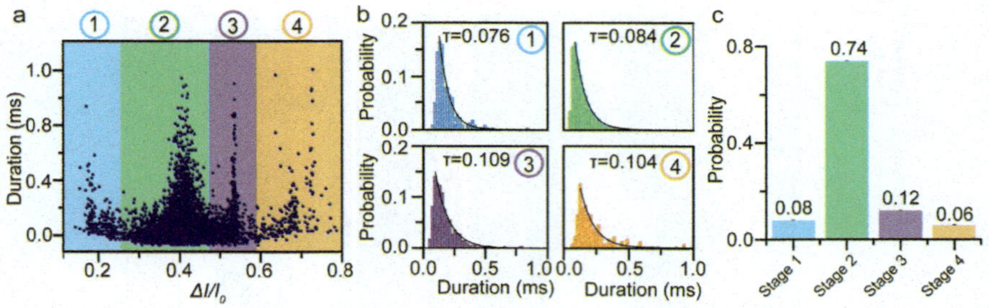

Fig. 5.11 Ionic current stage report on the dynamic folding state of a β-hairpin peptide. **a** Scatter plots of the spikes of a typical event. The current distribution is divided into four stages (color box) according to current histograms. **b** The duration histograms from the spikes fall into the stage 1, 2, 3 and 4, respectively. The lifetime of each stage was carried out by fitting the duration histogram with a single exponential function. **c** The time ratio of the summarized durations (tn, n = 1, 2, 3 and 4) at each stage to the total event lifetime for the characteristic event

increases as voltage increases from + 100 to + 150 mV. This is because the voltage-induced rupture of the hydrophobic cluster of the folded peptide to facilitates the occurrence of the unfolded conformation at stage 4. However, under the voltage of + 200 mV, the time probability at stage 2 increases while the other three stages all decrease, respectively. These results indicate that all the metastable conformations are less stable at a high voltage and experience a short lifetime. Further, we used a ~ 5 nm nanopore to perform this experiment. Similarly, the signal also produced four current stages and different pathway patterns (Figs. 5.12e–h and 5.13c–e). However, the time probability for the partially unfolded conformation and the total unfolded conformation is increased, while the compact conformation is decreased. So, we could deduce that the partially unfolded and total unfolded conformation is more stable in a relatively large nanopore. This indicates that the confinement effect has diminished in a large pore and the peptide has more space to form the conformation with a large excluded volume.

In summary, these results exhibit the ability of the solid-state nanopore to measure the dynamics and transition pathways of a single protein. The confined effect inside the nanopore is capable of prompting the protein folding into its stabilized conformation and has a high sensitivity to monitor the dynamic conformations in the protein folding/unfolding process.

5.4 Nanopore Combined Measurement Platform for Single Protein Molecules

Owing to the fast transition and subtle conformational changes of proteins, it has a high requirement for nanopore measurement instruments to acquire the ionic current with low noise, high sensitivity, and high bandwidth. Besides, new methods are

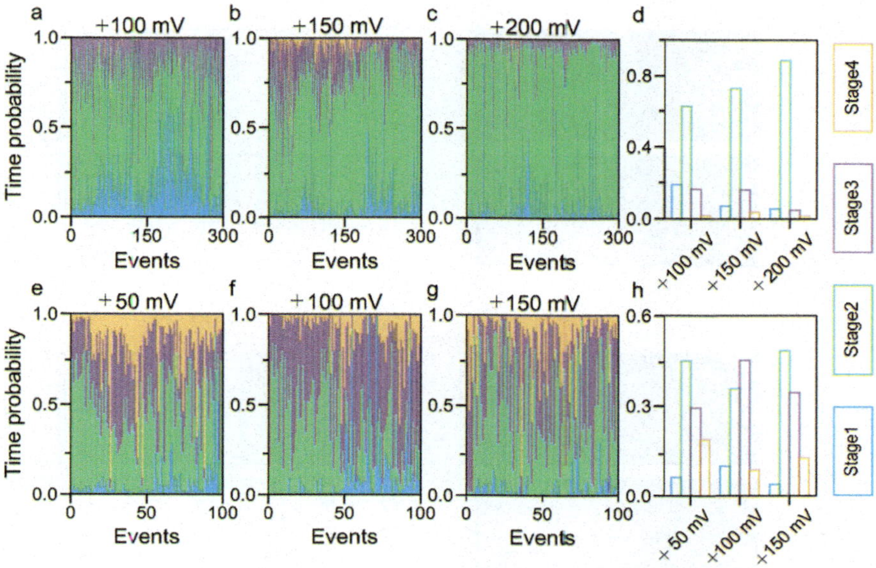

Fig. 5.12 The confined effects on the time probability of single β-hairpin for each current stage. The stacked bar charts for the time probability of each event under **a** + 100 mV, **b** + 150 mV and **c** + 200 mV using d ~ 4 nm nanopore. **d** The voltage-dependent average time probability from all the event of each current stage using d ~ 4 nm nanopore. The stacked bar charts for the time probability of each event under **e** + 50 mV, **f** + 100 mV and **g** + 150 mV using d ~ 5 nm nanopore. **h** The voltage-dependent average time probability from all the event of each current stage using d ~ 5 nm nanopore. Stage 1–4 are colored by blue, green, purple and orange. "Reprinted with permission from [65]. Copyright [65] Royal Society of Chemistry"

also necessary to integrate into this system to further assist in the identification of transition conformations. Generally, the key feature of the nanopore-based technique is providing a transient time and space window in which only the interested molecule existed by applying the bias to the nanopore. Therefore, the integrateion of the advantages of the nanopore method with other techniques, such as fluorescence and Raman spectroscopy, is realizable to acquire complementary information about proteins in this transient window. So, to further improve the applicability of the nanopore measurement system for the investigation of the mechanisms of protein dynamics, an optical acquisition system is necessary for solid-state nanopore analysis in protein sensing. In this section, we present our recent progress to build an integrated optical and electrochemical system for solid-state nanopores of single protein measurement [73].

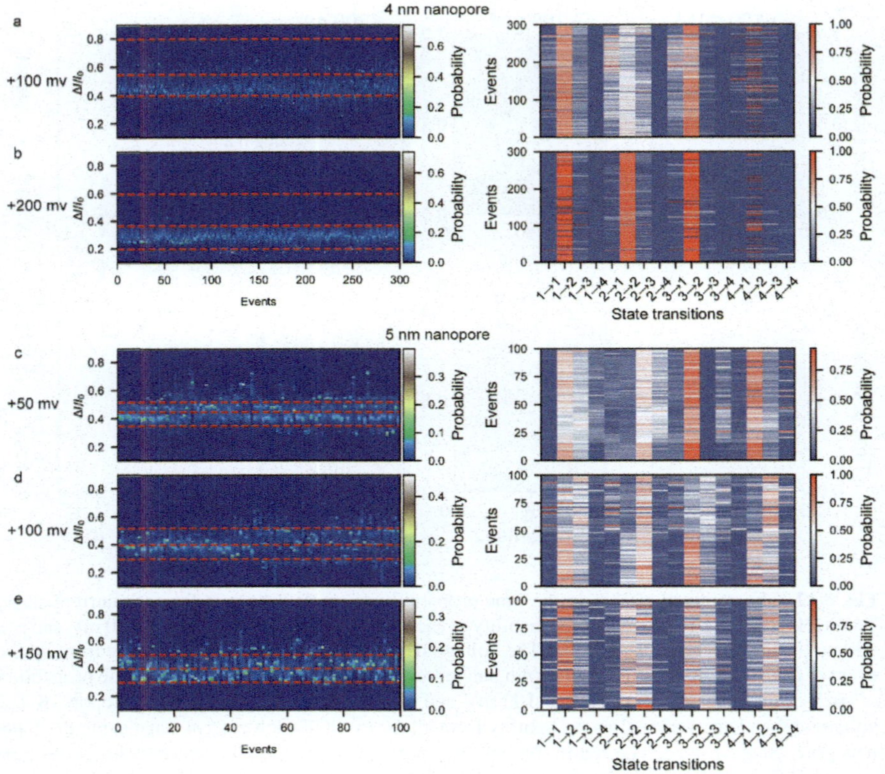

Fig. 5.13 Collection of the current histograms from 300 events and the corresponding heat map of folding/unfolding pathways for each β-hairpin peptide in ~ 4 nm nanopore at + 100 mV (**a**) and + 200 mV (**b**). Collection of the current histograms from 100 events and the corresponding heat map of folding/unfolding pathways for each β-hairpin peptide in ~ 5 nm nanopore at + 50 mV (**c**), + 100 mV (**d**) and + 150 mV (**e**). "Reprinted with permission from [65]. Copyright [65] Royal Society of Chemistry"

5.4.1 System Composition

The nanopore combined measurement system consists of a low-current amplifier, a nanopore chip with holders, and an optical system if necessary. As shown in Fig. 5.14, the amplifier is used to acquire the ionic current of the nanopore with a high signal-to-noise ratio. The inverted dark-field microscope is the best choice for nanopore optical analysis. The optical system consists of three parts: a light source, an excitation module, and an optical signal acquisition module. The light source could include a laser source with different wavelengths for fluorescent or Raman excitation. The collimated laser passes a series of mirrors and lenses to the microscope. According to the measurement requirements, Both the standard epi-illumination and total internal reflection (TIR) illumination had been integrated into this system. In the optical signal acquisition module, there are three kinds of optical information that should be focused on for nanopore analysis: imaging, spectrum, and optical intensity detection. As the example we provided in Fig. 5.14, the optical signal from a single nanopore was collected by a 40 × objective from the bottom of the nanopore chip. After filtering the source and background by filter sets equipped, the light with desired wavelength range was split into two beams by a splitter. One was transmitted through the lens to a camera for imagination or the spectrograph for spectrum analysis. Then the other light beam would be used for high-speed intensity monitoring with two photomultiplier tubes (PMT) or avalanche photodiodes (APD) for fluorescence with different emission wavelengths. In this system, the pinhole mounted onto 3-μm screws was used for spatial filterings, such as in confocal microscopy, to block the extra focal signal. All the components could be mounted along with a common optical axis with the 30 mm optical cage system. This flexible system could satisfy the requirements for diverse nanopore applications both the SiN_x solid-state nanopore and glass nanopipettes.

The ionic current detection is the kernel of the nanopore analysis. Generally, the patch-clamp amplifier system for electrophysiology is the most widely used for nanopore analysis, such as Axopatch 200B and the HEKA EPC-8. But these systems

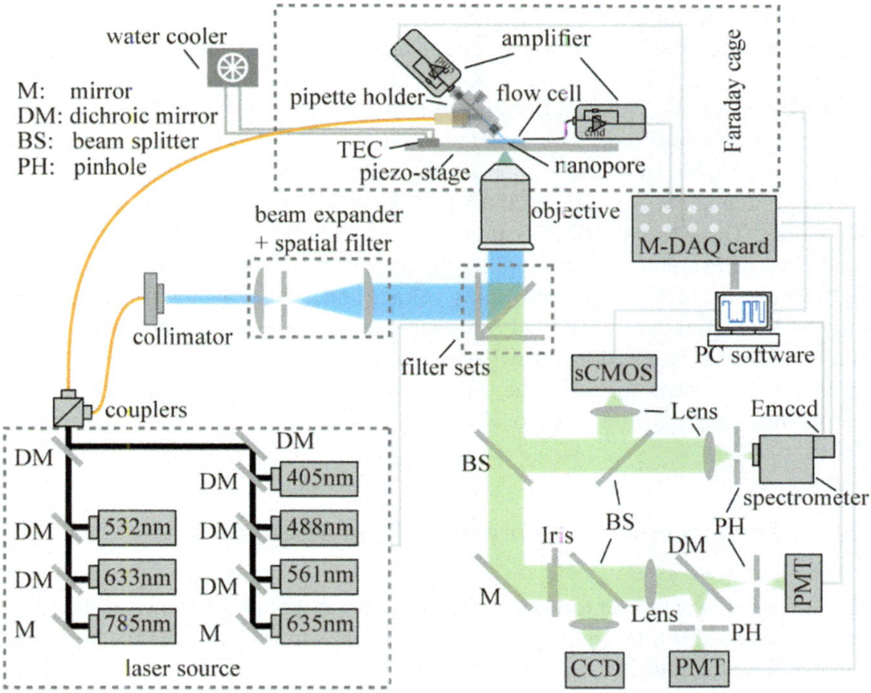

Fig. 5.14 Schematic of an integrated system for the optical and electrical detection of nanopore analysis. "Reprinted with permission from [73]. Copyright [73] American Institute of Physics"

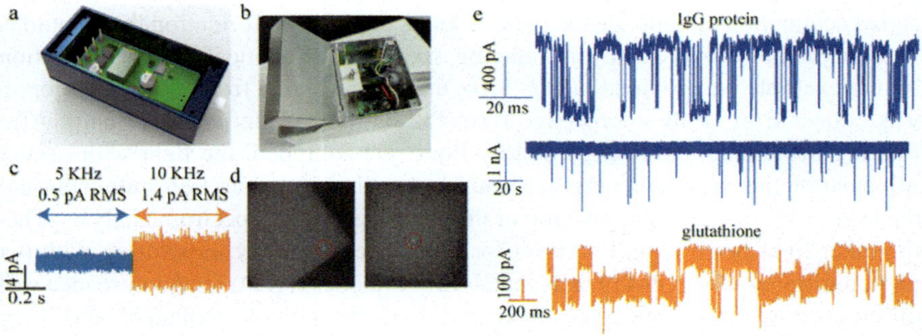

Fig. 5.15 **a** The renderings of the amplifier unit. **b** An image of the amplifier has been equipped. **c** Current trance of the amplifier at 5 and 10 kHz filter frequency with a sample rate of 100 kHz. **d** The calcium fluorescence images for confirmation and location of the solid-state nanopore fabricated on the SiN_x membrane. **e** Examples events of continuous current trace events for IgG protein and glutathione detected with the solid-state nanopore. "Reprinted with permission from [73]. Copyright [73] American Institute of Physics"

usually are expensive, bulky, and functional redundant. Miniaturization and Specialization is the luciferous future for nanopore amplifier. To satisfy this requirement, an ultralow-current amplifier was designed as shown in Fig. 5.15a and b. This amplifier includes a trans-impedance amplifier circuit to convert an input nanopore current into an output voltage, a 4-pole Bessel low-pass filter to eliminate noise, and a low-noise linear power supply. This amplifier is built on a circuit board with a size of 3 cm × 7 cm as a minimized unit. The potential range is ± 1 V and the current measuring range is adjustable by changing the feedback resistance, which is typically ± 2.5 nA for protein nanopores and ± 20 nA for solid-state nanopores and glass capillary nanopores. The low-pass filter is programmed with 5, 10, and 100 kHz for high bandwidth measurement. This amplifier has high compatibility with the commercial digitizer to control the amplifier to apply voltage and acquire the ionic current signal, such as Axon Digidata 1550B and National Instruments multifunction data acquisition DAQ board. In addition, a shielding cage with thermoelectric cooling (TEC) is applied in this system to reduce the various noises and keep the constant experiment temperature ranging from − 10 to 70 °C.

5.4.2 Performance Test for Single Protein Analysis

To test the performance of the amplifier, the open-circuit current trance of the amplifier at 5 and 10 kHz filter frequency was acquired, as shown in Fig. 5.15c, the average noise of the amplifier is 0.5 pA rms at 5 kHz and 1.4 pA rms at 10 kHz. Further, a protein detection experiment was conducted using SiN_x solid-state nanopore. Then the calcium fluorescence test was used to confirm and locate the nanopore on the SiN_x membrane (Fig. 5.15d). Further, these nanopores were used to detect the IgG protein, and glutathione, respectively. As shown in Fig. 5.15e, the IgG protein experienced a fast translocation in a 6.2 nm nanopore. When the pore size diminished to 1.8 nm, only the collision signals could be detected. We also used the nanopore to detect the glutathione, and distinct blockage events were acquired within a long duration time. These results confirmed that this platform for solid-state nanopores could be used for the measurement of single proteins. In further, by combining advanced spectrum analysis methods, the identification of the interaction and transition conformations in protein dynamics could be achieved in this platform.

5.5 Conclusion

In this chapter, we summarized our recent studies on solid-state nanopores for protein analysis. The nanopore technique opens a new paradigm of single molecule analysis both in vitro and in vivo. The confined space of the nanopore provides an isolated venue where only a single molecule or single molecule event existed with a very simple operation, repeatably and reliably. This feature makes nanopores become an ideal device for single molecule capture, transport, and detection all in one chip. We successfully used nanopores to study single proteins whether in vitro or complex cell systems. Regardless of the simple protein identification or the complicated protein dynamics problems, solid-state nanopores are qualified and could provide new insights into understanding the mechanism of protein and its dynamics. Furthermore, the nanopore technology also promotes the development of a series of combined single molecule technology and boost the improvement of the corresponding instrument system. Nanopore combined analysis is an exciting and relatively new area of research both for fundamental and applied science.

Acknowledgements This research was supported by the National Natural Science Foundation of China (22027806, 22090050). The authors thank all who contributed in this work from YTL group.

References

1. Bu Z, Callaway DJE (2011) Proteins MOVE! Protein dynamics and long-range allostery in cell signaling, pp 163–221
2. Orozco M (2014) A theoretical view of protein dynamics. Chem Soc Rev 43:5051–5066. https://doi.org/10.1039/C3CS60474H
3. Rueda M, Ferrer-Costa C, Meyer T et al (2007) A consensus view of protein dynamics. Proc Natl Acad Sci 104:796–801. https://doi.org/10.1073/pnas.0605534104
4. Surpeta B, Sequeiros-Borja C, Brezovsky J (2020) Dynamics, a powerful component of current and future in silico approaches for protein design and engineering. Int J Mol Sci 21:2713. https://doi.org/10.3390/ijms21082713
5. Anfinsen CB (1973) Principles that govern the folding of protein chains. Science 181:223–230. https://doi.org/10.1126/science.181.4096.223

6. Kendrew JC, Bodo G, Dintzis HM et al (1958) A Three-dimensional model of the myoglobin molecule obtained by X-ray analysis. Nature 181:662–666. https://doi.org/10.1038/181662a0
7. Shea J-E, Brooks CL III (2001) From folding theories to folding proteins: a review and assessment of simulation studies of protein folding and unfolding. Annu Rev Phys Chem 52:499–535. https://doi.org/10.1146/annurev.physchem.52.1.499
8. Dill KA, MacCallum JL (2012) The protein-folding problem, 50 years on. Science 338:1042–1046. https://doi.org/10.1126/science.1219021
9. Carrell RW, Lomas DA (1997) Conformational disease. Lancet 350:134–138. https://doi.org/10.1016/S0140-6736(97)02073-4
10. Kopito RR, Ron D (2000) Conformational disease. Nat Cell Biol 2:E207–E209. https://doi.org/10.1038/35041139
11. Dobson CM (2001) The structural basis of protein folding and its links with human disease. Philos Trans R Soc London Ser B Biol Sci 356:133–145. https://doi.org/10.1098/rstb.2000.0758
12. Horwich A (2002) Protein aggregation in disease: a role for folding intermediates forming specific multimeric interactions. J Clin Invest 110:1221–1232. https://doi.org/10.1172/JCI16781
13. Ross CA, Poirier MA (2004) Protein aggregation and neurodegenerative disease. Nat Med 10:S10–S17. https://doi.org/10.1038/nm1066
14. Costa CHS, Oliveira ARS, dos Santos AM et al (2019) Computational study of conformational changes in human 3-hydroxy-3-methylglutaryl coenzyme reductase induced by substrate binding. J Biomol Struct Dyn 37:4374–4383. https://doi.org/10.1080/07391102.2018.1549508
15. Fraser JS, van den Bedem H, Samelson AJ et al (2011) Accessing protein conformational ensembles using room-temperature X-ray crystallography. Proc Natl Acad Sci 108:16247–16252. https://doi.org/10.1073/pnas.1111325108
16. Davis IW, Arendall WB, Richardson DC, Richardson JS (2006) The backrub motion: how protein backbone shrugs when a sidechain dances. Structure 14:265–274. https://doi.org/10.1016/j.str.2005.10.007
17. Callaway DJE, Matsui T, Weiss T et al (2017) Controllable activation of nanoscale dynamics in a disordered protein alters binding kinetics. J Mol Biol 429:987–998. https://doi.org/10.1016/j.jmb.2017.03.003
18. Farago B, Li J, Cornilescu G et al (2010) Activation of nanoscale allosteric protein domain motion revealed by neutron spin echo spectroscopy. Biophys J 99:3473–3482. https://doi.org/10.1016/j.bpj.2010.09.058
19. Turton DA, Senn HM, Harwood T et al (2014) Terahertz underdamped vibrational motion governs protein-ligand binding in solution. Nat Commun 5:3999. https://doi.org/10.1038/ncomms4999
20. Chodera JD, Noé F (2014) Markov state models of biomolecular conformational dynamics. Curr Opin Struct Biol 25:135–144. https://doi.org/10.1016/j.sbi.2014.04.002
21. Lane TJ, Shukla D, Beauchamp KA, Pande VS (2013) To milliseconds and beyond: challenges in the simulation of protein folding. Curr Opin Struct Biol 23:58–65. https://doi.org/10.1016/j.sbi.2012.11.002
22. Varongchayakul N, Song J, Meller A, Grinstaff MW (2018) Single-molecule protein sensing in a nanopore: a tutorial. Chem Soc Rev 47:8512–8524. https://doi.org/10.1039/C8CS00106E
23. Waduge P, Hu R, Bandarkar P et al (2017) Nanopore-based measurements of protein size, fluctuations, and conformational changes. ACS Nano 11:5706–5716. https://doi.org/10.1021/acsnano.7b01212
24. Sutherland TC, Long Y-T, Stefureac R-I et al (2004) Structure of peptides investigated by nanopore analysis. Nano Lett 4:1273–1277. https://doi.org/10.1021/nl049413e
25. Wang H-Y, Ying Y-L, Li Y et al (2011) Nanopore analysis of β-amyloid peptide aggregation transition induced by small molecules. Anal Chem 83:1746–1752. https://doi.org/10.1021/ac1029874
26. Van Meervelt V, Soskine M, Singh S et al (2017) Real-time conformational changes and controlled orientation of native proteins inside a protein nanoreactor. J Am Chem Soc 139:18640–18646. https://doi.org/10.1021/jacs.7b10106

27. Cressiot B, Oukhaled A, Patriarche G et al (2012) Protein transport through a narrow solid-state nanopore at high voltage: experiments and theory. ACS Nano 6:6236–6243. https://doi.org/10.1021/nn301672g
28. Luchian T, Park Y, Asandei A et al (2019) Nanoscale probing of informational polymers with nanopores. Applications to amyloidogenic fragments, peptides, and DNA-PNA Hybrids. Acc Chem Res 52:267–276. https://doi.org/10.1021/acs.accounts.8b00565
29. Movileanu L, Schmittschmitt JP, Martin Scholtz J, Bayley H (2005) Interactions of peptides with a protein pore. Biophys J 89:1030–1045. https://doi.org/10.1529/biophysj.104.057406
30. Hu R, Rodrigues JV, Waduge P et al (2018) Differential enzyme flexibility probed using solid-state nanopores. ACS Nano 12:4494–4502. https://doi.org/10.1021/acsnano.8b00734
31. Mohammad MM, Prakash S, Matouschek A, Movileanu L (2008) Controlling a single protein in a nanopore through electrostatic traps. J Am Chem Soc 130:4081–4088. https://doi.org/10.1021/ja710787a
32. Shi W, Friedman AK, Baker LA (2017) Nanopore sensing. Anal Chem 89:157–188. https://doi.org/10.1021/acs.analchem.6b04260
33. Xue L, Yamazaki H, Ren R et al (2020) Solid-state nanopore sensors. Nat Rev Mater 5:931–951. https://doi.org/10.1038/s41578-020-0229-6
34. Zhang S, Li M, Su B, Shao Y (2018) Fabrication and use of nanopipettes in chemical analysis. Annu Rev Anal Chem 11:265–286. https://doi.org/10.1146/annurev-anchem-061417-125840
35. Stanley J, Pourmand N (2020) Nanopipettes—the past and the present. APL Mater 8. https://doi.org/10.1063/5.0020011
36. Yu R-JJ, Ying Y-LL, Gao R, Long Y-TT (2019) Confined nanopipette sensing: from single molecules, single nanoparticles, to single cells. Angew Chemie Int Ed 58:3706–3714. https://doi.org/10.1002/anie.201803229
37. Yi W, Li X, He XX et al (2022) Glass nanopipette sensing of single entities. J Electroanal Chem 909:116106. https://doi.org/10.1016/j.jelechem.2022.116106
38. Altschuler SJ, Wu LF (2010) Cellular heterogeneity: do differences make a difference? Cell 141:559–563. https://doi.org/10.1016/j.cell.2010.04.033
39. Actis P, Mak AC, Pourmand N (2010) Functionalized nanopipettes: toward label-free, single cell biosensors. Bioanal Rev 1:177–185. https://doi.org/10 1007/s12566-010-0013-y
40. Bulbul G, Chaves G, Olivier J et al (2018) Nanopipettes as monitoring probes for the single living cell: state of the art and future directions in molecular biology. Cells 7:55. https://doi.org/10.3390/cells7060055
41. Lv J, Qian RC, Hu YX et al (2016) A precise pointing nanopipette for single-cell imaging via electroosmotic injection. Chem Commun 52:13909–13911. https://doi.org/10.1039/c6cc08125h
42. Qian RC, Lv J, Long YT (2018) Ultrafast mapping of subcellular domains via nanopipette-based electroosmotically modulated delivery into a single living cell. Anal Chem 90:13744–13750. https://doi.org/10.1021/acs.analchem.8b04159
43. Ying YL, Li YJ, Mei J et al (2018) Manipulating and visualizing the dynamic aggregation-induced emission within a confined quartz nanopore. Nat Commun 9. https://doi.org/10.1038/s41467-018-05832-y
44. Chen H, Lin Y, Long YT et al (2021) Nanopore-based measurement of the interaction of P450cam monooxygenase and putidaredoxin at the single-molecule level. Faraday Discuss 233:295–302. https://doi.org/10.1039/d1fd00042j
45. Gao R, Lin Y, Ying Y-LL et al (2019) Wireless nanopore electrodes for analysis of single entities. Nat Protoc 14:2015–2035. https://doi.org/10.1038/s41596-019-0171-5
46. Gao R, Ying Y-L, Hu Y-X et al (2017) Wireless bipolar nanopore electrode for single small molecule detection. Anal Chem 89:7382–7387. https://doi.org/10.1021/acs.analchem.7b00729
47. Yu R-J, Ma W, Liu X-Y et al (2016) Metal-linked immunosorbent assay (MeLISA): the enzyme-free alternative to ELISA for biomarker detection in serum Theranostics 6:1732–1739. https://doi.org/10.7150/thno.16129
48. Gao R, Lin Y, Ying Y et al (2017) Dynamic self-assembly of homogenous microcyclic structures controlled by a silver-coated nanopore. Small 13:1700234. https://doi.org/10.1002/smll.201700234

49. Ying YL, Hu YX, Gao R et al (2018) Asymmetric nanopore electrode-based amplification for electron transfer imaging in live cells. J Am Chem Soc 140:5385–5392. https://doi.org/10.1021/jacs.7b12106
50. Lu SM, Li YJ, Zhang JF et al (2019) Monitoring hydrogen evolution reaction catalyzed by MoS_2 quantum dots on a single nanoparticle electrode. Anal Chem 91:10361–10365. https://doi.org/10.1021/acs.analchem.9b02364
51. McKelvey K, German SR, Zhang Y et al (2017) Nanopipettes as a tool for single nanoparticle electrochemistry. Curr Opin Electrochem 6:4–9. https://doi.org/10.1016/j.coelec.2017.06.006
52. Gao R, Ying Y-LL, Li Y-JJ et al (2018) A 30 nm nanopore electrode: facile fabrication and direct insights into the intrinsic feature of single nanoparticle collisions. Angew Chemie Int Ed 57:1011–1015. https://doi.org/10.1002/anie.201710201
53. Yu R-J, Xu S-W, Paul S et al (2021) Nanoconfined electrochemical sensing of single silver nanoparticles with a wireless nanopore electrode. ACS Sens 6:335–339. https://doi.org/10.1021/acssensors.0c02327
54. Sha JJ, Si W, Xu W et al (2015) Glass capillary nanopore for single molecule detection. Sci China Technol Sci 58:803–812. https://doi.org/10.1007/s11431-015-5779-2
55. Zhang S, Chen W, Song L et al (2021) Recent advances in ionic current rectification based nanopore sensing: a mini-review. Sens Actuat Rep 3:100042. https://doi.org/10.1016/j.snr.2021.100042
56. Gao R, Ying YL, Yan BY et al (2016) Ultrasensitive determination of mercury(II) using glass nanopores functionalized with macrocyclic dioxotetraamines. Microchim Acta 183:491–495. https://doi.org/10.1007/s00604-015-1634-1
57. Yu R-J, Chen K-L, Ying Y-L, Long Y-T (2022) Nanopore electrochemical measurement for single molecular interactions and beyond. Curr Opin Electrochem 101063. https://doi.org/10.1016/j.coelec.2022.101063
58. Lu S-M, Yu R-J, Long Y-T (2019) Real-time monitoring of click reaction based on confined glass nanopipette. Chinese J Anal Chem 47:1309–1313. https://doi.org/10.19756/j.issn.0253-3820.191333
59. Ying YL, Yu RJ, Hu YX et al (2017) Single antibody-antigen interactions monitored: via transient ionic current recording using nanopore sensors. Chem Commun 53:8620–8623. https://doi.org/10.1039/c7cc03927a
60. Karawdeniya BI, Bandara YMNDY, Nichols JW et al (2019) Challenging nanopores with analyte scope and environment. J Anal Test 3:61–79. https://doi.org/10.1007/s41664-019-00092-1
61. Hu YX, Ying YL, Gao R et al (2018) Characterization of the dynamic growth of the nanobubble within the confined glass nanopore. Anal Chem 90:12352–12355. https://doi.org/10.1021/acs.analchem.8b03923
62. Yu RJ, Ying YL, Hu YX et al (2017) Label-free monitoring of single molecule immunoreaction with a nanopipette. Anal Chem 89:8203–8206. https://doi.org/10.1021/acs.analchem.7b01921
63. Yu RJ, Lu SM, Xu SW et al (2019) Single molecule sensing of amyloid-β aggregation by confined glass nanopores. Chem Sci 10:10728–10732. https://doi.org/10.1039/c9sc03260f
64. Ma H, Yu RJ, Ying YL, Long YT (2022) Electrochemically confined effects on single enzyme detection with nanopipettes. J Electroanal Chem 908:116086. https://doi.org/10.1016/j.jelechem.2022.116086
65. Liu S-C, Ying Y-L, Li W-H et al (2021) Snapshotting the transient conformations and tracing the multiple pathways of single peptide folding using a solid-state nanopore. Chem Sci 12:3282–3289. https://doi.org/10.1039/D0SC06106A
66. Roccatano D, Amadei A, Di NA, Berendsen HJC (2009) A molecular dynamics study of the 41–56 β-hairpin from B1 domain of protein G. Protein Sci 8:2130–2143. https://doi.org/10.1110/ps.8.10.2130
67. Muñoz V, Thompson PA, Hofrichter J, Eaton WA (1997) Folding dynamics and mechanism of β-hairpin formation. Nature 390:196–199. https://doi.org/10.1038/36626
68. Dinner AR, Lazaridis T, Karplus M (1999) Understanding β-hairpin formation. Proc Natl Acad Sci 96:9068–9073. https://doi.org/10.1073/pnas.96.16.9068

69. Howarth M, Chinnapen DJ-F, Gerrow K et al (2006) A monovalent streptavidin with a single femtomolar biotin binding site. Nat Methods 3:267–273. https://doi.org/10.1038/nmeth861
70. Tsai J, Levitt M (2002) Evidence of turn and salt bridge contributions to β-hairpin stability: MD simulations of C-terminal fragment from the B1 domain of protein G. Biophys Chem 101–102:187–201. https://doi.org/10.1016/S0301-4622(02)00198-9
71. Si W, Aksimentiev A (2017) Nanopore sensing of protein folding. ACS Nano 11:7091–7100. https://doi.org/10.1021/acsnano.7b02718
72. Johnson RP, Fleming AM, Beuth LR et al (2016) Base flipping within the α-hemolysin latch allows single-molecule identification of mismatches in DNA. J Am Chem Soc 138:594–603. https://doi.org/10.1021/jacs.5b10710
73. Liu S-C, Xie B-K, Zhong C-B et al (2021) An advanced optical–electrochemical nanopore measurement system for single-molecule analysis. Rev Sci Instrum 92:121301. https://doi.org/10.1063/5.0067185

Chapter 6
Self-consistent Brownian Dynamics Simulations of the Ionic Current Blockade in Solid State Nanopores

Dmitriy V. Melnikov and Maria E. Gracheva

Abstract When a nanoparticle (or a biomolecule) translocates through a nanopore in a solid-state membrane submereged in an electrolyte solution, the ionic current changes and the ionic current blockade appears. In this work we elucidate how the concentration polarization, the surface charge on the nanopore and nanoparticle as well as the applied voltage affect the ionic current blockade. We use Brownian dynamics approach in conjunction with a full three-dimensional self-consistent solution of the Poisson-Nernst-Planck and Navier-Stockes system of equations to describe realistic ionic current response arising due to the random motion of a nanoparticle through a nanopore. We then derive a closed form expression for the ionic current blockade ratio and apply it to analyze the computed and experimentally measured ionic current blockades.

Keywords Ionic current blockade · Nanoparticle · Brownian dynamics · Nanopore · Multiscale modeling · Surface charge

6.1 Introduction

In recent years the detection, characterization and filtering of nanoparticles and biomolecules [1–6] has become a large and popular research area. In particular, a large and still growing number of research studies focuses on nanopore-based applications for the advancement of biosensing technologies [7–10]. In such a system, a nanoporous membrane separates two chambers filled with electrolyte solution, one where the objects of interest (such as nanoparticles or biomolecules) initially reside in (the *cis* chamber) and the other to which they are encouraged to translocate into (the *trans* chamber) via the nanopore(s) in a membrane. While the translocation

D. V. Melnikov · M. E. Gracheva (✉)
Department of Physics, Clarkson University, Potsdam, NY 13699, US
e-mail: gracheva@clarkson.edu

J.-P. Leburton (ed.), *Solid State Nanopores*, Nanostructure Science and Technology,
https://doi.org/10.1007/978-3-031-42336-9_6

of the object of interest through a nanopore can be achieved by a variety of means (such as applied pressure, for example), here we focus on a particular mechanism when these events are facilitated by means of the voltage bias applied between the chambers. In this case, owing to the presence of the electrolyte ions in the solution, the ionic current through the nanopore appears (the open pore current, I_o). When a nanoparticle or a biomolecule passes through a nanopore under the influence of the applied voltage, the ionic current changes, usually decreasing in value, from the open pore current value I_o to the blocked pore current I, $I < I_o$. This leads to the phenomenon of "the ionic current blockade“ which depends on the properties of a translocating object and its environment in a non-trivial way. The method of identifying and characterizing a passing object of interest from the ionic current blockade features [11–18] is the resisitive-pulse technique, and it is one the most popular and widely available tools for studying the ionic transport response to the passage of nanoparticles or biomolecules through a nanopore [11, 18–25] and performing DNA sequencing [26–29].

By evaluating the effective resistance of a conducting medium with small spherical particles of radius R_s made of a different material dispersed at large distances from each other, it can be shown that the current blockade ratio for a very long cylindrical pore of radius R_p, $R_p \gg R_s$, is proportional to the particle/pore volume ratio v_p/V_p [30–32]:

$$\frac{I - I_o}{I_o} = -\frac{3}{2}\frac{v_p}{V_p}. \tag{6.1}$$

To account for distortion of the electric field in a confined volume of the nanopore around the particle which become signifcant for larger particles, the above result has to be modified. This is usually accomplished in an *ad hoc* manner introducing the large particle correcting factor $F_1(R_s/R_p)$ [31, 33]. The values of this factor are numerically computed [34], and the results for spherical particles can be fitted as, for example,

$$F_1(R_s/R_p) = \frac{1}{1 - 0.8(R_s/R_p)^3} \tag{6.2}$$

for $R_s \lesssim 0.9R_p$ [35].

Furthermore, real nanopores have finite lengths L_p, and redistribution of the current density and electric field near the pore ends affects the ionic current through the pore. These effects can be captured by introducing the access resistance [36] in series with the nanopore resistance so that the applied voltage is then drops over the nanopore length and the two access regions. In principle, the access resistance also has different values for the blocked and open nanopores but for long nanopores with $L_p > R_p$, we can take it to be unaffected by the presence of the particle inside the nanopore away from its ends [37], in which case its effect on the ionic current blockade can be described with help of another factor $F_2(R_p/L_p) = [1 + \pi R_p/(2L_p)]^{-1}$. For a spherical particle, keeping only the leading terms $\sim v_p/V_p$, the expression for current blockade ratio (6.1) thus becomes

$$\frac{I - I_o}{I_o} = -\frac{3}{2}\frac{v_p}{V_p} F_1\left(\frac{R_s}{R_p}\right) F_2\left(\frac{R_p}{L_p}\right). \tag{6.3}$$

The above discussion makes no mention of any effects the surface charges have on the ionic current but in most experimental situations, however, both a nanoparticle (or a biomolecule) and the membrane with a nanopore carry a (usually) negative surface charge. The experimental measurements revealed that when a charged particle passes through such a pore (1) the current blockade ratio $(I - I_o)/I_o$ can be actually smaller than given by Eq. (6.1), particulalrly when the translocating object is just entering or about to exit the nanopore, and (2) the current blockade ratio decreases with applied voltage, that is, it is voltage dependent [38, 39]. On the other hand, other experiments also demonstrated [12, 39, 40] that for large driving voltages, the ionic current can increase above the open pore value, that is, $I > I_o$, leading to a positive current blockade ratios. This was found for both the nanopores with dimensions in the nanometer range as well as fairly large nanopore systems (with dimensions in hundreds of nm) for translocating DNA molecules [12] and nanoparticles [39–41]. For smaller nanopore systems [40], this current increase was tentatively attributed to the overlap of the double layers formed on the charged nanopore and nanoparticle surfaces leading to the increase in the ionic concentration and pore conductivity which at appropriate conditions can overcome the decrease in the current associated with the particle blocking the pore. However, in very large nanopores [39, 41], the increase in the ionic current due to the enhanced conductivity from ions accumulated in the double layer on the surface of the nanoparticle, should be small compared to the current decrease due to the volume excluded by the nanoparticle, and yet, the blocked pore current still exceeded the open pore current value.

Continuum electrostatic and hydrodynamic models [42] are used extensively to investigate and unravel the factors controlling nanoparticle (or even DNA if represented by a rod shaped rigid object) translocation through a pore along its central axis [39, 43–45]. A variety of models was used to explain the ionic current rectification in nanopores [46, 47] while other works investigated ionic current traces and hydrodynamic effects near the membrane surface [48, 49]. Such models allowed to identify conditions for the applied potential bias, membrane charge, and electrolyte concentration necessary for the postive current blockade ratio (see, for example, [38, 39, 41, 43, 50] and references therein).

Previously, we developed several models that account for the electric and dielectrophoretic effects in nanoporous membranes of different compositions, such as the dielectric and semiconductor membranes, in close contact with either a DNA molecule, protein, or a nanoparticle translocating through it [51–54]. With these models we were able to understand the effects of the membrane surface charge, electrolyte concentration, the applied membrane bias, etc. on the motion of the translocating molecule which was modeled with a help of Brownian dynamics. In particular, we determined how the charge on the membrane surface affects the nanoparticle motion in the electric landscape of the nanopore. The electric and other effects ultimately contribute to the ability of a nanoporous membrane system to detect a certain molecule via the ionic current change, and thus, it is important to study and understand these effects.

In this Chapter, we first discuss a sophisticated computational model that allows us to accurately simulate the random (Brownian) motion of a nanoparticle through a solid state nanoporous membrane and offers a significant improvement in the way the dynamic particle-pore interactions are taken into account. Our model describes the forces affecting the random nanoparticle motion in a self-consistent manner so that we are able to calculate the ionic current traces and observe blockade events as the particle moves in a random fashion through the pore. Based on the calculations, we then propose a comprehensive explanation of the above, seemingly contradictory, observations of the ionic blockade which accounts for both the enhanced conductivity in the nanopore volume, and not just around the surface of the nanoparticle, as well as its voltage dependence.

Our idea is rooted in the concentration polarization (CP) effect occuring in charged nanopores [41, 55]. It appears because the charge on the nanopore surface leads to the accumulation of the ions of the opposite sign inside the pore. When an electic bias is applied, not only the ionic current through the pore appears, but also a net ionic flux. When the nanopore surface is negatively charged, this flux creates an ionic build-up around the low electric potential end and depletes the ionic concentration near the other end of the nanopore. To put it in other words, the CP appears because the charged nanopore represents a constriction to the flow of ions, and when a particle is present inside the nanopore, it also represents an obstacle to the ionic flux, localized inside the pore. As such, one could expect the CP around the particle to appear in addition to the concentration polarization across the pore. Because the net ionic flux in the nanopore remains constant, this additional CP across the particle will thus affect the ionic concentration everywhere inside the nanopore, not just in the immediate vicinity of the particle itself. And while the idea of the concentration polarization affecting the ionic current was used previously to interpret the experimental data [39, 41], here we take it a step further and derive a closed form expression for the ionic current in a nanopore taking this effect into account.

The Chapter is organized as follows. In the following Sect. 6.2 we describe our simulated nanopore system and our computational model. Section 6.3 discusses behavior of the translocating particle in the nanopore as well as the numerically calculated ionic current traces using Brownian dynamics model. Section 6.4 develops an analytical model and provides a comprehensive analysis of the computed and experimental results in terms of the concentration polarization induced current. Section 6.5 provides concluding remarks and briefly discusses future work.

6.2 Continuum Model for the Ionic Current Calculations in Nanopores

6.2.1 Simulated System

Our three-dimensional system consists of a cylindrical nanopore in a solid-state membrane surrounded by two reservoirs (*cis* and *trans* filled with an electrolyte) with a nanoparticle allowed to move randomly throughout the whole volume of the

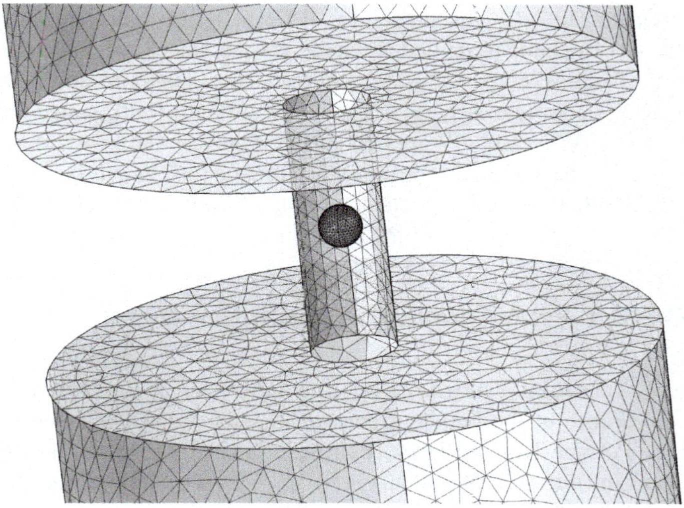

Fig. 6.1 Visual representation of the simulated structure with superimposed triangular mesh. Reproduced with permission from Ref. [58] Copyright (2018) (IOP Science)

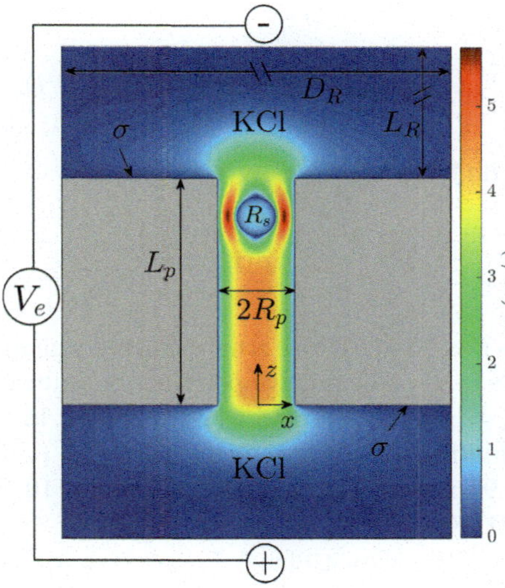

Fig. 6.2 A 2D profile of the modeled structure with a spherical nanoparticle of radius R_s inside the nanopore. The magnitude of the computed electrosmotic fluid velocty v_f is used as the background. Membrane is shown in grey. Reproduced with permission from Ref. [58] Copyright (2018) (IOP Science)

system (Fig. 6.1). The system is similar to a 2D axisymmetric model used in many works (see, for example, Ref. [56] and references therein) but being 3D makes it more realistic at the expense of greatly increased computational effort.

Specifically, in this Section we perform calculations for the system schematically shown in Fig. 6.2 in which the nanopore has radius of $R_p = 2$ nm and length of $L_p = 12$ nm while the reservoirs above ($z > L_p$) and below ($z < 0$) the nanopore have diameters of $D_R = 28$ nm and heights of $L_R = 18$ nm. Constant surface charge density $\sigma = -0.16$ e/nm^2 is applied at the membrane-electrolyte interface while the nanoparticle of radius $R_s = 1$ nm is uncharged. The nanopore and reservoirs are filled with KCl electrolyte at $c_b = 0.1$ M ionic strength (bulk concentration). An electrolyte bias of $V_e = 100$ mV applied between the *cis* and *trans* compartments. This bias induces the ionic current through the nanopore which is affected by the presence of the spherical nanoparticle at various positions in the electrolyte.

6.2.2 Poisson-Nernst-Planck-Navier-Stokes Model

As the membrane carries a surface charge and there is an applied voltage V_e responsible for appearance of the ionic current and translocation of the nanoparticle, we first solve the Poisson-Nernst-Planck (PNP) equations to determine the electrostatic potential $\Phi(\mathbf{r})$ and the ionic current fluxes $\mathbf{N}_{K^+(\mathrm{Cl}^-)}$ for the whole simulated structure:

$$\nabla^2 \Phi = -\frac{\rho(\mathbf{r})}{\varepsilon_0 \varepsilon_r}, \tag{6.4}$$

$$\nabla \cdot \mathbf{N}_i = \nabla \cdot \left[z_i \frac{e D_i}{k_B T} C_i \nabla \Phi + D_i \nabla C_i - \mathbf{v} C_i \right] = 0, \quad i = K^+, \mathrm{Cl}^- \tag{6.5}$$

with the charge density $\rho(\mathbf{r})$ given by

$$\rho(\mathbf{r}) = \begin{cases} 0, & \text{nanoparticle}, \\ e\left[C_{K^+} - C_{\mathrm{Cl}^-}\right], & \text{electrolyte}. \end{cases} \tag{6.6}$$

In Eqs. (6.4)–(6.6), e is the elementary charge, ε_0 is the permittivity of free space, $\varepsilon_r = 78$ (3.9) is the relative permittivity of the electrolyte solution (a SiO_2 nanoparticle), $z_i = \pm 1$, $C_{K^-}(\mathbf{r})$ and $C_{\mathrm{Cl}^+}(\mathbf{r})$ are the local ionic concentrations of potassium and chlorine ions, respectively. Also, D_i is the diffusion coefficient of ions, $D_{K^+} = 1.95 \times 10^{-9}$ m^2/s and $D_{\mathrm{Cl}^-} = 2.03 \times 10^{-9}$ m^2/s, and the ambient temperature $T = 300$ K.

In the Nernst-Planck Eq. (6.5), $\mathbf{v}$ is the velocity field of the electroosmotic flow (EOF) in the electrolyte fluid [57]. As was already mentioned, the ionic solution within the nanopore attains a non-zero charge, and when the electric bias along the axis of the pore is applied, the ions begin to move so that the fluid flow (EOF) appears. As the fluid is viscous, in the steady state regime, the bulk motion of the solution in the nanopore sets in; as a result, the EOF is present throughout the whole cross sectional area of the pore. To calculate the EOF velocity $\mathbf{v}(\mathbf{r})$, we solve the Navier-Stokes (NS) equation together with the continuity equation assuming an incompressible fluid and low Reynolds number $Re \sim 10^{-3}$:

$$\eta\nabla^2\mathbf{v} = \nabla p - e(C_{K^+} - C_{Cl^-})\nabla\Phi, \tag{6.7}$$

$$\nabla \cdot \mathbf{v} = 0, \tag{6.8}$$

where p is the local pressure and $\eta = 10^{-3}$ Pa·s is the viscosity of water.

The above PNP-NS system of equations is solved with the normal components of the ionic flux set to zero at the electrolyte-membrane, electrolyte-particle interface, and sides of the reservoirs. Normal components of the electric field were set to zero at the sides of the reservoirs while on the surface of the membrane (particle), in general, the normal component of the electric field

$$-\hat{n} \cdot \nabla\Phi = \frac{e\sigma_{(p)}}{\varepsilon_0\varepsilon_r k_B T}, \tag{6.9}$$

which is approximately valid when the dielectric constant of the membrane (or the particle) is much smaller than that of the solution [58]. Here $\sigma_{(p)}$ is the nanopore, σ, (nanoparticle, σ_p) surface charge density, and $\hat{n}$ is the unit vector normal. The electric potential was set to 100 mV at the lower boundary of the bottom reservoir and to zero at the upper boundary of the top reservoir to create the applied electric bias V_e. For the fluid flow simulations, a no-slip condition ($\mathbf{v} = 0$) was imposed at the membrane-electrolyte interface and at the nanoparticle-electrolyte interface (except for the drag coefficient calculations when we assume a moving particle with a velocity $\mathbf{v}_p$, see Sect. 6.3.2 below for details), and a slip boundary condition was used at the side walls of reservoirs. The pressure was set to zero at the top and bottom of the reservoirs.

Note that in the above equation, surface charge densities σ and σ_p are assumed to be given constants that enter the model through boundary conditions. In reality, both the dielectric membrane and the nanoparticle are submerged in the electrolyte solution with a given pH value. In this situation, acidic groups on the membrane (and nanoparticle) surface ionize, and in a simplest case [59]

$$AH \rightleftarrows A^- + H^+. \tag{6.10}$$

If the total surface density of ionizable groups is $n = [A^-] + [AH]$, at equilibrium we can find for the surface charge density on the membrane (nanoparticle):

$$\sigma_{(p)} = -e[A^-] = \frac{-enK}{K + [H^+]_o}, \tag{6.11}$$

where $K = [A^-][H^+]_o/[AH]$ with $[AH]$, $[A^-]$ being surface concentrations of AH and A^-, and $[H^+]_o$ is the concentration of hydrogen ions in the solution on the surface of the membrane (particle). As both OH^- and H^+ ions are present in the solution together with K^+ and Cl^- ions, this means that the surface charge density is now pH-dependent and should be determined in the self-consistent way from the solution of Eqs. (6.4)–(6.6) where $\rho = e[C_{K^+} - C_{Cl^-} + C_{H^+} - C_{OH^-}]$ and Nernst-Planck equations for H^+ and OH^- ions are added. The result is that $\sigma\,(\sigma_p)$ will depend on the pH of the solution. However, for a fixed bulk pH value, the computed surface charge densities [60] were found to be approximately constant over the nanopore length for $pH = 6 - 10$, and as such, we set here σ (σ_p) to fixed values, to simplify calculations and perform a general analysis irrespective of a particular membrane and particle composition.

A nanoparticle translocating through a nanopore induces a change in the ionic current. To obtain the ionic current I for different particle positions, the ionic fluxes $\mathbf{N_i}$, $i = K^+, \mathrm{Cl}^-$, from Eq. (6.5) are integrated over the cross-sectional area of the nanopore:

$$I = eN_A \int \left(z_{K^+} N_{z,K^+} + z_{\mathrm{Cl}^-} N_{z,\mathrm{Cl}^-}\right) dA. \tag{6.12}$$

From the above system of the PNP-NS equtions, we can also compute the electric and hydrodynamic drag forces acting on a particle in the nanopore and outside near the membrane. By computing these forces as well as the ionic current at various nanoparticle locations in the electrolyte, we thus create a 3D force and current maps that is later used to interpolate their values at the intermediate locations between the stored positions to be used in the Brownian Dynamics simulations (Sect. 6.3) in order to produce realistic particle trajectories and the corresponding ionic current traces. As the system has axial symmetry, in cylindrical coordinates, we only need to compute the axial and radial components of the electric force, F_z^E and F_r^E, and the hydrodynamic force, F_z^H and F_r^H [59]:

$$F_z^E = \varepsilon_0 \varepsilon_w \int \left[\frac{E_z^2 - E_r^2}{2} n_z + E_z E_r n_r \right] dA, \tag{6.13}$$

$$F_r^E = \varepsilon_0 \varepsilon_w \int \left[\frac{E_r^2 - E_z^2}{2} n_r + E_z E_r n_z \right] dA, \tag{6.14}$$

$$F_z^H = \int \left[\left(-p + 2\eta \frac{\partial v_z}{\partial z} \right) n_z + \eta \left(\frac{\partial v_z}{\partial r} + \frac{\partial v_r}{\partial z} \right) n_r \right] dA, \tag{6.15}$$

$$F_r^H = \int \left[\left(-p + 2\eta \frac{\partial v_r}{\partial r} \right) n_r + \eta \left(\frac{\partial v_z}{\partial r} + \frac{\partial v_r}{\partial z} \right) n_z \right] dA, \tag{6.16}$$

where E_z, E_r, v_z, and v_r are the z- and r-components of the electric field $\mathbf{E} = -\nabla\Phi$ and the fluid velocity $\mathbf{v}$, respectively. In Eqs. (6.13)–(6.16), the integration is performed over the surface of the nanoparticle, and n_z and n_r are, respectively, the z- and r-components of the unit vector normal to that surface.

Numerical solution of the system of PNP-NS equations with corresponding boundary conditions is accomplished self-consistently with COMSOL® Multiphysics 5.1 using segregated solver where PNP and NS equations are solved sequentially with the global relative tolerance of calculations set to 10^{-6}. A 3D triangular mesh was implemented everywhere and adaptive mesh refinement was applied resulting in the increase of the number of degrees of freedom from 3×10^5 to 1.5×10^6 to generate mesh with the maximum size of 1 Å on the nanoparticle and $\sim$ 8 Å in other regions. We also verified our 3D calculations against 2D results when the particle motion is restricted along the pore axis and found that the ionic current values are within 0.5 % of each other. The simulations were executed for different particle positions in the xz-plane with 2 Å increments for particle displacements in both the x and z directions.

6.3 Brownian Dynamics Simulations of the Ionic Current

6.3.1 Brownian Dynamics Model

Theoretical studies of the nanoparticle dynamics through numerical simulations offers valuable insights on the motion of particles and their interaction with the nanopore-membrane system. Molecular dynamics (MD) provides a detailed atomic level description of particle-pore-electrolyte interactions [51, 61], but it is computationally expensive, limited in time scale, and not practical for gathering statistical information over long time periods. On the other hand, numerical simulations using the Brownian dynamics (BD) method provide a practical means of modeling the stochastic motion of nanoparticles within a nanopore on time scales of the order of 1 μs or larger [54].

To describe the translational motion of the spherical nanoparticle within the BD model, we use the over-damped (no acceleration term) Langevin equation [62] in which the velocity $\dot{\mathbf{r}}_p$ and the position of the spherical nanoparticle, $\mathbf{r}_p(t)$, at time t are determined from the solution of the following equation:

$$\dot{\mathbf{r}}_p(t) = \boldsymbol{D}_p \left(\mathbf{F}^H[\mathbf{r}_p(t)] + \mathbf{F}^E[\mathbf{r}_p(t)] - \nabla U_m[\mathbf{r}_p(t)]\right) - k_B T \nabla \boldsymbol{D}_p + \sqrt{\frac{6}{\delta t}} \boldsymbol{D}_p^{1/2} \hat{\mathbf{n}}. \tag{6.17}$$

where $\delta t = 0.8$ ps is the time step used in simulations, $\boldsymbol{D}_p$ is the particle's translational diffusion tensor [63] (the mobility matrix) and $\boldsymbol{D}_p^{1/2} [(\boldsymbol{D^T}_p)^{1/2}]$ is the (transposed) square root of it, $\boldsymbol{D}_p = \boldsymbol{D}_p^{1/2} (\boldsymbol{D^T}_p)^{1/2}$. Only translational diffusion is included in $\boldsymbol{D}_p$ owing to the spherical symmetry of the nanoparticle, so that $\boldsymbol{D}_p$ is given as a 3×3 matrix with diagonal terms being the diffusion coefficients along the coordinate axes in Fig. 6.2.

The first term on the right of Eq. (6.17) gives contribution to the particle's displacement due to the deterministic forces acting on it which is the sum of the electric and hydrodynamic forces, $\mathbf{F}^E(\mathbf{r})$ and $\mathbf{F}^H(\mathbf{r})$ as well as the short-range Lennard-Jones interaction $-\nabla U_m$ between the nanoparticle and the membrane surface [52–54]. Both hydrodynamic and electric forces are computed as a result of a self-consistent solution of the PNP-NS system of equations as described in the previous Section.

In writing of the above Langevin equation, we also assumed the position dependent diffusion tensor $\boldsymbol{D}_p$ [64, 65] responsible for the second term in the above equation which is required for a system to reach equilibrium. The last term on the right of Eq. (6.17) accounts for the stochastic motion of a nanoparticle (the Brownian force) with $\hat{\mathbf{n}}$ being the three-dimensional random unit vector.

In the course of a BD simulation, the particle's positions were determined, and at those positions, the ionic current values were determined by linear interpolation of the data computed as described in Sect. 6.2.2. Thus, an ionic current trace was obtained for each particle translocation through the nanopore.

6.3.2 Drag Coefficient in Nanopores

When the nanoparticle moves through the nanopore, its diffusion speed decreases compared to the bulk due to the constriction of the (incompressible) fluid flow between the nanopore and nanoparticle surfaces, this effect is sometimes called "the hindered diffusion" [64]. As the particle is allowed to move randomly in our 3D simulations, in principle, we need three independent diffusion coefficients along the principal directions [48, 66]. To this end, we performed a set of additional 3D PNP-NS computations for different positions of the particle in the radial direction and along the pore's axis assuming a moving particle and setting the EOF fluid velocity on its surface to $\mathbf{v} = v_p\hat{z}(\hat{r})$ where v_p is the particle's velocty along a particular direction given by the $\hat{z}(\hat{r})$ unit vectors. We found that the diffusion coefficient remains approximately the same in both directions and weakly depends on the position of the particle in the radial direction as long as the particle is within $\sim R_p/2$ distance from the central axis. These conclusions were later corroborated by a more comprehensive calculations [67] for an infinitely long cylindrical pores. As such, we use only one value of the diffusion coefficient, D_p, for all three directions of particle's motion, and set it equal to its value along the pore's z-axis. This is an approximation which was required by the overall complexity of calculations and justified by the fact that the diagonal coefficents of the resistance matrix [67] (the inverse of the diffusion tensor $\boldsymbol{D}_p$) only weakly depend on the radial position of the particle (they change by $\sim 20\ \%$ in both axial and radial directions over the whole diameter of the cylinder).

To find the corresponding values of the enhanced drag coefficient $\xi = k_b T/D_p$, we performed an additional set of the PNP-NS simulations for a 2D axisymmetric model assuming particle's translocation only along the central z-axis through the nanopore. To this end, we evaluated the hydrodynamic force [Eq. (6.15)] acting on the nanoparticle at different positions of the particle along the z-axis and different values of the particle's velocity v_p in the z-direction. Figure 6.3 shows the dependence of F_z^H vs v_p for three different z-positions of the nanoparticle. As can be seen from this plot, this force depends linearly on the particle's velocity. The condition of the zero force, $F_z^H = 0$, determines the actual hydrodynamic velocity of the particle which is found to be within 20 % of the open pore EOF fluid velocity v_f (see inset in Fig. 6.3).

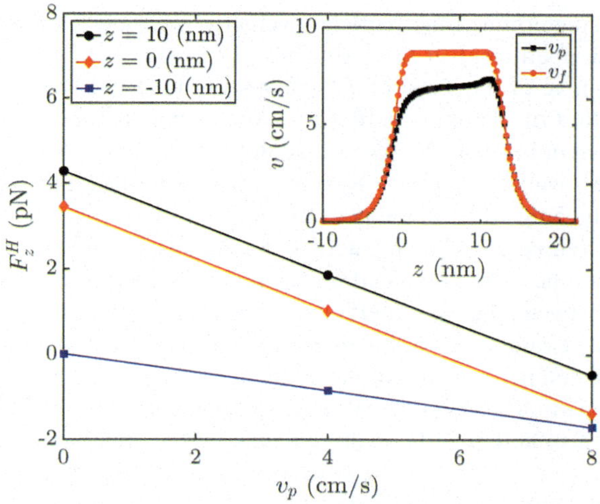

Fig. 6.3 z-component of the hydrodynamic force F_z^H versus particle's velocity v_p for the particle's location at $x_p = 0$ nm and $z_p = 11$ (black), 0 (red), and -10 (blue) nm. The inset compares the fluid velocity v_f (red curve) without the nanoparticle with the particle's velocity v_p determined from the condition $F_z^H = 0$ (black curve). Reproduced with permission from Ref. [58] Copyright (2018) (IOP Science)

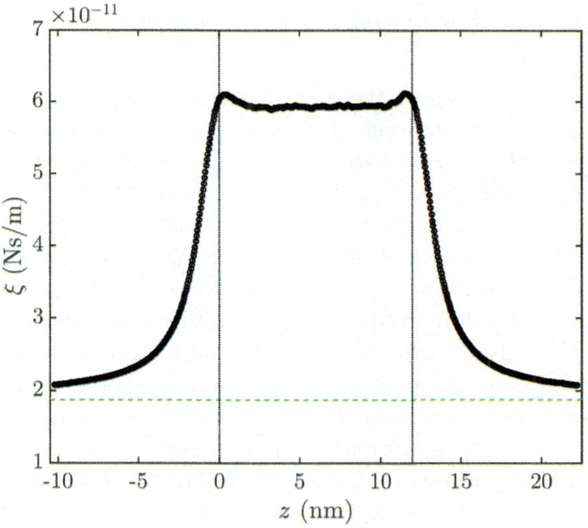

Fig. 6.4 Drag coefficient ξ along the central pore z-axis (black circles connected with lines). The green horizontal dashed line gives the bulk (Stokes) drag coefficient $\xi_{bulk} = 1.88 \times 10^{-11}$ Ns/m, and the vertical black dotted lines show the boundaries of the nanopore. Reproduced with permission from Ref. [58] Copyright (2018) (IOP Science)

Having thus determined the particle's velocity, we then extracted the position-dependent drag coefficient in the z-direction as $\xi(z) = -F_z^H(z)|_{v_p=0}/v_p$, that is, $\xi(z)$ is the negative of the slope of F_z^H versus v_p curve in Fig. 6.3. The computed in this way values of the drag coefficient ξ versus nanoparticle position along the z-axis are shown in Fig. 6.4. We see in this plot that the drag coefficient approaches the bulk (Stokes) value $\xi_{bulk} = 6\pi\eta R_s$ (the horizontal dashed line) when the particle is away from the pore while in the pore the drag coefficient is enhanced by a factor of three. We also note here that the increase in the drag coefficient is still a factor of two smaller than the values obtained for the infinitely long pores [64]. Recovery of an agreement between the drag coefficient computed in the above way and the previously computed values [64, 67], requires increasing the length of the pore beyond 100 nm; a more detailed analysis of the effect of the nanopore dimensions and other nanopore-nanoparticle properties on the drag coefficient tensor will be performed elsewhere. We also note that the hydrodynamic particle relaxation time [68, 69] in the nanopore m/ξ is ~ 0.2 ps, where m is the mass of the SiO_2 nanoparticle, which is much smaller then the time step used in calculations justifying the use of the steady state approximation for solving PNP-NS equations in our dynamic problem. Also, the Peclet number $Pe = v_f L_p/D_p \sim 6$ in the nanopore, so that the effects on the particle's motion from the random and the deterministic forces are both important.

6.3.3 Force and Current Mapping

Figure 6.5 shows the map of the z-component of the electric force, F_z^E, and the hydrodynamic force, F_z^H, acting on the nanoparticle at its different positions. It is seen that the hydrodynamic force is the dominant force when the particle is in the pore. This is to be expected for a neutral particle, because in this case the electric force is solely due to the dielectrophoretic effects [70–72], and hence it is largest near the bottom of the pore where the electric field changes the most [56]. Both of these forces are positive everywhere in the pore "pushing" the nanoparticle along the fluid flow from the *cis* to the *trans* reservoirs in the positive z-direction (up). Note also that the hydrodynamic force reaches its largest value near the top of the nanopore due to the maximum in the local pressure experienced by the particle [56]; this leads to the largest particle velocity in that region (Fig. 6.4).

The calculated ionic current map for different particle positions inside the pore and in the reservoirs is shown in Fig. 6.6. One can immediately see that the ionic current is lower than its value in the bulk for all nanoparticle positions inside the pore (the ionic current blockade) and also next to the pore in the reservoirs within the distance of $\sim R_p$ away from its entrance/exit. It is also visible that the current depends on the particle's location in both z- and x-directions along the pore axis and perpendicular to it: The lowest current values are observed when the particle is on the central axis of the pore at $z \sim 10$ nm and the current increases by a $1-2$ pAs when the particle edges towards the nanopore wall.

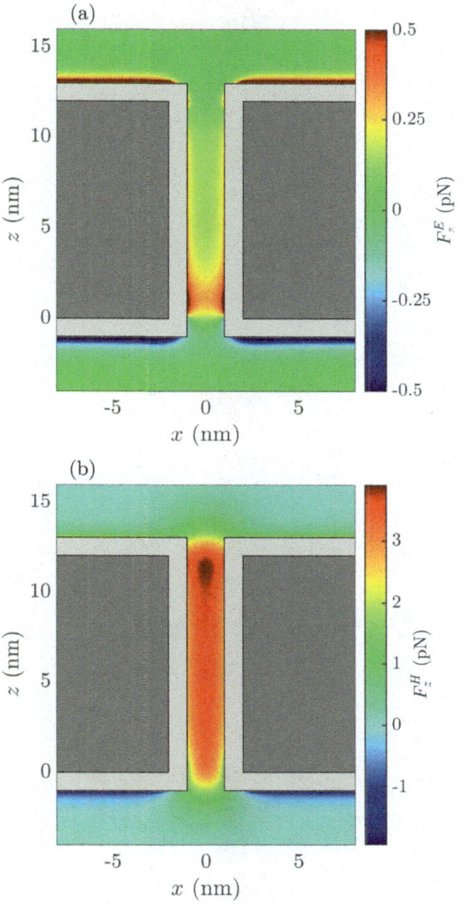

Fig. 6.5 z-component of (a) the electric force, F_z^E, and (b) the hydrodynamic force, F_z^H, for different particle positions. The membrane is dark gray while light gray color shows the regions inaccessible to the particle. Reproduced with permission from Ref. [58] Copyright (2018) (IOP Science)

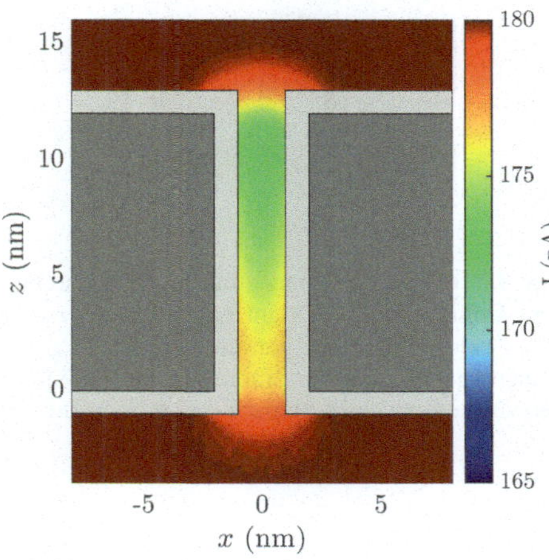

Fig. 6.6 The ionic current map at different particle positions

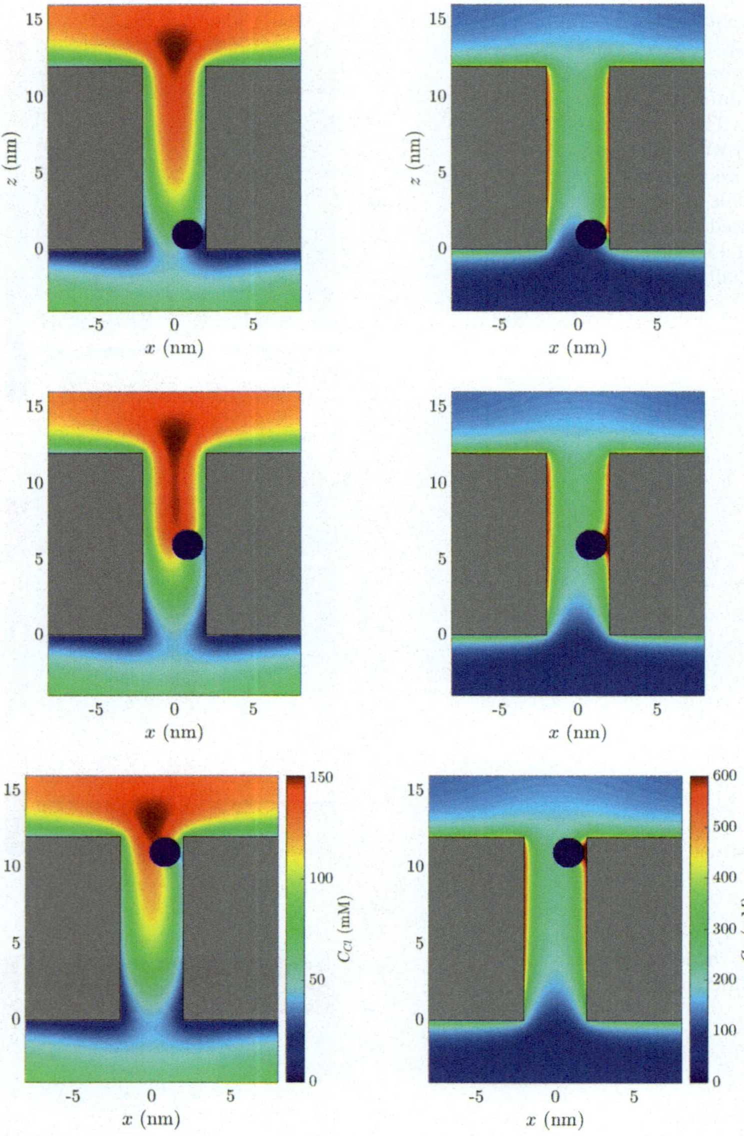

Fig. 6.7 $Cl^-$and K^+concentrations for different positions (x_p, z_p) of the nanoparticle (dark blue): Left Column is C_{Cl^-}, right column is C_{K^+}. Top row—$(x_p, z_p) = (0.8, 1)$ nm, middle row—$(x_p, z_p) = (0.8, 6)$ nm, bottom row—$(x_p, z_p) = (0.8, 11)$ nm. In order to show the concentration changes in the nanopore more clearly, only portions of reservoirs adjacent to the nanopore are shown. Reproduced with permission from Ref. [58] Copyright (2018) (IOP Science)

To qualitatively understand these changes in the ionic current, in Fig. 6.7 we plot the ionic concentrations in our system for different nanoparticle positions (to save space, we only show results for the off-center particle positions but the variation of the concentration along the z-direction discussed in this paragraph remains the same when the particle is located on the central pore axis). When the particle enters the nanopore, the current drops as expected. Then, as the particle translocates through the pore from bottom to top, one can see in the plots that regions in the pore with large Cl^- and K^+ concentrations formed above the particle due to the concentration polarization [39, 55] across the particle decrease while the volume of the pore below the particle filled with Cl^- and K^+ ions with concentrations lower than the bulk value increases (note also enhanced (depleted) Cl^- and K^+ concentrations relative to the bulk values in the *trans* (*cis*) reservoirs above (below) the nanopore due to the concentration polarization across the membrane). This indicates that the conductance of the electrolyte solution in the pore which is proportional to the net ionic concentration progressively decreases as the particle moves through it, and consequently, the current also becomess smaller, that is, the ionic current blockade magnitude increases reaching its maximum value (the smallest blocked current) when the particle is located near the top of the pore. The effect when the ionic current (blockade) becomes a function of the particle's position inside the pore, rather than staying at a constant level, is not captured by the basic theory outlined in Sect. 6.1, and the quantitative understanding of this behavior will be provided in the next Section.

We also observe in Fig. 6.7 that when the particle is located close to the negatively charged membrane, there is a very large increase in K^+ concentration (up to six time larger than the bulk value) in the region between the particle and the nanopore surface. At the same time, concentration of Cl^- ions in the same region is depleted. This is because the nanoparticle located close to the negatively charged nanopore wall (the distance between the nanoparticle's side close to the nanopore surface and the nanopore is less than Debye screening length $\lambda_D = (\varepsilon_0 \varepsilon_r k_B T / 2e^2 c_b)^{1/2} \sim$ 1 nm for our parameters) effectively blocks Cl^- ions and thus precludes formation of the double layer in the "contact region" between the particle and pore, so that only K^+ ions accumulate there. Overall this means a local increase in the ionic concentration and conductivity, but because the "contact area" between the particle and the membrane is much smaller than the surface area of the nanopore, it results in a rather small increase in the ionic current compared to the overall current blockade drop due to the particle blocking the pore (Fig. 6.6). This is somewhat similar to the already mentioned in Sect. 6.1 ionic current responses observed [40] when a long cylindrical rod translocated the nanopore, except that in our case the particle is uncharged and the contact area is so small that we find only a slight current increase. A qualitatively similar behavior in the computed ionic current blockade was also recently attributed to the change in the ionic mobility next to the protein translocating through the nanopore [61].

6.3.4 Ionic Current Traces

We next perform a statistical analysis of the ionic current traces to obtain insights on how random motion of a particle as it translocates through the nanopore, affects the ionic current. As it was noted above, the hydrodynamic force is the main factor determining the translocation direction which means that since there is no sizable electrostatic force acting on the particle, it will translocate the pore only along the fluid flow [73], subject to the random Brownian motion.

Figure 6.8, left, shows the particle's track from a single BD simulation as the particle successfully finds its way from the *cis* reservoir into the pore, and then moves through the pore to the *trans* side of the membrane. On the right of Fig. 6.8, the ionic current blockade corresponding to this successful particle translocation through the pore is shown, zoomed in on to the time of the translocation event. In this particular example, upon entering the pore, the particle spends most of its time in the top half of the pore, where the current is at its lowest level (Fig. (6.6)), before exiting the pore. Shorter current spikes before and after the large current drop in Fig. 6.8 indicate the unsuccessful particle's attempts to enter the pore from either the *cis* and *trans* reservoirs.

The concatenated ionic current traces (top plot) together with the corresponding particle's z-position along the pore axis (bottom plot) for several simulations during which the particle does translocate through the nanopore are shown in Fig. 6.9. We observe in these plots several distinct current levels in the traces. The open pore current $I_o \approx 180$ pA. When the particle first accesses the pore, the ionic current is reduced to a value $I \sim 176$ pA, and it stays around that level while the z-coordinate of the particle slightly increases. While the particle continues to move toward the

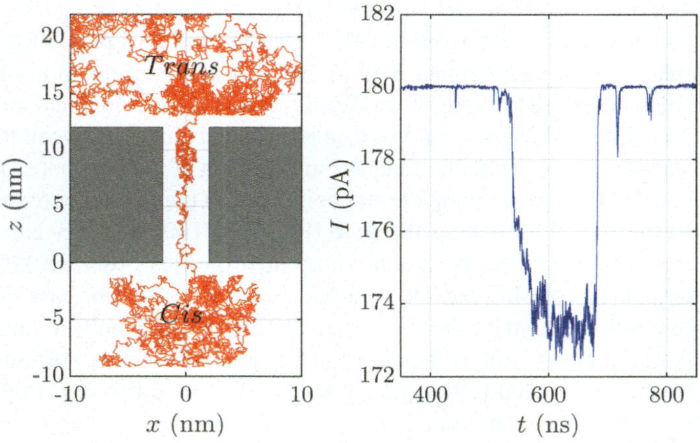

Fig. 6.8 The particle's track (left) as the particle translocates from the bottom compartment to the top compartment, and the corresponding ionic current trace (right). Reproduced with permission from Ref. [58] Copyright (2018) (IOP Science)

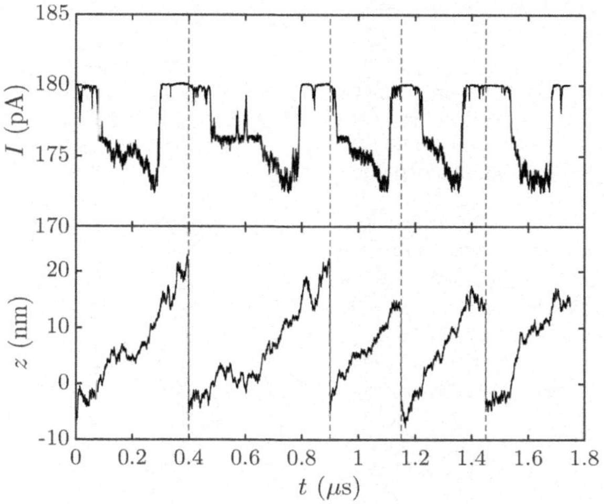

Fig. 6.9 Concatenated ionic current blockades (top) together with the corresponding particle's z-position (bottom). The vertical dashed lines separate individual simulations. Position $z = 0$ is at the bottom threshold of the membrane on the *cis* side of the membrane where the particle is initially placed. Reproduced with permission from Ref. [58] Copyright (2018) (IOP Science)

exit of the pore (z-coordinate further increases), the current drops even more until it reaches its lowest level of $\approx$ 173 pA near the top of the pore. Once the particle is pushed out into the top reservoir on the *trans* side of the membrane, the ionic current quickly returns back to the open pore current of $\sim$ 180 pA.

Figure 6.10 summarizes the results of about 10^3 translocation events. Figure 6.10a shows a scatter plot of the current values (blue dots) versus z-position of the particle. The spread of the current values is the result of the particle's random motion in the radial direction inside the pore (as discussed above) and in the access regions just above and below the nanopore. The bold (red) curve, given as a reference, is obtained for a particle that was restricted to move only along the central pore axis. Overall, we again observe that the largest current blockade occurs when the particle is near the top half of the pore, and the current drop is the smallest when the particle is near the entrance of the pore. This shows that the ionic current drop is strongly position dependent (specifically, along the pore axis), even for a neutral particle. This is result will be further explored and understood in the next Section.

Figure 6.10b shows a distribution of the calculated current drop values $\Delta I = I_o - I$ vs the translocation time of the nanoparticle through the nanopore. One can see that the current blockade values range from about 6.5–7.6 pA consistent with the data shown in Fig. 6.10a while the translocation time varies from about 0.1 up to 0.5 μs. The majority of the events, however, have $\Delta I \sim 7.4$ pA which is the median value of current drop for our system, see Fig. 6.10a, and the average translocation time (the conditional mean exit time [74] for $Pe \sim 6$) of $\approx 0.14(L_p^2\xi/2k_BT) \sim 140$ ns where $L_p = 12$ nm is the pore length and $\xi \approx 6 \times 10^{-11}$ Ns/m is the particle's drag coefficient in the nanopore (see Fig. 6.4). However, there is a sizable number of translocation events with smaller current drops, most of those correspond to faster translocation times. In general, one can observe from this plot that the larger the current drop, the larger the spread of the possible translocation times. This is because,

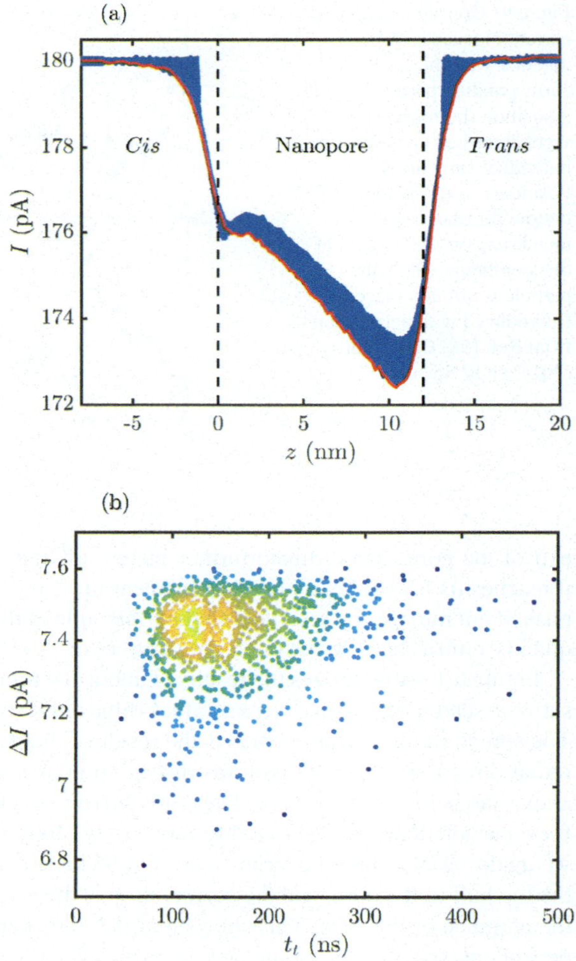

Fig. 6.10 **a** Ionic current values (blue dots) for 10^3 simulations as the particle moves from the *cis* side to the *trans* side of the membrane, along with the current trace of a particle translocating along the central pore axis only (bold red curve). **b** Scatter plot of the maximum values of the current drop $\Delta I = I_o - I$ versus nanoparticle's translocation times. Event frequency is indicated by the varying color from yellow (a lot of events) to dark blue (a few events). Reproduced with permission from Ref. [58] Copyright (2018) (IOP Science)

the more time the nanoparticle spends in the pore, the more opportunities it has to sample different current values [recall that the ionic current depends on the particle's position in the pore, Fig. (6.6)] and more likely it will find itself moving in the vicinity of the central axis of the pore where the largest current changes occur.

6.4 Concentration Polarization and the Ionic Current Blockade in Nanopores

6.4.1 Ionic Current Blockade Calculations

In order to better understand properties of the ionic current in nanopores, we now consider how the ionic concentration in the nanopore is affected in a general case when a nanoparticle with a surface charge density σ_F passes through it. When no nanoparticle is present (the open pore case), for sufficiently large nanopores with $R_p \gg \lambda_D$, and at zero applied bias, the averaged over the cross-sectional area of the pore net ionic concentration $\bar{c} = \bar{c}_{\mathrm{Cl}^-} + \bar{c}_{K^+}$ within the Gouy-Chapman model [59] and planar approximation for the electric potential Φ can be calculated as:

$$\bar{c} \approx 2c_b(1 + Du), \tag{6.18}$$

where the term $2c_b Du$ describes the accumulation of ions in the double layer on the nanopore surface due to its surface change σ (this increase in concentration results in the surface conductance [75]). This contribution is written with help of the so-called Dukhin number $Du = |\varepsilon|\gamma$ [75, 76] where $\varepsilon = \sigma/(ec_b R_p)$ is the nanopore charge parameter, $\gamma = -b/\lambda_D + [1 + (b/\lambda_D)^2]^{1/2}$ with $b = 2\varepsilon_0\varepsilon_r k_B T/(e|\sigma|)$ being the Gouy-Chapman length [59, 77].

Figure 6.11 shows the net ionic concentration $\bar{c}$ along the pore for different applied biases. One can see that outside the nanopore, $\bar{c}$ is above for $z < 0$ and below for

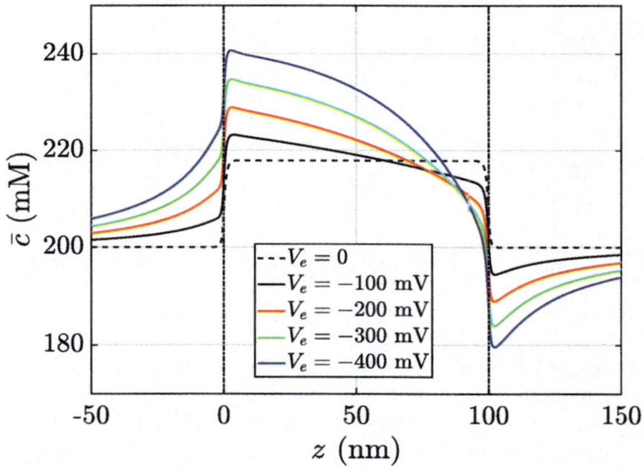

Fig. 6.11 The average net ionic concentration $\bar{c}$ along the nanopore axis for different applied voltages V_e. Dashed line is $\bar{c}$ at $V_e = 0$ [at equilibrium, Eq. (6.18)]. For these calculations, we took $L_p = 100$ nm, $R_p = 10$ nm, $c_b = 100$ mM, $\sigma = -0.16$ e/nm^2. For these parameters, the EOF contribution to the current is within 5 %, and as such, it is omitted. Here the numerical solution of the PNP equations [Eqs. (6.4)–(6.6)] was performed for the two-dimensional axio-symmetric model. Reproduced with permission from Ref. [79] Copyright (2020) (ACS Publications)

$z > L_p$ the bulk value of $2c_b$, and linearly increases with the applied voltage. It is also seen that the concentration decay to the bulk value is not exponential over a distance comparable to the Debye length $\sim \lambda_D$, as could be expected from the basic electrostatic considerations, but much slower (in fact, it decreases as 1/distance from the pore to be consistent with the net ionic flux conservation). At $z \sim 0$ and

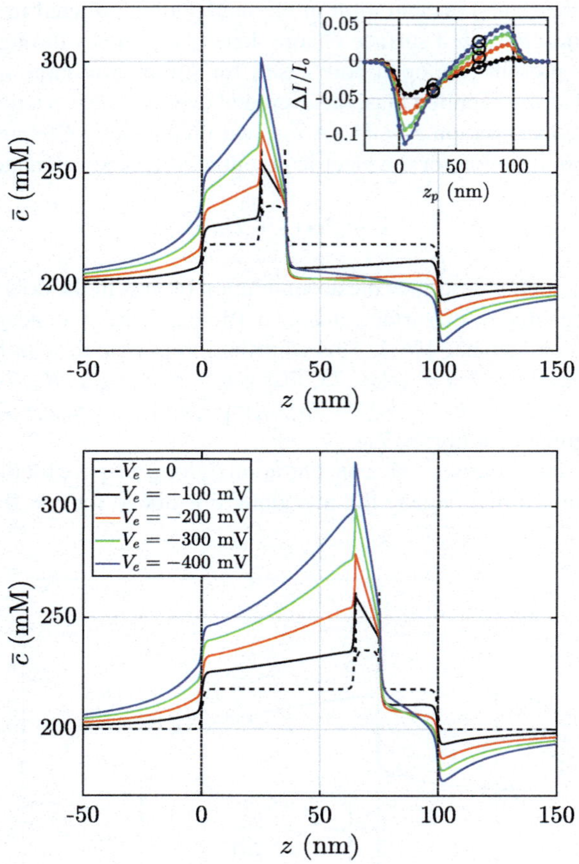

Fig. 6.12 The average net ionic concentration $\bar{c}$ along the nanopore axis for different applied voltages V_e with a cylindrical nanoparticle with diameter equal to its length $l_p = 10$ nm located at $z_p = 30$ nm (top) and at $z_p = 70$ nm (bottom). The dashed line is $\bar{c}$ at $V_e = 0$. Nanopore parameters are the same as in Fig. 6.11. Sharp changes in the average concentration next to the left and right edges of the particle ($z \sim z_p \pm l_p/2$) are due to the contribution from the ionic build up over its whole surface while diminished cross sectional area of the pore open to the current leads to the increase of the average concentration along its length for $z_p - l_p/2 < z < z_p + l_p/2$ [Eq. (6.22)]. Inset shows the computed current blockade ratios $\Delta I/I_0 = (I - I_o)/I_o$ with large circles marking particle positions for which concentrations are shown in the main plots. Reproduced with permission from Ref. [79] Copyright (2020) (ACS Publications)

$z \sim L_p$, the concentration sharply increases over a distance $\sim \lambda_D$ by $\approx 2c_b Du$ to reach its in-pore values [Eq. (6.18)]; this increase is independent of the applied bias V_e. While inside the pore, $0 < z < L_p$, the concentration changes linearly with for small biases (such as $V_e = -100$ mV) but deviations from the linearity appear with increasing $|V_e|$.

Figure 6.12 shows the average net ionic concentration $\bar{c}$ when a charged cylindrical nanoparticle with surface charge σ_p equal to that of a nanopore, $\sigma_p = \sigma = -0.16$ e/nm^2. is present in the pore. It is clearly seen that when the particle is close to the left (right) edge of the pore, most of the nanopore is filled with the solution at concentration lower (higher) than the equilibrium in-pore value at $V_e = 0$ (dashed curves in Fig. 6.12). This indicates that the pore's conductivity which is proportional to the ionic concentration and the blocked pore current I change when the particle moves along the pore, as was already demonstrated by our BD calculations described in the last Section. The ionic current can even overshoot the open pore current I_o when $z_p \gtrsim L_p/2$ and $|V_e| > 100$ mV (see the inset in Fig. 6.12). Note also that the strongest current blockade (the smallest I) also increases with the applied voltage as expected from a larger concentration variation along the pore's length.

The linear dependencies on the particle's position inside the nanopore visible in the average concentration plots at smaller biases suggest that it is possible to find a closed form solution for the ionic current blockade ratio. Averaging the ionic concentration over the cross-sectional area of the nanopore allows us to decouple products $c_{K^+(Cl^-)}\nabla\Phi$ in Eqs. (6.5) and (6.12) [78]. Further utilizing conservation of the net ionic current and flux in the nanopore, the above system of PNP equations can be solved in closed form for small nanoparticles [79]. For the general case of a small spherical charged nanoparticle translocating a charged cylindrical nanopore, the leading term of the resulting ionic current blockade ratio can be written as

$$\frac{I - I_o}{I_o} = -\frac{3}{2}\frac{v_p}{V_p}\left[1 + \Sigma\left(1 - \frac{2z_p}{L_p}\right)\right] F_3(R_p, L_p), \tag{6.19}$$

where

$$\Sigma = \frac{1}{3}\left(\varepsilon + \frac{3}{2}\varepsilon_p\right)\frac{e|V_e}{k_B T} \tag{6.20}$$

is the charge-voltage parameter, $\varepsilon_p = \sigma_p/(ec_b R_s)$ is the nanoparticle charge parameter, and factor $F_3(R_p, L_p) = \left[1 + 2\pi R_p(1 + Du)/(4L_p(1 + Du/4))\right]^{-1}$ accounts for the nanopore and the access resistances in finite-length nanopores with surface charge [56, 77], parametrised by the nanopore Dukhin number Du (similarly to factor F_2 in Sect. 6.1, we set F_3 to be the same for both open and blocked pores).

The first term in Eq. (6.19) is responsible for the current drop due to excluded volume of the nanopore occupied by the nanoparticle and is identital to the classical result of Sect. 6.1 when $\sigma = \sigma_p = 0$. The contribution to the ionic current blockade due to surface charge and the induced concentration polarization around the translocating object given by the second term is controlled by the charge-voltage parameter Σ which depends on both the surface charge densities, σ and σ_p, and the applied voltage V_e. Its contribution also varies with the particle position, so that while the particle moves through the pore, $|(I - I_o)/I_o|$ changes from the maximum value at $z_p \sim 0$ to the minimum at $z_p \sim L_p$, even for a perfectly cylindrical nanopore.

In other words, the maximum/minimum current blockade ratio occurs when the particle is located next to the nanopore entrance/exit. Whether those values correspond to the beginning or end of the translocation event, depends on which way the particle goes through the pore. As described in Sect. 6.3.3, the direction is determined by which force—the electric $\mathbf{F}_E$ or hydrodynamic drag $\mathbf{F}_H$ due to the EOF [given by Eqs. (6.13)–(6.16)]—is larger [73]. For $V_e < 0$ and both nanopore and nanoparticle being negatively charged, the drag force is directed down (see Fig. 6.2) while the electric force is up. In this situation, if $F_E > F_H$ and the nanoparticle translocates through the nanopore electrophoretically (that is, in the direction from $z = 0$ to $z = L_p$), the volume of the pore filled by high ionic concentration gradually increases, and since the conductivity is determined by the concentration value, so does the ionic current: $\Delta I/I_o$ will be the large and negative when the nanoparticle just enters the pore and decreases in magnitude or even become positive right before it exits. For the electroosmotic direction of particle's motion (when the drag force dominates and the particle can only enter the nanopore at $z = L_p$), just as the particle enters the pore, the pore becomes filled with the ionic solution at high concentration, and the decrease in the ionic current due to the volume excluded by the particle will be partially compensated by its increase due to the enhanced nanopore conductivity. In this case, if the concentration polarization induced addition to the ionic current is strong, I may even overshoot the open pore current.

As seen from Eq. (6.19), the blocked pore current will indeed exceed the open pore current ($I > I_o$, a positive current blockade ratio) when the charge-voltage parameter $\Sigma \gtrsim 1$ and $z_p \sim L_p$. This phenomenon was experimentally observed [38, 41] and is frequently found in numerical solutions of the PNP-NS equations. When $z_p \sim 0$, however, we find a stronger and voltage dependent blockade than given by Eq. (6.1):

$$-\left(\frac{I - I_o}{I_o}\right)_{\max} = \frac{3}{2}\frac{v_p}{V_p}(1 + \Sigma)\, F_3(R_p, L_p). \tag{6.21}$$

In practice, for a given nanopore-nanoparticle system, both the current increase above the baseline and stronger voltage-dependent current blockade can be achieved, for example, when a large electric bias is applied and/or a solution with small ionic concentration c_b is used (in which case both ε and ε_p become large). This behavior of the ionic current persists even when a neutral particle passes through a charged pore, as seen from the results of Sect. 6.3, but in general the effect will be strongest (and easiest to observe) when both the translocating object and the nanopore carry surface charges of the same sign. For example, for data in Fig. 6.12, the charge-voltage parameter $\Sigma = (\varepsilon + \varepsilon_p)e|V_e|/(3k_BT)$ varies from ≈ 1 at $V_e = -0.1$ V to ≈ 4 at $V_e = -0.4$ V so that at $V_e = -0.1$ V the current blockade ratio $\Delta I/I_o \approx 0$ and it becomes positive for other voltages when $z_p \sim L_p$, consistent with computed current traces.

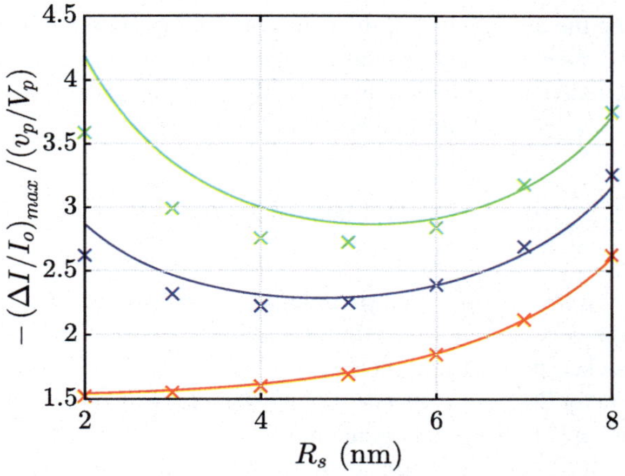

Fig. 6.13 Maximum ionic current blockade $-(\Delta I/I_0)_{\max} = -[(I - I_o)/I_o]_{\max}$ normalized by the ratio of the nanoparticle to the nanopore volume v_p/V_p versus radius of the particle R_s for for different charge densities on the particle: $\sigma_p = 0$ (red solid curve and symbols), -0.08 e/nm^2 (blue), -0.16 e/nm^2 (green). Other nanopore parameters are the same as in Fig. 6.11. Solid lines are from Eq. (6.21), multiplied by the factor $F_1(R_s/R_p)$, while symbols are the results of the numerical simulations. The deviations between the two sets at small radii are due to the breakdown of the linear approximation at large $\varepsilon_p \sim R_s^{-1}$ Reproduced with permission from Ref. [79] Copyright (2020) (ACS Publications)

In general, the ionic current is more sensitive to the charge on the particle than on the pore which can be traced back to the charge-voltage parameter in Eqs. (6.19) and (6.21). When $R_s \to 0$, the term $\propto \varepsilon_p$ becomes dominant so that the ratio $|(I - I_o)/I_o| \propto R_s^2$ for a charged nanoparticle as opposed to $\propto R_s^3$ for a neutral one. This indicates that when a small charged particle translocates through a nanopore, it produces a much larger current blockade than a neutral one of the same size. This is demonstrated in Fig. 6.13 where the computed values of the ionic current blockade (normalized by the volume ratio v_p/V_p) for an uncharged particle monotonically decrease with $R_s \to 0$ while they increase and diverge from each other when $\sigma_p \neq 0$. Eventually, for very small particles, the current signal will be washed out by the thermal noise as ΔI becomes too small for accurate measurement. Nevertheless, such sensitivity of the current to the particle charge may allow in principle to identify and characterize the translocating objects by their size *and* charge with help of Eq. (6.21) and using only resistive pulse current measurements without resorting to separate mobility measurements and analysis [20, 80].

When a charged nanoparticle is present inside the nanopore, ions of the sign opposite to the surface charge accumulate around it creating a double layer, in addition to the layer on the charged nanopore surface, resulting in a further increase of the ionic concentration around the particle, see Fig. 6.12. For small nanoparticles, we can estimate the effect of the enhanced ionic concentration on the ionic current on example of a small ($\lambda_D \ll l_p \ll L_p$) cylindrical nanoparticle of radius $r_p \gg \lambda_D$

positioned along the central axis of the nanopore. In the planar approximation, with non-overlappind Debye layers around the nanopore and nanoparticle, the average ionic charge density in the region between the particle and nanopore surface for $z_p - l_p/2 < z < z_p + l_p/2$ (dashed curve in Fig. 6.12) is

$$\bar{c} \approx 2c_b(1 + Du) + 2c_b Du_p \frac{r_p^2}{R_p^2}, \tag{6.22}$$

where $Du_p = |\varepsilon_p|\gamma_p$ is the nanoparticle Dukhin number. Approximating the nanopore resistance with that of a tube of varying cross section [31], one can then find that the blocked pore current increases by $\delta I \sim (v_p/V_p)Du_p/(1 + Du)I_o$ over the current value for the pore with a neutral particle, leading to the contribution to the current blockade $\delta I/I_o$ being independent of the particle's location in the pore and the applied bias. It is seen that for this increase to be comparable to the current change induced by the concentration polarization, the nanoparticle Dukhin number has to be large which means that the nanoparticle has to be either very small or the surface charge has to be very large which is not always possible: For example, for the 5-nm-radius particles with $|\sigma_p| < 0.16$ e/nm^2 discussed in this Section, $Du_p < 0.15$, and as such, this contribution to the ionic current is small and cannot be responsible for dramatic modifications of the ionic current blockade with the nanoparticle position and applied voltage (Fig. 6.12).

6.4.2 Analysis of Experimental Data

We next apply results of Sect. 6.4.1 to the analysis of a recent experimental study [39] of the ionic current blockade observed when the same neutral and charged nanoparticles of radius $r_p = 200$ nm were translocating through two nanopores—the polyethylene (PET) nanopore of nominal length $L_p = 11\ \mu$m and radius $R_p = 775$ nm and the poycarbonate (PC) one with $L_p = 29\ \mu$m and $R_p = 575$ nm. Numerical solution of the PNP-NS equations for such very large nanopores (or mesopores) is prohibitively expensive as it requires a huge number of matrix elements to accommodate fine mesh with sizes $\ll \lambda_D \sim 1$ nm over a μm sized domains to accurately compute minute (a fraction of a percent) changes in the ionic current. An alternative way to rescale the geometrical dimensions of the system has to be exercised with caution as not just the dimensions have to be scaled but also the applied voltage or the surface charge density to keep the value of the charge-voltage parameter and the respective ionic current contribution in Eq. (6.19) intact.

The measurements we analyze here were performed at $c_b = 100$ and 300 mM, $pH = 10$. For uncharged nanoparticles passing through the PET pore at 100 mM and for charged ones at 300 mM, the current blockade ratios were at or slightly above the theoretical level given by Eq. (6.1) for all applied biases. These results are consistent with expectations from Eq. (6.21): Larger values of the bulk concentration c_b lead

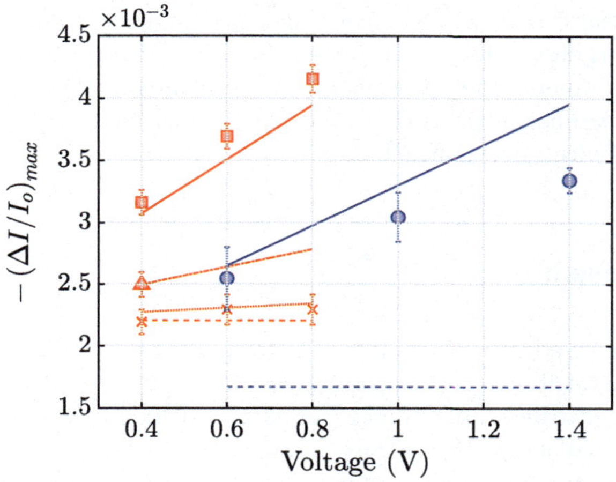

Fig. 6.14 Maximum ionic current blockade $-(\Delta I/I_o)_{\max}$ versus applied voltage for two types of nanopores and nanoparticles at different ionic concentrations from experimental work [39]. The symbols represent the experimental values while lines are their fit by Eq. (6.21) with $\sigma_p = -0.6$ e/nm^2 and $\sigma = -0.3$ and 0 e/nm^2 for the PET (red lines and symbols) and PC (blue lines and symbols) nanopores, respectively. For charged nanoparticles, red squares, blue circles and solid lines correspond to $c_b = 100$ mM, red triangle and dash-dotted line—to $c_b = 300$ mM; red crosses and dotted line are for uncharged nanoparticles and $c_b = 100$ mM. Horizontal dashed lines are the current blockade without the charge-voltage effects [Eq. (6 1)]. Note that for the PC pore, the experimental data at 1.4 V correspond to very large values of the charge-voltage parameter so that a linear fit is not very accurate. Reproduced with permission from Ref. [79] Copyright (2020) (ACS Publications)

to smaller contribution to the ionic current as charge parameters ε and ε_p decrease ($\Sigma \to 0$) and the current blockade ratio approaches the value given by Eq. (6.1). This can further be exacerbated by a decrease of the surface charge density with increasing electrolyte concentration [12] due to competition between $K^+$and H^+ ions near the membrane surface. Weak dependence of $|\Delta I/I_o|$ on the applied voltage observed for the neutral particles ($\varepsilon_p = 0$) at $c_b = 100$ mM also suggests a small surface charge on the nanopore so that $\Sigma \approx 0$.

However, the passage of charged nanoparticles at 100 mM was characterized by considerably larger values of $|(I - I_o)/I_o|$ which increased with voltage. The experimental results and their fits with Eq. (6.21) are summarized in Fig. 6.14. In order to fit data for both pores with the same value of the nanoparticle charge $\sigma_p = -0.6$ e/nm^2, we put surface charge density on the PC pore to zero and limit the charge on the PET pore to $\sigma = -0.3$ e/nm^2, that is, our analysis suggests that the PC pore carries a much lower charge than the PET pore while the nanoparticle charge density is consistent with other prior measurements on similar particles [14].

We also note here that one has to be cautious about inferring the shape of the nanopore directly from the shape of the current traces since, as it follows from our calculations, in addition to the geometric blockade, the current is also affected by the flux-induced ionic concentration redistribution, and this contribution may be nontrivial to uncover in pores with irregular cross-sectional geometry. This contribution can, in principle, be minimized by lowering the charge on both nanopore and nanoparticle which experimentally can be accomplished, for example, by lowering pH or increasing bulk electrolyte concentration.

6.5 Conclusion

In this Chapter we addressed the issue of how the ionic current through a nanopore changes when a small charged nanoparticle translocates through it. It turns out that the main contribution to the ionic current arises because of the concentration polarization induced in the nanopore when a particle blocks the ionic flow. This leads to the strong dependence of the relative current change (the ionic current blockade) on the nanopore/nanoparticle charge and the applied voltage as well as on the location of the nanoparticle in the pore. If the applied bias and/or surface charge densities are large enough, the induced polarization of the ionic solution in the nanopore will be strong, and the blocked pore current may even overshoot the open pore current as the particle enters/exits the nanopore. Quantitatively, this behavior of the ionic current is described by the derived Eqs. (6.19) and (6.21) and by numerical simulations of the nanoparticle's Brownian motion through the nanopore performed based on the self-consistent solution of the PNP-NS equations. The model is then applied to the analysis of experimental data to extract the surface charge densities directly from the resistive pulse measurements. Our calculations were performed for the simplest case of a cylindrical nanopore and do not account for irregularities in the nanopore shape, the presence of the Stern layer [75] on the nanopore surface or inhomogenuities in the distribution of ionizable groups on the surfaces in equilibrium with the ionic solution [59]; these effects can be incorporated in our model in furture works.

References

1. Gu L-Q, Shim JW (2010) Single molecule sensing by nanopores and nanopore devices. Analyst 135(3):441–451
2. Venkatesan BM, Bashir R (2011) Nanopore sensors for nucleic acid analysis. Nat Nanotechnol 6:615–624
3. de la Escosura-Muñiz A, Merkoçi A (2012) Nanochannels preparation and application in biosensing. ACS Nano 6(9):7556–7583
4. Stoloff DH, Wanunu M (2013) Recent trends in nanopores for biotechnology. Curr Opin Biotechnol 24(4):699–704
5. Feng Y, Zhang Y, Ying C, Wang D, Du C (2015) Nanopore-based fourth-generation DNA sequencing technology. Genomics, Proteomics Bioinf 13(1):4–16
6. Maugi R, Hauer P, Bowen J, Ashman E, Hunsicker E, Platt M (2020) A methodology for characterising nanoparticle size and shape using nanopores. Nanoscale 12(1):262–270

7. Movileanu L (2014) Watching single proteins using engineered nanopores. Protein Peptide Lett 21(3):235–246
8. Oukhaled A, Bacri L, Pastoriza-Gallego M, Betton J-M, Pelta J (2012) Sensing proteins through nanopores: fundamental to applications. ACS Chem Biol 7(12):1935–1949
9. Arjmandi-Tash H, Belyaeva LA, Schneider GF (2016) Single molecule detection with graphene and other two-dimensional materials: nanopores and beyond. Chem Soc Rev 45(3):476–493
10. Lee K, Park K-B, Kim H-J, Yu J-S, Chae H, Kim H-M, Kim K-B (2018) Recent progress in solid-state nanopores. Adv Mater 30(42):1704680
11. Kasianowicz JJ, Brandin E, Branton D, Deamer DW (1996) Characterization of individual polynucleotide molecules using a membrane channel. Proc National Acad Sci 93(24):13770–13773
12. Smeets RMM, Keyser UF, Krapf D, Wu M-Y, Dekker NH, Dekker C (2006) Salt dependence of Ion transport and DNA translocation through solid-state nanopores. Nano Lett 6:89–95
13. Waduge P, Hu R, Bandarkar P, Yamazaki H, Cressiot B, Zhao Q, Whitford PC, Wanunu M (2017) Nanopore-based measurements of protein size, fluctuations, and conformational changes. ACS Nano 11(6):5706–5716
14. Qiu Y, Hinkle P, Yang C, Bakker HE, Schiel M, Wang H, Melnikov D, Gracheva M, Toimil-Molares ME, Imhof A, Siwy ZS (2015) Pores with longitudinal irregularities distinguish objects by shape. ACS Nano 9(4):4390–4397
15. Zhang Y, Edwards MA, German SR, White HS (2016) Multipass resistive-pulse observations of the rotational tumbling of individual nanorods. J Phys Chem C 120(37):20781–20788
16. Talaga DS, Li J (2009) Single-molecule protein unfolding in solid state nanopores. J Am Chem Soc 131(26):9287–9297
17. Bacri L, Oukhaled AG, Schiedt B, Patriarche G, Bourhis E, Gierak J, Pelta J, Auvray L (2011) Dynamics of colloids in single solid-state nanopores. J Phys Chem B 115:2890–2898
18. Plesa C, Kowalczyk SW, Zinsmeester R, Grosberg AY, Rabin Y, Dekker C (2013) Fast translocation of proteins through solid state nanopores. Nano Lett 13(2):658–663
19. Bayley H, Martin CR (2000) Resistive-pulse sensing - from microbes to molecules. Chem Rev 100:2575–2594
20. Luo L, German SR, Lan W-J, Holden DA, Mega TL, White HS (2014) Resistive-pulse analysis of nanoparticles. Annu Rev Anal Chem 7:513–535
21. Willmott GR (2018) Tunable resistive pulse sensing: better size and charge measurements for submicrometer colloids. Anal Chem 90(5):2987–2995
22. Wanunu M, Dadosh T, Ray V, Jin J, McReynolds L, Drndić M (2010) Rapid electronic detection of probe-specific microRNAs using thin nanopore sensors. Nat Nanotechnol 5:807–814
23. Larkin J, Henley RY, Muthukumar M, Rosenstein JK, Wanunu M (2014) High-bandwidth protein analysis using solid-state nanopores. Biophys 106(3):696–704
24. Edwards MA, German SR, Dick JE, Bard AJ, White HS (2015) High-speed multipass coulter counter with ultrahigh resolution. ACS Nano 12(9):12274–12282
25. Blundell ELCJ, Mayne LJ, Lickorish M, Christie SDR, Platt M (2016) Protein detection using tunable pores: resistive pulses and current rectification. Faraday Discuss 193:487–505
26. Li J, Stein D, McMullan C, Branton D, Aziz MJ, Golovchenko JA (2003) Ion-beam sculpting at nanometre length scales. Nature 412:166–169
27. Storm AJ, Chen JH, Ling XS, Zandbergen HW, Dekker C (2003) Fabrication of solid-state nanopores with single-nanometre precision. Nat Mater 2:537–540
28. Aksimentiev A (2010) Deciphering ionic current signatures of DNA transport through a nanopore. Nanoscale 2:468–483
29. Feng J, Liu K, Bulushev RD, Khlybov S, Dumcenco D, Kis A, Radenovic A (2015) Identification of single nucleotides in MoS2 nanopores. Nat Nanotechnol 10:1070–1077
30. Maxwell JC (1954) A treatise on electricity and magnetism, vol 1, §314. Dover Publications
31. DeBlois RW, Bean CP (1970) Counting and sizing of submicron particles by the resistive pulse technique. Rev Sci Instrum 41:909
32. Hurley J (1970) Sizing particles with a coulter counter. Biophys J 10:74–79

33. Anderson JL, Quinn JA (1971) The relationship between particle size and signal in coulter-type counters. Rev Sci Instrum 42(8):1257–1258
34. Smythe WR (1964) Flow around a spheroid in a circular tube. Phys Fluids 7(5):633–638
35. DeBlois RW, Bean CP, Wesley RKA (1977) Electrokinetic measurements with submicron particles and pores by the resistive pulse technique. J Colloid Interface Sci 61:323–335
36. Hall JE (1975) Access resistance of a small circular pore. J Gen Physiol 66(4):531–532
37. Wang J, Ma J, Ni Z, Zhang L, Hu G (2014) Effects of access resistance on the resistive-pulse caused By translocating of a nanoparticle through a nanopore. RSC Adv 4:7601–7610
38. Lan W-J, Kubeil C, Xiong J-W, Bund A, White HS (2014) Effect of surface charge on the resistive pulse waveshape during particle translocation through glass nanopores. J Phys Chem C 118:2726–2734
39. Qiu Y, Lin C-Y, Hinkle P, Plett TS, Yang C, Chacko JV, Digman MA, Yeh L-H, Hsu J-P, Siwy ZS (2016) Highly charged particles cause a larger current blockage in micropores compared to neutral particles. ACS Nano 10(9):8413–8422
40. Venta KE, Zanjani MB, Ye X, Danda G, Murray CB, Lukes JR, Drndić M (2014) Gold nanorod translocations and charge measurement through solid-state nanopores. Nano Lett 14:5358–5364
41. Menestrina J, Yang C, Schiel M, Vlassiouk I, Siwy ZS (2014) Charged Part nodulate local ionic concentrations and cause formation of positive peaks in resistive-pulse-based detection. J Phys Chem C 118(5):2391–2398
42. Bocquet L, Charlaix E (2010) Nanofluidics, from bulk to interfaces. Chem Soc Rev 39:1073–1095
43. Liu H, Qian S, Bau HH (2007) The effect of translocating cylindrical particles on the ionic current through a nanopore. Biophys J 92(4):1164–1177
44. Kim SC, Kannam SK, Harrer S, Downton MT, Moore S, Wagner JM (2014) Geometric dependence of the conductance drop in a nanopore due to a particle. Phs Rev E 89:042702
45. Töws T, Reimann P (2017) Lateral trapping of DNA inside a voltage gated nanopore. Phys Rev E 95:062413 Jun
46. Gracheva ME, Vidal J, Leburton J-P (2007) PN semiconductor membrane for electrically tunable ion current rectification and filtering. Nano Lett 7(6):1717–1722
47. Laohakunakorn N, Keyser UF (2015) Electroosmotic flow rectification in conical nanopores. Nanotechnology 26(27):275202
48. Kim M-M, Zydney AL (2004) Effect of electrostatic, hydrodynamic, and brownian forces on particle trajectories and sieving in normal flow filtration. J Colloid Interface Sci 269(2):425–431
49. Craven TJ, Rees JM, Zimmerman WB (2008) On slip velocity boundary conditions for electroosmotic flow near sharp corners. Phys Fluids 20(4):043603
50. Kesselheim S, Müller W, Holm C (2014) Origin of current blockades in nanopore translocation experiments. Phys Rev Lett 112:018101
51. Gracheva ME, Xiong A, Aksimentiev A, Schulten K, Timp G, Leburton J-P (2006) Simulation of the electric response of DNA translocation through a semiconductor nanopore-capacitor. Nanotechnology 17:622–633
52. Melnikov D, Leburton J-P, Gracheva M (2012) Slowing down and stretching out DNA with an electrically tunable nanopore in a p-n semiconductor membrane. Nanotechnology 23:25501
53. Jou IA, Melnikov DV, McKinney CR, Gracheva ME (2012) DNA translocation through a nanopore in a single-layered doped semiconductor membrane. Phys Rev E 86(6):061906
54. Jou IA, Melnikov DV, Nadtochiy A, Gracheva ME (2014) Charged particle separation by an electrically tunable nanoporous membrane. Nanotechnology 25(14):145201
55. Pu Q, Yun J, Temkin H, Liu S (2004) Ion-enrichment and ion-depletion effect of nanochannel structures. Nano Lett 4(6):1099–1103
56. Melnikov DV, Hulings ZK, Gracheva ME (2017) Electro-osmotic flow through nanopores in thin and ultrathin membranes. Phys Rev E 95:063105
57. Rice CL, Whitehead R (1965) Electrokinetic flow in a narrow cylindrical capillary. J Phys Chem 69(11):4017–4024

58. Hulings ZK, Melnikov DV, Gracheva ME (2018) Brownian dynamics simulations of the ionic current traces for a neutral nanoparticle translocating through a nanopore. Nanotechnology 29(44):445204
59. Russell WB, Saville DA, Schowalter WR (1989) Colloidal dispersions. Cambridge University Press, Cambridge, United Kingdom
60. Yeh L-H, Zhang M, Qian S (2013) Ion transport in a pH-regulated nanopore. Anal Chem 85:7527–7534
61. Si W, Aksimentiev A (2017) Nanopore sensing of protein folding. ACS Nano 11(7):7091–7100
62. Sun X, Lin T, Gezelter JD (2008) Langevin dynamics for rigid bodies of arbitrary shape. J Chem Phys 128(23):234107
63. Fernandes MX, de la Torre JG (2002) Brownian dynamics simulation of rigid particles of arbitrary shape in external fields. Biophys J 83(6):3039–3048
64. Dechadilok P, Deen WM (2009) Electrostatic and electrokinetic effects on hindered diffusion in pores. J Membrane Sci 336(1):7–16
65. Doi M, Edwards SF (1986) The theory of polymer dynamics. Oxford University Press, Oxford, U.K.
66. Bhattacharya S, Mishra C, Bhattacharya S (2010) Analysis of general creeping motion of a sphere inside a cylinder. J Fluid Mech 642:295–328
67. Gubbiotti A, Chinappi M, Casciola CM (2019) Confinement effects on the dynamics of a rigid particle in a nanochannel. Phys Rev E 100(5):053307
68. Ermak DL, McCammon J (1978) Brownian dynamics with hydrodynamic interactions. J Chem Phys 69(4):1352–1360
69. Santiago JG (2001) Electroosmotic flows in microchannels with finite inertial and pressure forces. Anal Chem 73(10):2353–2365
70. Nakano A, Ros A (2013) Protein dielectrophoresis: advances, challenges, and applications. Electrophoresis 34(7):1085–1096
71. Freedman KJ, Otto LM, Ivanov AP, Barik A, Oh SH, Edel JB (2015) Nanopore sensing at ultra-low concentrations using single-molecule dielectrophoretic trapping. Nat Commun 7(10217)
72. Tian K, Decker K, Aksimentiev A, Gu L-Q (2017) Interference-free detection of genetic biomarkers using synthetic dipole-facilitated nanopore dielectrophoresis. ACS Nano 11(2):1204–1213
73. Firnkes M, Pedone D, Knezevic J, Doblinger M, Rant U (2010) Electrically facilitated translocations of proteins through silicon nitride nanopores: conjoint and competitive action of diffusion, electrophoresis, and electroosmosis. Nano Lett 10(6):2162–2167
74. Redner S (2001) A guide to first-passage processes. Cambridge University Press
75. Schoch RB, Han J, Renaud P (2008) Transport phenomena in nanofluidics. Rev Mod Phys 80:839–883
76. Delgado A, González-Caballero F, Hunter R, Koopal L, Lyklema J (2007) Measurement and interpretation of electrokinetic phenomena. J Colloid Interface Sci 309:194–224
77. Lee C, Joly L, Siria A, Biance A-L, Fulcrand R, Bocquet L (2012) Large apparent electric size of solid-state nanopores due to spatially extended surface conduction. Nano Lett 12(8):4037–4044
78. Yan Y, Wang L, Xue J, Chang H-C (2013) Ion current rectification inversion in conic nanopores: nonequilibrium ion transport biased by ion selectivity and spatial asymmetry. J Chem Phys 138:044706
79. Melnikov D, Hulings Z, Gracheva M (2020) Concentration polarization, surface charge, and ionic current blockade in nanopores. J Phys Chem C 124(36):19802–19808
80. Kozak D, Anderson W, Vogel R, Chen S, Antaw F, Trau M (2012) Simultaneous size and ζ potential measurements of individual nanoparticles in dispersion using size-tunable pore sensors. ACS Nano 6(8):6990–6997

Chapter 7
Modeling of Ionic and Electronic Bio-Sensing with Semiconductor Nanopores

Mingye Xiong, Rajat Chakraborty, and Jean-Pierre Leburton

Abstract In this chapter, we provide a detailed description of a comprehensive computational approach for simulating biomolecules translocating through an electrically active solid-state nanopore. Emphasis is placed on bio-detection via in-plane electronic current variations across semiconducting membranes made of two-dimensional (2D) materials, such as monolayer graphene and molybdenum disulfide. This approach that consists in coupling molecular dynamics (MD) with self-consistent electronic transport modeling is illustrated by examples of applications in the detection of single- (ssDNA) and double-strand (dsDNA), DNA methylations, epigenetic biomarkers, DNA nicks, and DNA with RNA tails, for which statistical signal processing techniques are implemented to improve the signal to noise ratio (SNR). Technical issues related to the use of advanced multilayer 2D solid-state nanopores for reducing the conformational noise by electrical tuning and compounded van der Waals interactions, sensitivity boosting by introducing back gate layer, and the homopolymer distinction via multi-probe nanopore field-effect transistor (FET) are also addressed. Finally, a proof-of-concept analysis for the large-scale parallel biomolecule detection that shows immunity to crosstalk of the in-plane electronic current signals is discussed.

Keywords Two-dimensional membranes · Molecular dynamics · Self-consistent modeling · Boltzmann transport · Green's function

M. Xiong · R. Chakraborty · J.-P. Leburton (✉)
Department of Electrical and Computer Engineering, Nick Holonyak, Jr. Micro and Nanotechnology Lab, University of Illinois at Urbana—Champaign, 61820 Champaign, IL, USA
e-mail: jleburto@illinois.edu

M. Xiong
e-mail: mxiong5@illinois.edu

R. Chakraborty
e-mail: rajatc2@illinois.edu

J.-P. Leburton (ed.), *Solid State Nanopores*, Nanostructure Science and Technology,
https://doi.org/10.1007/978-3-031-42336-9_7

7.1 Introduction

In 1990, the idea of using nanopores for biomolecule sensing and DNA sequencing was proposed by David Demear and Daniel Branton [8]. Years later, they demonstrated, for the first time, the ssDNA translocation through an α-hemolysin nanopore [21], suggesting the possibility of nanopore sequencing. Nowadays, this approach is the most promising candidate for the third-generation DNA sequencing methodology that can achieve the "$1000, 24 hr" genome sequencing platform challenge to realize reliable, low-cost, and rapid single molecule detection without functionalization or labeling.

In the nanopore sensing scenario, a membrane containing a nano-scale ion channel is placed in the electrolyte solution, separating the solution into two chambers and allowing the ions and target molecules to flow only through the nanopore. When an electric field is applied to the two chambers, ions and charged biomolecules (such as DNAs) are driven to migrate through the nanopore generating an ionic current signal. Since the nanopore diameter is comparable to the diameter of the biomolecules, the translocation of the targeted biomolecule temporarily blocks the ion flow, reflected as a dip in the ionic current. The amplitude of the blocking current (dip) and its width provide information on the size (diameter) of the biomolecule and its dwell time in the pore, related to the molecule length, respectively. Correspondingly, the separation between two current dips is identified as the waiting time between two translocation events [37]. In the DNA nanopore sequencing process, each of the four nucleotides on the ssDNA is blocking the ionic current with a specific amplitude, which enables its identification.

While the use of biological membranes such as α-hemolysin or MspA nanopores achieved some success in DNA and RNA sequencing in the past years, they carry drawbacks in their fixed sizes and their sensitivity to chemical, mechanical, electrical, and thermal conditions [23]. With the development of nanotechnology fabrication techniques, nanopores on solid-state materials have emerged as alternatives to their biological counterparts for their robustness to experimental conditions. Moreover, with the advances in current solid-state technologies, solid-state nanopores (SSNPs) can be fabricated into tunable sizes and shapes [24], with the potential to achieve massive sensing platform production [3, 13, 14, 16].

Earliest solid state nanopores were fabricated from Si_3N_4 [7, 23] and SiO_2 [35] dielectric materials with membrane thicknesses ranging from 5 to 100 nm, which is $10 - 30$ times DNA base pair separation in solution ($\sim$ 0.33 nm [25]). Because of their oversized dimensions, these early SSNPs suffered from low spatial resolution. However, with the 2004 emergence of graphene and two-dimensional (2D) materials like transition metal dichalcogenides (TMD) with a thickness ranging from 0.3 to 0.7 nm [9, 33], 2D SSNPs improve their intrinsic spatial biomolecule detection and appear well suited for single nucleotide sensing.

Although DNA translocation has been successfully observed in 2D SSNPs via ionic blocking current, actual sequencing remains challenging, because the resolution of ionic blockage current is too low compared to the translocation speed of

DNAs (0.01–1 μs per base) [18], which is an intrinsic drawback of the ionic current measurement approach.

In this respect, 2D solid-state membranes made of semiconducting materials such as graphene or MoS_2 offer additional advantages in their ability to carry in-plane electronic currents, providing electronic signals, for bio-detection and DNA sequencing. Prior works have shown the in-plane current can respond to high gigahertz frequency in both graphene and MoS_2 [17, 22]. Consequently, added to conventional sensing of ionic current blocking mechanism, this approach can overcome the difficulty in the aforementioned ionic current measurements to distinguish biomolecule structural details at the speed of its translocation in SSNPs. Among the various proposals for advanced structures based on electronic current sensing is the quantum point contact nanopore transistor that uses a lateral constriction of the membrane and a back gate to increase the current sensitivity [11]. It is anticipated that the realization of SSNPs with in-plane current sensing could increase the DNA sequencing rate by 2–3 orders of magnitude compared to ionic current sequencing [5].

Owing to the wide variety of available materials with versatile geometric, electronic, and structural properties in SSNPs in-plane current sensors, a comprehensive and accurate modeling scheme aimed at understanding the sensing mechanisms and optimizing their designs is desirable.

This chapter focuses on a computational methodology that combines molecular dynamics with advanced electronic device modeling to explore various structures and nanopore designs for optimum sensitivity of SSNPs via in-plane electronic signals. It is organized as follows: Sect. 7.2 describes the structure of a SSNP in-plane sensing device. Section 7.3 introduces the comprehensive computational model to simulate the ionic and in-plane current response of biomolecules translocated with examples. Section 7.4 discusses the advantages of using stacked SSNP. Section 7.5 introduces the scalability and parallel detection of SSNPs, followed by Sect. 7.6, the conclusion.

7.2 Semiconductor Nanopore Electronic Device Structure

A generic semiconductor nanopore sensor that simultaneously gauges the ionic and in-plane current with controllable electronic sensitivity by using a back gate is schematically illustrated in Fig. 7.1. It consists of a 2D semiconducting membrane sandwiched between two thin oxide layers. An aperture with a diameter of around 6 nm or less is etched through the membrane layers, making it a SSNP device. The pore diameter is usually determined by the target biomolecule to allow the biomolecule to pass through and maximize the sensing activity, which will be discussed in later sections. The device is immersed in the ionic water solution and separates the electrolyte into two chambers, in which the ions can only translocate between the two chambers through the nanopore forming the ionic current. A source and drain electrode are connected to two ends of the 2D SSN to measure the in-plane electronic current flow at the same time. A back gate made of metal or heavily doped semiconductor membrane is fabricated underneath the lower oxide substrate

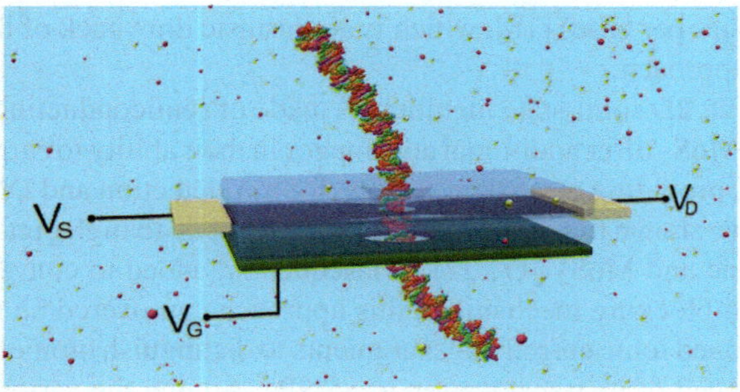

Fig. 7.1 Schematic diagram of a prototypical multilayer device containing a 2D solid-state layer (black) with a nanopore, sandwiched between two oxides (transparent). The DNA is translocated through the pore, and the current is measured with the source (V_s) and drain leads (V_d). Reprinted from [11] with permission

to control the charge carrier concentration in the 2D nanopore membrane as in a FET configuration [11]. The back gate can be tuned and enhance the electrical sensitivity to biomolecule translocation.

7.3 Comprehensive Modeling

In order to achieve deep physical insight into biomolecule interactions with the 2D solid-state nanopore and the resulting ionic and electronic current signals, a comprehensive computational model should take into consideration the interaction between solid-state nanopore and the biomolecule, conformations of the biomolecule, ion screening, electronic transport in the 2D nanopore membranes, etc. As shown in Fig. 7.2 the modeling scheme starts with the molecular dynamics (MD) simulation to achieve the biomolecule trajectories translocating through the nanopore, then solves for the potential induced at the vicinity of the pore using a self-consistent Poisson-Boltzmann solver, and obtains the ionic current and the in-plane electronic current with electronic transport calculation. The aforementioned simulation techniques are discussed in detail in the following sections.

7.3.1 Molecular Dynamics and Ionic Current Modeling

As a powerful computational tool to investigate the interactions between biomolecules and nanopore membranes at an atomic scale, molecular dynamics (MD) provides detailed insight of the dynamics of a bio-system, not achievable by experiments alone. Both interpretations and predictions of experimental outcomes can be made by tuning the system parameters of the MD simulation, for which the funda-

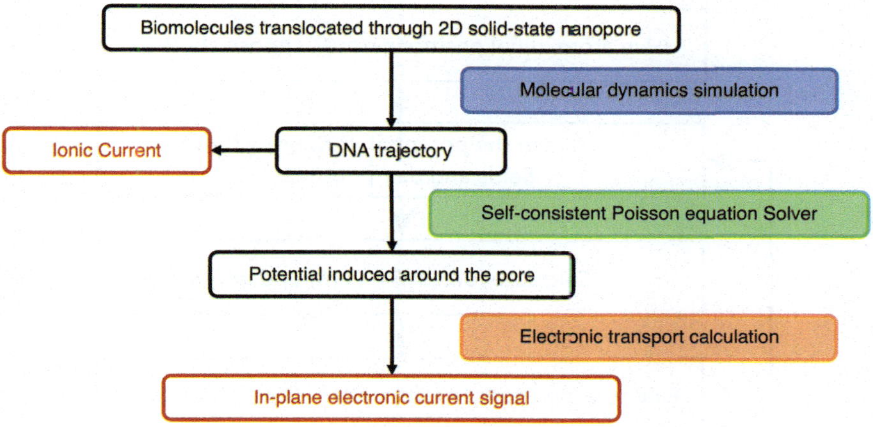

Fig. 7.2 A flowchart illustrating the comprehensive modeling scheme of the 2D solid-state nanopore

mental steps are shown in Fig. 7.3. Initially, the positions and velocities of atoms are assigned to a system. The system can be built either by manually creating a position file for the atoms or by using visualization software, such as VMD [19]. Next, the interaction energy or potential (V(**R**)) among all the atoms at any given coordinate **R** is given by the following basic equation:

$$V(\mathbf{R}) = \sum_{bonds} k_{ij}^{bond}(r_{ij} - r_{ij,0})^2 + \sum_{angles} k_{ijk}^{angle}(\theta_{ijk} - \theta_{ijk,0})^2 + \sum_{dihedrals} k_{ijkl}^{dihe}[1 + \cos(n_{ijkl}\phi_{ijkl} + \delta_{ijkl})] + \sum_{i}\sum_{i \neq j} \epsilon_{ij}\left[\left(\frac{\sigma_{ij}}{r_{ij}}\right)^{12} - \left(\frac{\sigma_{ij}}{r_{ij}}\right)^{6}\right] + \sum_{i}\sum_{i \neq j} \frac{q_i q_j}{\epsilon r_{ij}} \tag{7.1}$$

where the first three terms on the right-hand side of this equation represent bonded potential among atoms, whereas the last two terms denote non-bonded potentials corresponding to the Van der Walls and Coulombic interactions, respectively. Bonded potential includes interactions within molecular structures e.g., atoms within the same DNA strand or oxygen-hydrogen bond in a water molecule, whereas non-bonded potential computes interaction between atoms which do not share a bond e.g., potential between oxygen atom in water and potassium atom of KCl salt. In Eq. 7.1, i, j, k and l denote the index of a specific atom, while q_i and $\boldsymbol{r}_i$ represent the charge and position vector of the ith atom, respectively. The three terms in the bonded potential represent 2-, 3-, and 4-body interactions of covalently bonded atoms, respectively. Moreover, r_{ij} represents the length of $\boldsymbol{r}_{ij}$ where $\boldsymbol{r}_{ij} = \boldsymbol{r}_j - \boldsymbol{r}_i$,

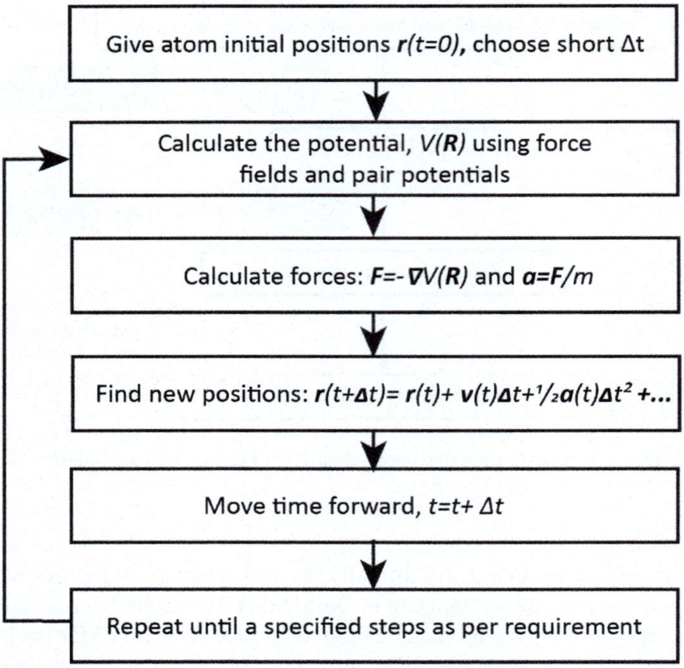

Fig. 7.3 Flowchart demonstrating basic steps of a molecular dynamics simulation

θ_{ijk} is the angle between $\boldsymbol{r}_{ij}$ and $\boldsymbol{r}_{jk}$, and ϕ_{ijkl} is the angle between (i, j, k)- and (j, k, l)-plane in a dihedral. $r_{ij,0}$, and $\theta_{ijk,0}$ represent the equilibrium value of r_{ij} and θ_{ijk}, respectively. n_{ijkl}, δ_{ijkl} and ϵ_{ij} denote a non-negative integer constant for dihedral periodicity, phase shift angle between dihedral planes, and Van der Walls unitless scaling factor, respectively. Furthermore, k_{ij}^{bond}, k_{ijk}^{angle}, and k_{ijkl}^{dihe} are internal molecular bond constant, angle constant and dihedral constant, respectively. Calculating this interaction potential is the most critical step in the MD simulation process. k's, $r_{ij,0}$, $\theta_{ijk,0}$, δ_{ijkl} and n_{ijkl} in Eq. 7.1 are empirical parameters, described in force fields designed for potential calculations and are obtained through extensive research and verified against experimental data. An excellent example of such a force field is CHARMM27 [10], which can describe the underlying mechanisms of biological structures. Pair potentials, such as the Lennard-Jones (LJ) potentials [20], are used to describe a portion of the non-bonded potentials. The first part of the non-bonded potential in Eq. 7.2 is the standard form of the LJ potential, while the second part accounts for Coulombic interactions. Using this potential, the force acting on a molecule can be calculated by:

$$m_i \frac{d^2\mathbf{r_i}}{dt^2} = \mathbf{F_i} = -\nabla V(\mathbf{R}) \tag{7.2}$$

Acceleration and velocity can be calculated from the force using simple Newtonian physics. The subsequent positions of all atoms after a short time interval Δt can be determined using:

$$\mathbf{r}(t + \Delta t) = \mathbf{r}(t) + \mathbf{v}(t)\Delta t + \frac{1}{2}\mathbf{a}\Delta t^2 + \cdots \tag{7.3}$$

Velocity and other relevant information can be obtained from the positions if required. These steps can be applied recursively until a specific time to obtain a complete trajectory of all atoms. This trajectory can then be used to calculate any subsequent signal traces, such as ionic current and in-plane conductance. Various MD packages, such as NAMD [26], Anton [34], and others, are available to apply these methods using numerical techniques.

In MD simulations for biomolecule detection with solid-state nanopores, the system is initially minimized to obtain a stable configuration. Equilibration is then performed to achieve the volume and pressure of the minimized system at a given temperature. This is commonly achieved by equilibrating the system first as an NPT ensemble by varying the volume of the system at a constant pressure, and then as an NVT ensemble by varying the pressure at a constant volume. A constant temperature (usually 300 K) is maintained using a Langevin piston. After minimization and equilibration, an electric field is applied, specified by the corresponding transmembrane voltage, to obtain the trajectory of the system under the applied bias. In special cases, steered molecular dynamics (SMD) is used to apply forces to a protein or a biomolecule and manually manipulate the translocation of the structure.

The instantaneous ionic current I(t) through the nanopores for every trajectory frame is calculated using the following equation [1]:

$$I(t) = \frac{1}{\Delta t L_z} \sum_{i=1}^{N} q_i (z_i(t + \Delta t) - z_i(t)) \tag{7.4}$$

where q_i and z_i are the charge and z coordinate of the ion i respectively, N being the number of ions, L_z is the z-coordinate dimension of the specified region and Δt is the interval between the trajectory frames. The idea here is to find the average change in position for all the charges in the cell in a short period Δt and divide it by the period itself to get the ionic current.

7.3.1.1 DNA Detection with Ionic Current in Graphene Nanopore

Here, for the purpose of illustration, we analyze the DNA translocation kinetics and detection with the ionic current in graphene nanopores by MD simulations using the instantaneous ionic current model [32].

All-atom MD simulations were conducted for graphene nanopores under various bias voltages in 1M KCl offering an atomic-level description of DNA translocation through graphene nanopores. The open pore resistance was first obtained for graphene nanopores with diameters ranging from 2 to 7 nm. The resulting dependence of ionic

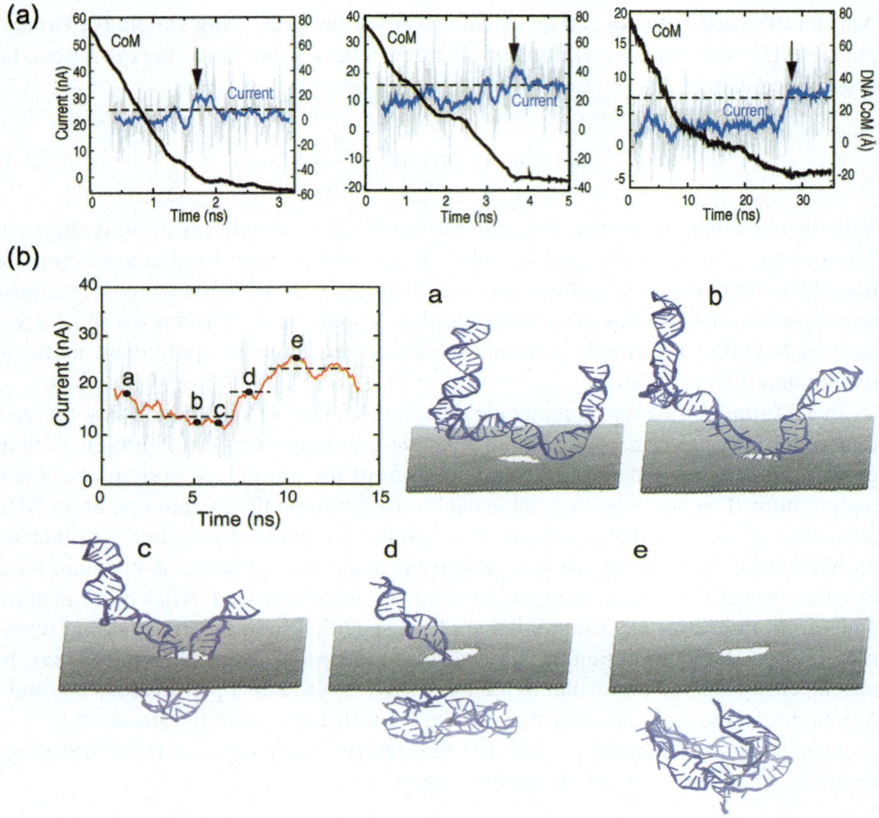

Fig. 7.4 **a** The ionic current (blue line) and the position of DNA center of mass (black line) during electrophoresis of dsDNA through graphene nanopores for bias voltages of 4.3, 2.5, and 0.8 V from left to right respectively. **b** The time evolution of the ionic current during the translocation of partially folded dsDNA associated with snapshots of DNA conformation. The three dotted lines correspond to plateaus in the ionic current signature. Adapted with permission from [32]. Copyright 2011 American Chemical Society

resistance on the pore diameters follows closely to $R \sim 1/d^2$, which agrees well with experiments [33]. Then dsDNA was translocated through a 2.4 nm diameter graphene nanopore driven by different applied bias (4.3, 2.5, and 0.8 V), and the time evolution of the ionic current was calculated as shown in Fig. 7.4a. The ionic blocking current during DNA translocation events is observed across all cases, among which ion blockade due to DNA is most obvious at a lower applied bias (0.8 V), resulting from the less stretched DNA at low bias and more blockage of the ion flow. During DNA translocation events, the attractive hydrophobic force between base pairs and graphene leads to the adhesion of DNA to the graphene membrane and slows down the translocation speed of DNA. Under high bias voltage, the electric force is much stronger than the attractive hydrophobic force, which keeps the DNA relatively vertical inside the pore, resulting in less ion blockage and shorter translocation time.

The translocation of folded dsDNA molecules was also simulated with MD and the resulting ionic current is shown in Fig. 7.4b. The observed ionic currents are attributed to different stages of translocation events, namely (a) the capture of DNA onto the pore rim, (b) DNA entering the pore in folded conformation, (c) translocation of the folded dsDNA, (d) translocation of the remaining single dsDNA chain, and (e) dsDNA exiting the pore. In addition, MD simulations of dsDNA permeation through a p-charged pore and an n-charged pore were also performed to analyze the kinetics of dsDNA translocation in graphene nanopores. The translocation of dsDNA through a p-charged pore is shown to be faster than its counterpart, due to the attraction between the negative charge on DNA backbones and the positive charge on a p-charged pore rim. In this work on graphene nanopores, we also demonstrate the discrimination of poly(A-T)$_{20}$ and poly(G-C)$_{20}$ using graphene nanopores.

Therefore, this initial work illustrates the possibility of dsDNA base pair identification using ionic blocking current at an atomic level, and can provide guidance in the design of graphene-based biomolecule sensing devices.

7.3.2 *Electronic Current Modeling*

As introduced briefly above, the modeling of electronic current across the semiconducting membrane (Fig. 7.1) involves the self-consistent solution of the Poisson-Boltzmann scheme to evaluate the electrostatic potential induced around the pore by the biomolecule translocation and its influence on the electronic transport in the membrane. In this section, we discuss this modeling approach in detail.

7.3.2.1 Self-Consistent Poisson-Boltzmann Solver

Here, one first obtains the trajectory of the target biomolecule translocation through the solid-state nanopore from MD simulations. The charge distribution of the DNAs (ρ_{DNA}) is then extracted in each frame of the trajectory, from which the potential profile is calculated by solving the self-consistent Poisson-Boltzmann equation.

Let us assume that the surrounding electrolyte is KCl. In the solution, the total charge distribution in the space consists of DNAs, positive potassium ions, and negative chloride ions. Thus, we obtain the Poisson equation

$$\nabla \cdot [\epsilon(\mathbf{r})\nabla\phi(\mathbf{r})] = -e[C_{K^+}(\mathbf{r}) - C_{Cl^-}(\mathbf{r})] - \rho_{DNA}(\mathbf{r}) \tag{7.5}$$

where C_{K^+} and C_{Cl^-} are the corresponding ion concentration. The ion concentrations follow Boltzmann distribution with virtual solid-state parameters [11]:

$$C_{K^+}(\mathbf{r}) = C_0 \exp \frac{-e\phi(\mathbf{r})}{k_B T} \tag{7.6}$$

$$C_{Cl^-}(\mathbf{r}) = C_0 \exp \frac{e\phi(\mathbf{r})}{k_B T} \tag{7.7}$$

where C_0 is the nominal concentration in solution, which is usually between 10 mM to 1 M.

In the solid-state material regions, the carrier concentrations follow the Fermi-Dirac distribution:

$$n(\mathbf{r}) = N_c \frac{2}{\sqrt{\pi}} F_{1/2}(\eta_c(\mathbf{r})) \tag{7.8}$$

$$p(\mathbf{r}) = N_v \frac{2}{\sqrt{\pi}} F_{1/2}(\eta_v(\mathbf{r})) \tag{7.9}$$

where N_c and N_v are the density of states in the conduction and valence bands of the solid-state material; $F_{1/2}$ is the 1/2 order Fermi-Dirac function. $\eta_c(\mathbf{r})$ and $\eta_v(\mathbf{r})$ are:

$$\eta_c(\mathbf{r}) = \frac{E_f - E_c(\mathbf{r})}{kT} \tag{7.10}$$

$$\eta_v(\mathbf{r}) = \frac{E_v(\mathbf{r}) - E_f}{kT} \tag{7.11}$$

where $E_c(\mathbf{r})$ is the solid-state material conduction band edge and $E_v(\mathbf{r})$ is the valence band edge. Similar to the case in the buffer solution, the charge density in the solid-state regions is given by:

$$\rho_{ss}(\mathbf{r}) = q(N_d^+(\mathbf{r}) - N_a^-(\mathbf{r}) + p(\mathbf{r}) - n(\mathbf{r})) \tag{7.12}$$

In both the solid-state region and the solution region, Poisson Equation and the charge distribution can be solved simultaneously using iterations until convergence.

7.3.2.2 Electron Transport Modeling of Graphene Nanopores

The electronic transport properties of the graphene nanopore are calculated using the tight-binding approximation [12]. The Hamiltonian can be expressed as

$$H = \sum_{i,\mu} [\epsilon_\mu - e\phi(\boldsymbol{r_i})] a_i^{\mu\dagger} a_i^\mu + \sum_{<ij>\mu} V_\mu(\mathbf{n}) a_i^{\mu\dagger} b_j^\nu + V_\nu(\mathbf{n}) b_j^{\nu\dagger} a_i^\nu \tag{7.13}$$

where ϵ_μ is the on-site energy of an electron in state μ located at position i, $\phi(\boldsymbol{r_i})$ is the electrostatic potential at position i, and $a_i^{\mu\dagger}/b_i^{\mu\dagger}$ and a_i^μ/b_i^μ are creation and annihilation operators for an electron in state μ at position i for the graphene A/B sublattice. The states μ, ν are the p_z, d_{yz}, and d_{zx} orbitals of atomic carbon. The transfer integrals $V(\mathbf{n})$ are determined by fitting to *ab initio* calculations where $\mathbf{n}$ is the unit displacement between i and j. Non-equilibrium Green's function (NEGF) formalism is used after acquiring the Hamiltonian ($\mathbf{H}$) with the Green's function

$$G(E) = [E - \mathbf{H}]^{-1} \tag{7.14}$$

Hence, the transmission $T(E)$ between two leads of the graphene-QPC is

$$T_{12} = -T_r[(\Sigma_1 - \Sigma_1^\dagger)G_c(\Sigma_2 - \Sigma_2^\dagger)G_C^\dagger] \tag{7.15}$$

where Σ_α is the self energy of lead α. From the transmission, one can find the conductance at a source-drain bias V_{DS}

$$G = \frac{2e}{V_{DS}h}\int_{-\infty}^{\infty} T(E)[f_1(E) - f_2(E)]dE \tag{7.16}$$

where $f_\alpha(E)$ is the probability of an electron occupying a state at energy E in the lead α. So far, we can identify the conductance and hence the current in a graphene nanopore from the potential profile achieved in the previous step.

7.3.2.3 Electrically Tunable Graphene Nanopore Sensor

The use of the aforementioned NEGF formalism is illustrated on a multilayer graphene quantum point contact (QPC) proposed to detect DNA molecule translocation, for which the electronic carrier concentration in the membrane is controlled by a back-gate voltage [11]. Here a constriction around the pore (QPC) increases the detection sensibility by forcing the flow of charge carriers in the vicinity of the nanopore. The simulation follows the modeling scheme discussed in Sect. 7.3 with the device setup shown in Fig. 7.1.

Here, we consider the detection of biomolecules with the graphene QPCs of four edge geometries and four pore configurations, giving 16 situations in total. Specifically, the four edge geometries are: 5-nm-wide and 15-nm-wide pure armchair-edge GNR, and 8-nm-wide and 23-nm-wide QPC edge. For simplicity, they will be referred as 5-GNR, 15-GNR, 8-QPC, and 23-QPC in the following discussions. The QPCs have pinch widths of 5 and 15 nm in 8-QPC and 23-QPC respectively, the same as the widths of the 5-GNR and 15-GNR for the purpose of comparison. The four pore configurations are (i) pristine (no pore), (ii) 2 nm pore at the center (point P) (iii) 2 nm pore centered at 75% of the total (pinch) width in GNR (QPC) (at point Q), and (iv) 4 nm pore at the center.

The transmission probabilities for these configurations are simulated and summarized for all 16 situations shown in Fig. 7.5 top. The Fermi level is chosen to be ranging from 0 to 0.5 eV, corresponding to carrier concentrations of approximately

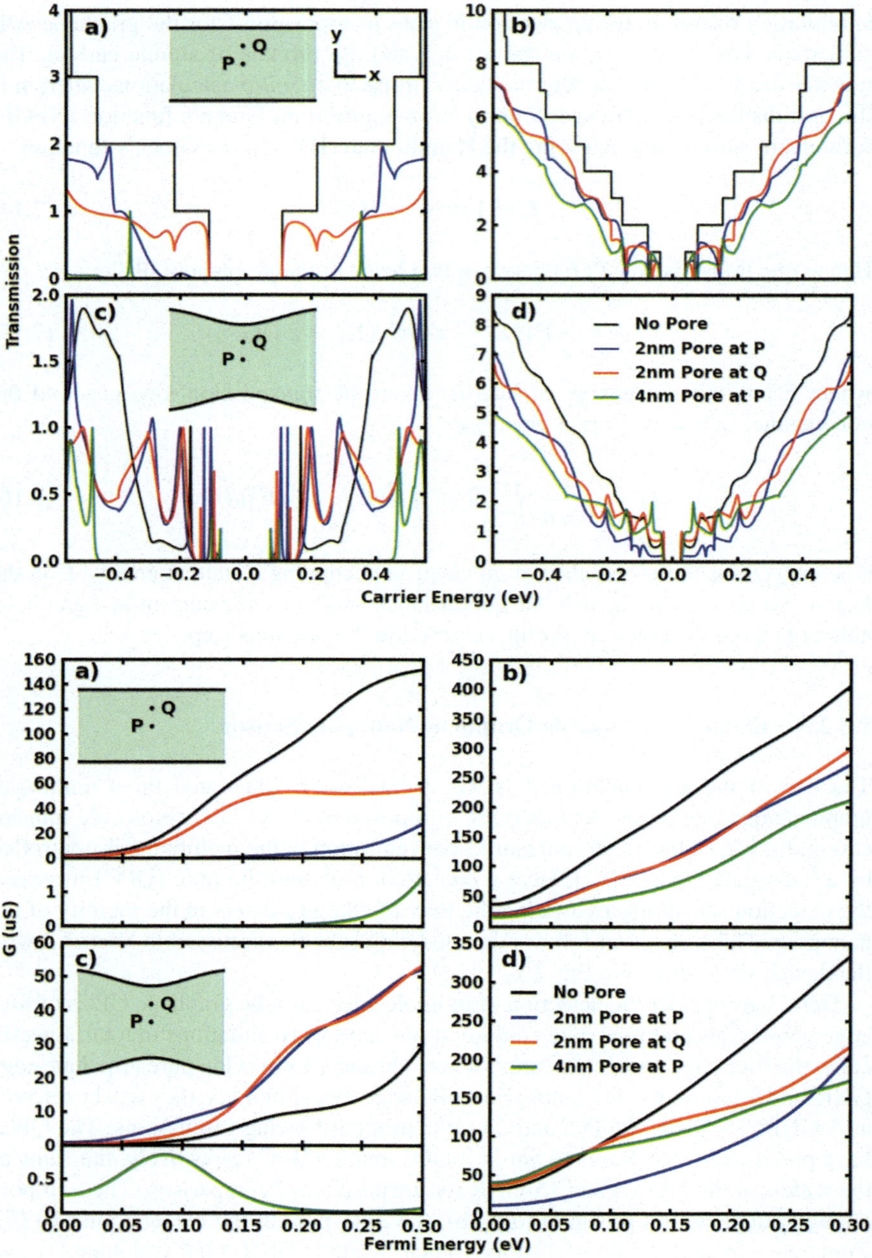

Fig. 7.5 Top: transmission functions for **a** 5-GNR, **b** 15-GNR, **c** 8-QPC, and **d** 23-QPCA; bottom: Conductance versus Fermi energy for **a** 5-GNR, **b** 15-GNR, **c** 8-QPC, and **d** 23-QPC. Reprinted from [11] with permission. Copyright (2013) National Academy of Sciences

10^{11} cm^{-2} to $5 \sim 7 \times 10^{12}$ cm^{-2} at 300 K, which is a reasonable level for a conventional graphene FET. The pristine case in 5-GNR shows a classic staircase shape as a result of the armchair boundary conditions. When the nanopore is introduced to the nanoribbon, the pore serves as a scattering center in the GNR and acts as additional boundary conditions at the pore edge. As a result, the number of the allowed electronic states reduces, while the probability currents also decrease as a result of the scattering from the nanopore, as shown in Fig. 7.5 top (a). Increasing pore sizes favors the scattering, and therefore less transmission is observed as shown in Fig. 7.5 top (a). Altering the pore locations results in the change of boundary conditions, and hence the transmission probability also varies accordingly. In the 5-GNR case, the transmission with the pore located at P is lower compared to that when the pore is located at Q.

In the 15-GNR case shown in Fig. 7.5 top (b), which exhibits a larger width than the 5-GNR, a similar trend case can be observed with several differences. First, because of the increased dimensions, more electronic states within a particular carrier energy become available, increasing the transmission probability for all four pore configurations. Second, because the density of allowed electronic states in the 15-GNR is greater at high energies, the 2-nm pore at both P and Q shows a similar number of allowed electronic states within a specific energy range.

For QPC cases, shown in Fig. 7.5 top (c) and (d), very different trends are observed mainly because of more strict boundary conditions introduced by the non-uniform QPC edges. As a result, resonant peaks throughout the Fermi energy range are observed. A decrease in transmission can still be observed with the increase of pore size as shown in Fig. 7.5 top (c). When going from 8-QPC to 23-QPC, the increase of electronic states still occurs, therefore smoothing out the transmission at higher carrier energies.

After obtaining the transmission coefficient, the electronic conductance as a function of the Fermi energy is calculated as the average of the transmission around such carrier energy weighted by the Fermi-Dirac distributions. Shown in Fig. 7.5 bottom (a), the conductance significantly depends on nanopore sizes and positions in narrow GNR. Specifically, the pristine 5-GNR has the largest conductance and increases monotonically over the Fermi energy up to 0.3 eV. For 2 nm pore at P, the conductance is much lower compared to the pristine case. The 2 nm pore at Q case is one order of magnitude higher than the P case with a plateau beyond 0.15 eV. The 4 nm pore case is negligibly small because of the suppressed transmission probability. For the 15-GNR case, all four situations show a similar monotonic trend and exhibit higher conductance compared to the 5-GNR case, which is expected because of the scaling of the GNR width. The pore location effect is no longer obvious, but the the conductance of the 4 nm pore case is still the lowest among all cases studied.

Very different trends are seen for the 8-QPC case in Fig. 7.5 bottom (c). One significant difference is that a 2 nm pore at either P or Q shows an enhancement of the conductance compared to the 8-QPC pristine case. Although seeming to be contradicting to the fact that the pore is a scattering barrier, it can be explained by the rich interaction of the electronic states with the edge and pore boundaries. Another significant difference is that the 4-nm pore case features a negative differential transconductance (NDTC) behavior, a feature that is not observed for the GNR systems. The conductance in the 23-QPC case, as shown in Fig. 7.5 bottom (d), is less sensitive to pore geometry and no NDTC behavior is observed. The 4 nm pore case in 23-QPC has larger conductance than the 2 nm pore case at low Fermi energies, which is a result of the enhanced transmission probability at low Fermi energies in the 4 nm pore geometry. The key message from the conductance study is that the conductance changes significantly with Fermi energy, which can be further exploited to build sensitive charge-sensing devices.

In this discussion, the shape of the edge, the carrier concentration, and the nanpore position and size in graphene nanoribbon can strongly affect its electronic conductance as well as its sensitivity to external charges. The added back-gate layer is used to tune the QPC carrier concentration to enhance the detection sensitivity of the graphene nanopore biosensor.

7.3.2.4 Stepwise Translocation of Stretched ssDNA

In the following, we further investigate the stretching conformation of single-stranded DNA (ssDNA) in narrow graphene nanopores and predict the stepwise translocation of ssDNA [27]. For this purpose, an ssDNA is placed at the center of the graphene nanopore, and within a short period of 1 ns, it spontaneously moves away from the center and adheres to the pore rim without the application of any external forces. This spontaneous adhesion of DNA bases arises due to the hydrophobic interaction between the biomolecule and the graphene membrane. The adhesion is found to be robust, as the DNA remains near the pore rim without any external biases while still being able to freely diffuse along the pore edge. The longitudinal diffusion of the two bases in contact with the membrane is observed to be 100 times lower than those that are not in contact with the membrane. This denotes a significant reduction in fluctuations normal to the membrane due to the base-graphene adhesion.

In order to gain insight into the translocation process, steered molecular dynamics (SMD) is then used to pull the ssDNA at the $5'$ end with a harmonic spring and a constant velocity of 2 Å/ns. Figure 7.6a shows the stepwise translocation of the ssDNA as one base moves through the pore. The variation in force shown in Fig. 7.6b originates from the requirement to overcome the base-graphene adhesion, which slows the translocation process. Figure 7.6c further clarifies the mechanism of each base permeation. Insets 1 and 2 illustrate the approach and adhesion of a base to the lower end of the graphene nanopore. During the adhesion process, the DNA requires a maximum force to be pulled. As the base suddenly flips about the backbone, it is translocated through the pore (inset 3) and rebinds with the upper end of the graphene. Repetition of these steps leads to the periodic nature of the force and ensures a stepwise translocation. This process is insensitive to the shape and size of nanopores, base type, and pulling speed. However, if the DNA is pulled in the direction of the $3'$ end, the translocation changes to a steady sliding with a significantly lower pulling force. Pulling in the original direction (the $5'$ end) creates a small angle between the bases and the pulling direction, causing them to get stuck and providing

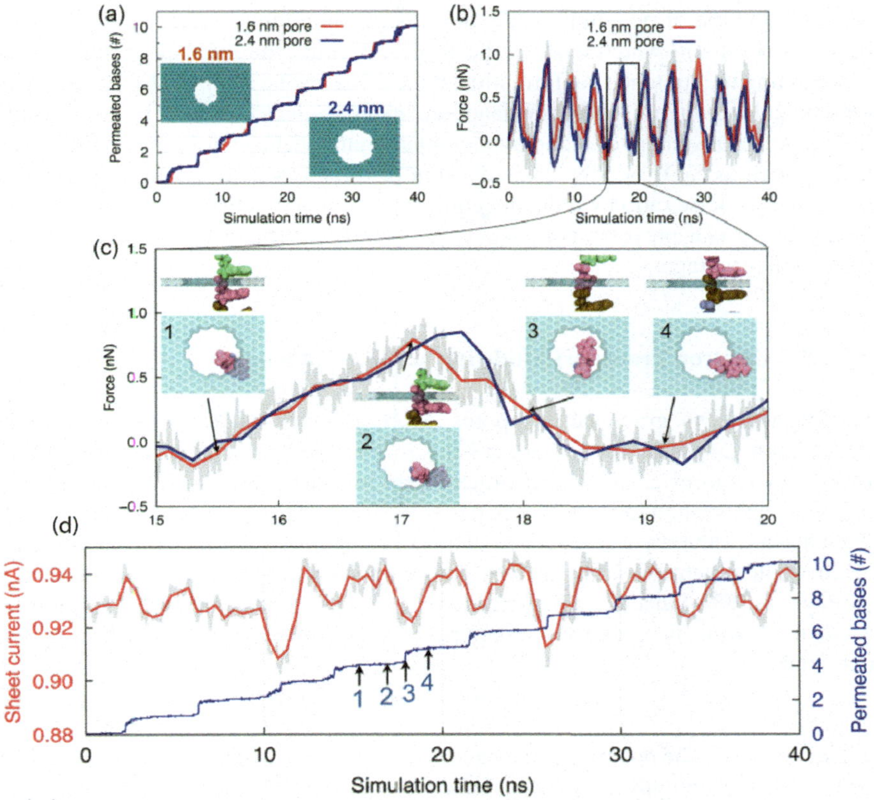

Fig. 7.6 **a** Number of permeated nucleotides and **b** the required pulling force during the translocation of poly (dA) ssDNA through 1.6 nm (red line) and 2.4 nm (blue line) diameter graphene nanopores at a constant velocity of 2 Å/ns. **c** The pulling force during a short phase of translocation (15–20 ns) with insets showing side and top views of the ssDNA and nanopore membrane at different stages during the permeation of a single base. **d** Calculated transverse sheet current (red line) and the number of permeated bases (blue line) during the translocation of poly (dA) ssDNA through a 1.6 nm diameter graphene nanopore. Reprinted from [27] with permission

a stepwise translocation. Such angle is not observed for pulling in the 3′ direction. For MoS_2 nanopores, stepwise translocation is observed with skips as more than one base permeates at a time because of the membrane's low surface hydrophobicity.

Next, a bias voltage is applied instead of the pulling force, and a similar stepwise translocation is observed for 1.5 V with no skips. However, a significant skip is observed for high voltages. The permeation process is relatively short-lived compared to the adhesion conformation, as expected. The transverse sheet current during the translocation is shown in Fig. 7.6d. During the adhesion conformation, as shown in steps 1, 2, and 4 of Fig. 7.6c, the electric potential is localized near the pore rim, whereas during permeation (step 3), the potential spreads around the pore.

This distribution of potential results in a distinct minimum in current during each base translocation, which enables the device to count the number of nucleotides. Maintaining the stretched conformation of the DNA can be achieved by threading the biomolecule through a solid-state nanopore under an electric field or by employing a multilayer arrangement of monolayer graphene sandwiched between two solid-state layers, such as SiO_2 or SiN. The adhesion of the bases can significantly enhance the signal-to-noise ratio by reducing conformational fluctuations and stabilizing the bases, and it can prevent spontaneous backward movement of the bases for low applied bias voltage.

7.3.2.5 Electron Transport Modeling of MoS_2 Nanopores

Unlike monolayer graphene nanopores, MoS_2 is much thicker, which makes the tight binding approach prohibited. For this reason, the electronic transport in MoS_2 nanopores requires a different treatment than in graphene nanopores. As a MoS_2 nanopore nanoribbons are much wider, usually hundreds of nanometers ($L_y = 100 - 200$ nm) in width [15], a semi-classical model using Fermi's golden rule, treating the electrostatic potential variations as a scattering source to the transverse current, is appropriate [38]. Since the nanopore diameter is much smaller than the width of the MoS_2 membrane, the perturbative nanopores can be modeled as Dirac delta function potential

$$V(\mathbf{r}) = V_{tot}\delta(x - x_0)\delta(y - y_0)A_{np} \tag{7.17}$$

where (x_0, y_0) is the position of the nanopore center, A_{np} is the area of the nanopore, and V_{tot} is the strength of the perturbation energy that includes the combined electrostatic effects of DNA, ions, and nanopore onto the conducting electrons. Because of the narrow width of the 2D MoS_2 membrane, we assume that conducting electrons are confined in the y-direction, and moves freely along the x-direction of the nanoribbon. The conducting electrons can be represented by their quantum mechanical wave function $\psi = Ce^{ikx}\sin(\frac{n\pi y}{L_y})$ where $C = \sqrt{\frac{2}{L_x L_y}}$ is the normalization constant. Hence, from Fermi's golden rule, the scattering rate from the (n, k) state to the (n', k') state is

$$S(\alpha, k, k', n, n') = \frac{8\pi V_{tot}^2 A_{np}^2}{\hbar L_x^2 L_y^2}\sin^2\left(\frac{n\pi y_0}{L_y}\right)\sin^2\left(\frac{n'\pi y_0}{L_y}\right)\delta(E_{k',n',\alpha} - E_{k,n,\alpha}) \tag{7.18}$$

where L_x is the length of the nanoribbon and α denotes the valley in the MoS_2 band structure. From the scattering rate and the Boltzmann transport equation, the membrane conductance due to the scattering of the nanopore potential is

$$G_{pore} = \frac{F}{V_{tot}^2 A_{np}^2} \tag{7.19}$$

where

$$F_\alpha(L_y, y_0, E_f^\alpha) = -\frac{e^2}{2\pi}\frac{\hbar^2 L_y^2}{2\sqrt{2m_\alpha^3}}\sum_n \sin^{-2}\left(\frac{n\pi y_0}{L_y}\right)\int \frac{\sqrt{E-E_n}dE}{\sum_{n'}\frac{\sin^2(\frac{n'\pi y_0}{L_y})}{k'(k,n,n')}}\frac{\partial f_{\alpha(E)}}{\partial E} \quad (7.20)$$

is a form factor as a function of the dimension of the nanoribbon, nanopore position, material properties, and carrier concentration.

From this model, we were able to achieve great agreement with the experiment measurement of the electronic current sensing signal in the MoS_2 nanopore. The sensitivity of the MoS_2 nanopore varies with the nanopore size and the nanoribbon geometric ratio.

7.3.2.6 Detection of DNA Methylation

In this section, we analyze the detection of methylated DNA (mDNA) using MoS_2 [29] and graphene [30] nanopore membranes. DNA methylation is an epigenetic modification involving the addition of a methyl group to DNA, which plays a crucial role in gene expression and regulation. DNA methylation is associated with various biological processes such as cell development, aging, and diseases including cancer. Methylation usually occurs in DNA sequences with a relatively high content of CpG dinucleotides (5′-CG-3′). The methyl-binding protein (MBD) serves as an excellent marker for methylation detection. Both MoS_2 and graphene nanopores can be used to detect the presence of methylation using DNA-MBD1 complexes. Figure 7.7a shows a single CpG site in an mDNA-MBD1 complex along with the chemical structures of regular and methylated cytosine. Qiu et al. utilized a 5.2 nm diameter MoS_2 membrane because the cross-sectional area of the mDNA-MBD1 complex is significantly larger (diameter of 5nm) than that of a dsDNA [29].

The ionic current trace and the transverse electronic conductance during the translocation of a mDNA-MBD1 complex are shown in Fig. 7.7b, c, respectively. Ionic current decreases and electronic differential conductance increases as the number of atoms reaches the maximum, denoting the presence of the protein in the pore. Therefore, it is possible to detect methylation using both ionic current and electronic conductance. For practical purposes, the detection of multiple methylation sites is necessary. The ionic current can detect two methylation sites with a separation of 15 base pairs (bp), as shown in Fig. 7.7d, but such distinction cannot be achieved for a separation of 10 bp (Fig. 7.7e). Hence, the resolution of methylation detection using ionic current is approximately 15 base pairs. On the other hand, Fig. 7.7f shows that two sites with a 10 bp separation can be distinguished with transverse conductance, providing better resolution for the detection of methylation.

In a similar study, Sarathy et al. employed a graphene nanopore to detect methylated DNA [30]. The conductance trace of the graphene pore for DNA without methylations shows two distinct levels before and after 19 ns, as illustrated by the blue line in Fig. 7.7g. This phenomenon can be attributed to the screening of the potential on

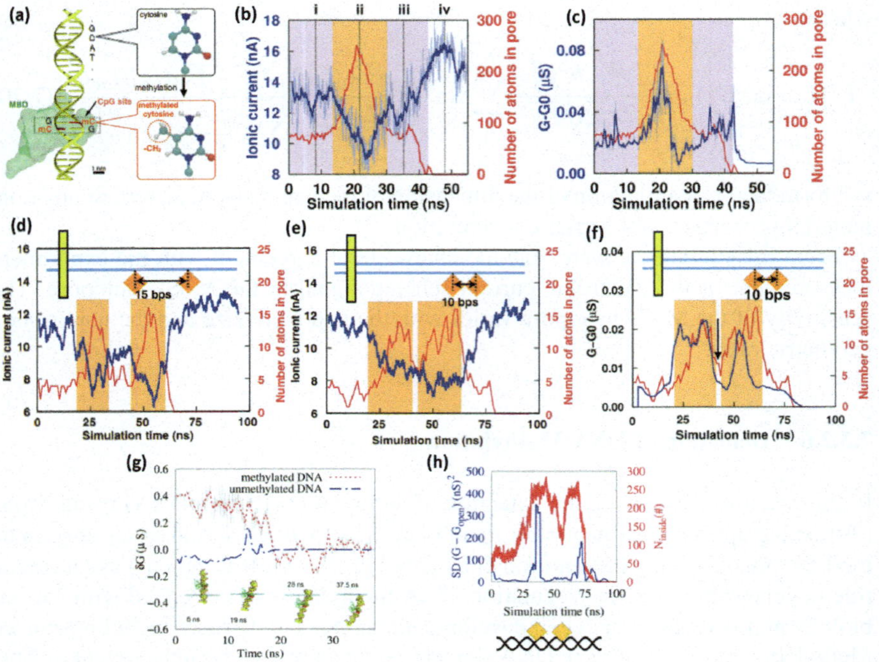

Fig. 7.7 **a** Schematic diagram of a CpG dinucleotide site in an mDNA molecule in complex with a MBD1 protein. Insets show the chemical structures of cytosine and methylated cytosine. **b–c** Ionic current (**b**) and differential electronic conductance (**c**) together with the number of DNA/protein atoms residing in a MoS_2 nanopore during the translocation of mDNA-MBD1 complex. **d–e** Ionic current and atom number in the pore during the translocation of mDNA-MBD1 complexes through MoS_2 nanopore with two methylation sites for separation of **d** 15 bp and **e** 10 bp. **f** Differential electronic conductance along with atom number for the same translocation as (**e**). **g** Conductance current trace with respect to open pore conductance during the translocation of the mDNA-MBD1 complex through a graphene nanopore with snapshots of important events at the inset. **h** Square deviation of sheet conductance with the number of atoms in the pore for translocation of mDNA-MBD1 complexes with two methylation sites with a 10 bp separation through the graphene nanopore. Reprinted from [29] with permission

the edges of the graphene membrane by the solute. However, for methylated DNA, three different regions are observed in the conductance trace (red line), corresponding to the translocation of DNA above and below the protein and the MBD protein itself. Although the conductance trace alone does not directly indicate the presence of the protein, a sudden dip in the trace can be used as an indicator. Thus, the square deviation (SD) of the conductance profile in successive time windows can be used to detect such a large deviation within a time window. Figure 7.7h demonstrates that the SD of the conductance profile, displaying two peaks, can distinguish between two methylated sites with a separation of 10 bp.

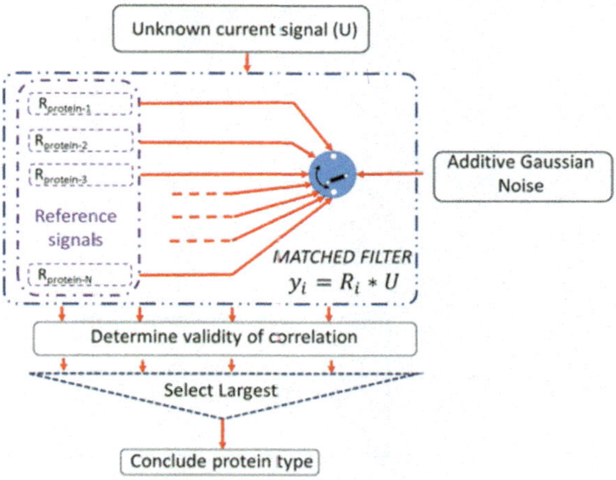

Fig. 7.8 Flowchart of a matched filter-based detection for the classification of epigenetic biomarkers using the correlation between the noisy input signal (test signal) and a dictionary of the precomputed reference current signatures from "noise-free" translocation of DNA-marker complexes. Reprinted with permission from [31]. Copyright 2018 American Chemical Society

7.3.2.7 An Example: Classification of Epigenetic Biomarkers

In this subsection, we discuss a versatile and general approach developed for the classification of epigenetic biomarkers using solid-state nanopore [31]. Three epigenetic biomarkers, MBD1, MeCP2, and γ-cyclodextrin (GCD), attached to double-stranded DNA (dsDNA) are analyzed. MBD1 and MeCP2 indicate hypermethylation, while GCD is a biomarker that denotes hypomethylation. The match-filtered process involves obtaining noise-free reference signals for each biomarker by artificially translocating the biomolecule through a solid-state nanopore using all-atom molecular dynamics (MD). This reference signal will contain a unique trace for the corresponding biomarker due to the unique charge distribution across the proteins. A set of these reference signals constitutes a dictionary for the detection process. The detection technique has been outlined in the flowchart in Fig. 7.8. The reference signals are denoted by $r_i(t)$, where i is the marker protein. MD simulations are then used to obtain a noisy signal ($u(t)$) with an unknown protein marker. The filter bank is then used to detect the protein type based on correlations between the unknown signal and reference signals using the following equation:

$$\rho(t) = \int_{-\infty}^{\infty} r_i(\tau) u(t+\tau)\, \mathrm{d}\tau \tag{7.21}$$

The matched filter technique is employed due to its optimality in detecting pulses that have been distorted by additive white Gaussian noise. To ensure uniformity during detection, all correlated signals are normalized to a range between -1 and 1. The Q factor can be calculated from the correlated signal using the following equations:

$$\rho_{\text{norm}}(t_i) = \frac{\rho(t_i) - \rho(t_{\text{open}})}{\max(\rho(t)) - \min(\rho(t))} \tag{7.22}$$

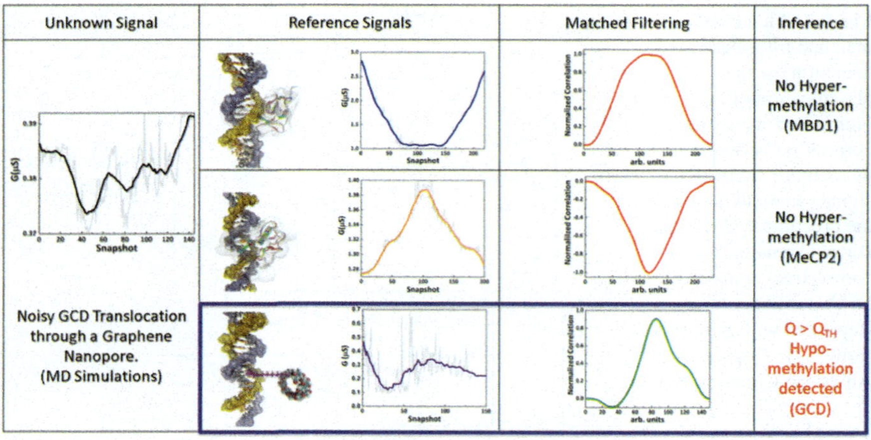

Fig. 7.9 A noisy current trace (first column) is correlated with the reference signals (second column). The decision (fourth column) is made based on the obtained correlations (third column). Reprinted with permission from [31]. Copyright 2018 American Chemical Society

$$Q = \frac{1}{BW_{corr}} \tag{7.23}$$

where $\rho(t_i)$ is the correlation trace at a sampled time t_i, and BW_{corr} is the bandwidth of the correlation at a particular value of correlation (ρ_i). The hyperparameter ρ_i is a user-defined value that depends on the specific application. In this study, a value of $\rho_i = 0.7$ is selected. The Q factor serves as the metric to distinguish between the biomarkers. A detailed illustration of the entire process can be found in Fig. 7.9. In this process, an unknown signal (first column) is cross-correlated with a set of reference signals (second column) to generate a normalized correlation profile (third column). Since no correlation value exceeds the threshold of 0.7, MeCP2 has a Q value of 0. MBD1 and GCD exhibit a bandwidth for $\rho_i = 0.7$, but the bandwidth of MBD1 is wider, resulting in a smaller Q value (0.017) as compared to GCD (0.04). Since the Q value of GCD surpasses a predetermined threshold value (Q_{TH}), the presence of this biomarker is confirmed.

This technique can be further extended to detect multiple biomarkers of the same type on a DNA molecule, as outlined in Fig. 7.10. The presence of the biomarker protein is detected based on the Q factor of the peak exhibiting the maximum correlation (ρ_{first}). Once the presence of the biomarker is confirmed, the correlation value for the second peak (ρ_{second}) is computed. If this value exceeds a predefined threshold (ρ_{TH}), it signifies the existence of another biomarker. The normalized correlations between the frozen single DNA-marker complex and the noisy signals from the translocation of the DNA with a single complex (known as reference correlation) and multiple complexes of the same type are measured. By aligning the correlation peaks, the presence of a peak can be identified using the difference between

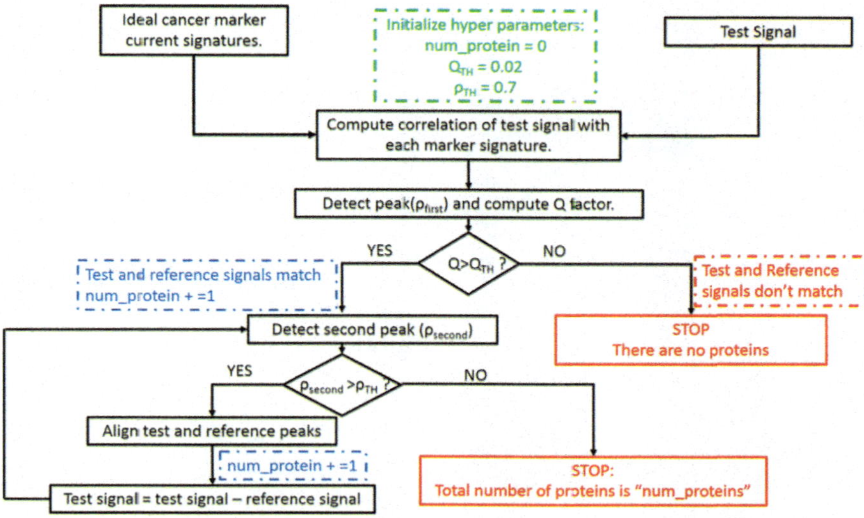

Fig. 7.10 Flowchart for the recursive algorithm to detect the type of marker protein and the number of markers in the vicinity of the protein. Reprinted with permission from [31]. Copyright 2018 American Chemical Society

the signals. These procedures can be recursively applied until the hypothesis is no longer valid. This study demonstrates a technique that can be employed for various applications and can further be implemented in hardware for real-time detection.

7.3.2.8 Detection of DNA Nicks

Another example of application of MD simulation combined with self-consistent electronic transport modeling is the detection of DNA nicks in solid-state nanopores. Nicks are defined as single-stranded breaks in the DNA backbone that can be caused by both endogenous and exogenous agents. While nicks are a common occurrence in cells, they are typically repaired by cellular mechanisms. However, an unusual increase in the number of nicks can result in double-stranded breaks, leading to genomic instability, neurological disorders, and other diseases. Additionally, nicks have been utilized for data storage on native DNA sequences [36]. One common type of nick involves the missing phosphodiester bond and the phosphate group at the 5′ end, which can be caused by hydroxyl and other DNA-damaging radicals.

Athreya et al. have conducted an investigation to detect nicks in DNA using both graphene and MoS_2 membranes with a 2.6 nm diameter pore [2]. As illustrated in Fig. 7.11a, there is an increase in the velocity of the biomolecule at the beginning of the translocation through a graphene nanopore. When the nicked backbone approaches the pore around 5 ns, the DNA strand experiences a decrease in velocity to 0 nÅ/ns, and comes to a stop. However, the velocity increases again after 8 ns, and

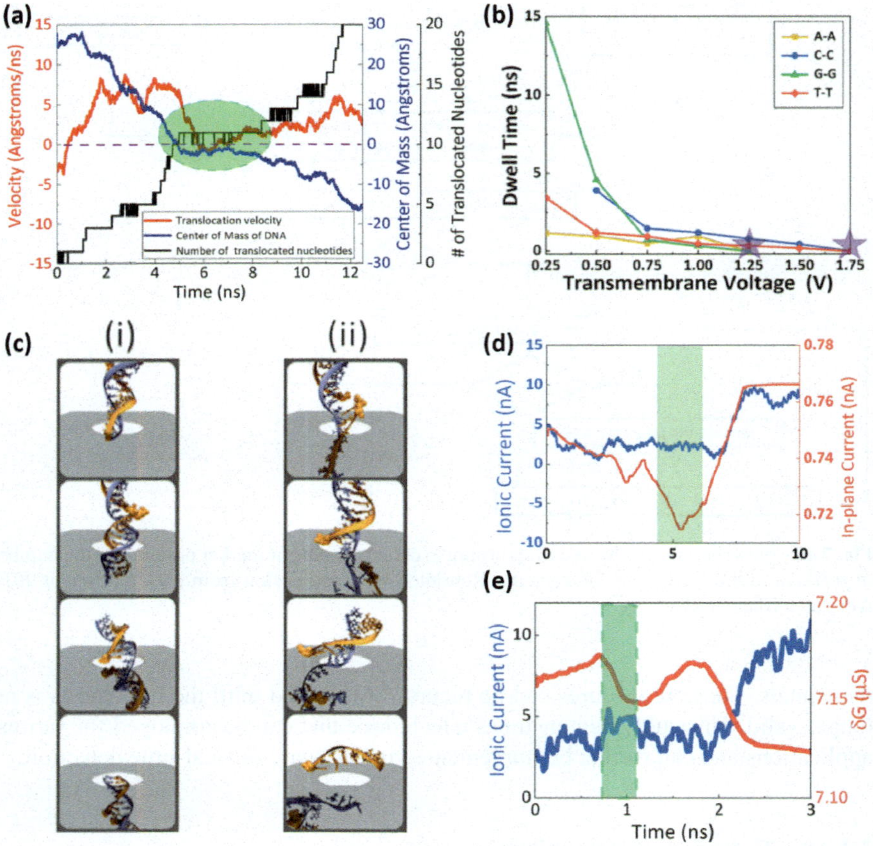

Fig. 7.11 **a** Velocity (left axis, red curve), the center of mass (CoM) of the nicked DNA (right axis, blue curve), and the number of translocated nucleotides (far right axis, black curve) during the entire translocation time of the nicked DNA through a graphene nanopore. The dotted line in the center of the plot indicates the location of the pore. **b** Dwell time of the nick in the pore as a function of transmembrane voltage. Star symbols denote the threshold voltages at which nicked DNA unzips. **c** Snapshots of important events showing conformational dynamics during (i) normal translocation and (ii) unzipping of the nicked DNA. **d** Ionic current and in-plane electronic current signals for translocation of DNA strand with nick between A-A through a graphene nanopore. **e** Ionic current and in-plane conductance variation with MoS_2 nanopore for the same case in (**d**). Reprinted from [2] with permission

the nick site leaves the pore. The plot of the center of mass (CoM) and the number of translocated nucleotides also verify the halt of the DNA as it takes almost a constant value during this period. The stop results from the attraction between the nick and the graphene membrane. Due to the absence of the hydrophilic phosphate group at the nick site, it becomes hydrophobic and induces a strong interaction with the membrane which is verified by the van der Walls (vdW) energies. MoS_2 also shows a similar behavior with a weaker interaction.

The dwell time of the nick site inside the pore depends on the neighboring atoms and also on the applied bias voltages (Fig. 7.11b). As expected, the dwell time monotonically decreases as the voltage increases. For low voltages, the DNA passes through the pore in an expected manner as shown by Fig. 7.11c(i). However, for high applied bias, the DNA unzips from the nick site after coming to a halt as shown in 7.11c(ii) (second and third snapshots) thereby putting a limitation on the voltage applied across cis- and trans- chamber. Here, the nucleotide in the cis- chamber breaks the hydrogen bond with its complementary nucleotide and slides onto the top of the graphene membrane due to strong hydrophobic interaction, thereby stopping the translocation. The high electric field present due to the applied voltage than unzips the DNA at the nick sites.

This limit for applied voltage is denoted by stars in Fig. 7.11b. This unzipping is absent in MoS_2 due to the weak interaction with the membrane. Ionic current and in-plane electronic current for translocation through graphene nanopore with a nick between A-A is shown in Fig. 7.11d. The ionic current does not show any significant change when the nick is in the pore due to the lack of conformational fluctuations arising from the vdW interaction. However, electrostatic scattering of charge carriers in the membrane around the pore due to the missing bonds and atoms of the DNA-nick in the pore, results in a clear dip in the in-plane transverse current. Hence, in-plane current signals can be used to detect nicks with graphene nanopores. For the MoS_2 nanopore, a dip in the in-plane conductance coincides with a rise of ionic current during the stay of the nick site in the pore. As MoS_2 is less hydrophobic than graphene, the two nucleotides on each side of the nick are seen to widen and open up, moving the DNA molecule closer to the edge of the membrane. This makes room for ions to drift through the pore which increases the ionic current. Therefore, a combination of ionic current and in-plane conductance can be used to analyze nicked DNAs with MoS_2 nanopore. The absence of unzipping in MoS_2 can be attributed to its weak interaction with the membrane.

7.3.2.9 Detection of DNA with Tails

As a final implementation of MD-self-consistent electronic transport, we describe, we describe a technique to detect RNA tails grown on a dsDNA by using a MoS_2 nanopore [6]. Figure 7.12a shows the differential conductance (ΔG) profile for the translocation of a dsDNA with a 20 nucleotides (nt) RNA tail through a 3-nm pore diameter MoS_2 membrane. Although the ΔG exhibits a fall when the tail enters the pore, there is no evident rise as the tail exits the pore, as can be observed from the key events (acquired from MD simulation) presented in Fig. 7.12a. To eliminate the dsDNA contribution from the signal, relative conductance, r_g, is defined as follows:

$$r_g = \frac{\Delta G - \Delta G_{DNA}}{|\Delta G_{DNA-i}|} \tag{7.24}$$

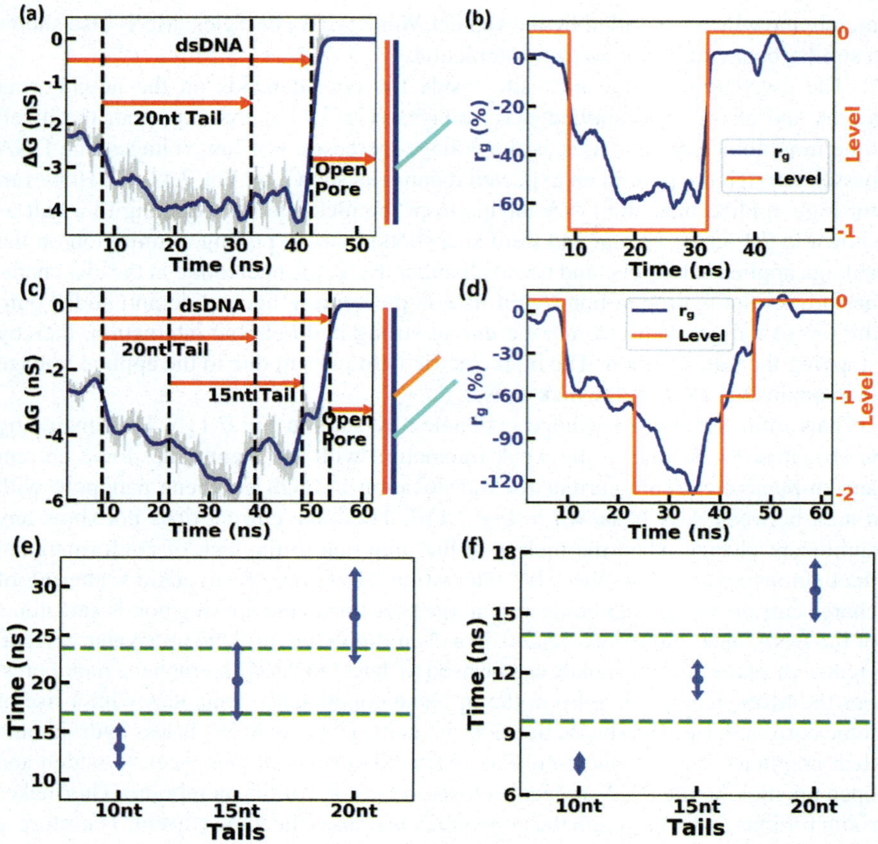

Fig. 7.12 Conductance signals for a 70 bp dsDNA with **a**, **b** a 20 nt RNA tail, and **c**, **d** a 20 and a 15 nt RNA tail placed 10 bp apart. **a**, **c** Differential conductance (ΔG) as a function of time where key events during translocation of the biomolecule (obtained from MD) are shown by red arrows and dotted vertical lines. Schematic of the individual biomolecule is also shown where the blue and red sticks denote the dsDNA and the orange and cyan sticks denote the 15 and 20 nt tails respectively. **b**, **d** Relative conductance, r_g (left axis, blue curve), and calculated levels (right axis, red curve) for the respective biomolecules. Here, level -1 denotes the presence of a single non-overlapping tail, while level-2 denotes the presence of two overlapping tails in the nanopore. The duration of different levels can be used to calculate dwell times. **e** Error bar plot of dwell times for different tails for original dwell times. **f** Error bar plot of dwell times for different tails after velocity modification for the same cases as in (**e**). Green dashed lines represent the decision boundaries in both plots (**e**, **f**). Adapted with permission from [6]. Copyright 2023 American Chemical Society

where ΔG_{DNA} is the contribution of the dsDNA in ΔG and ΔG_{DNA-i} is the value of the ΔG_{DNA} in the initial phase of the translocation. The relative conductance signal (r_g) shown in Fig. 7.12b exhibits a noticeable fall and rise as the RNA tail enters and exits the pore, respectively. A specific threshold value, L_1 (− 20% for this study), is selected for r_g below which the presence of RNA tail can be identified. Level -1 is defined as the state where r_g is below L_1, and the absence of any tail is indicated by level 0, as shown in Fig. 7.12b. The duration of level -1 provides the dwell time of the RNA tail.

Figure 7.12c, d demonstrates the ability of the proposed technique to detect overlapping tails. Specifically, a 15 nt RNA tail is placed 10 bp away from a 20 nt tail, resulting in the 20 nt tail signal overlapping with the 15 nt RNA tail as the separation distance is shorter than the length of the first (20 nt) tail. Despite this overlap, the first (20 nt) tail can still be detected using the same threshold value L_1 as before. To detect the overlapping tail, a new threshold value L_2 is defined as -70% in this study below which level -2 is denoted in Fig. 7.12d. When the first tail exits the pore, r_g rises above L_2, and eventually returns to level 0 as all tails exit the pore. The dwell times for each tail can be calculated from the duration of different levels. Dwell times can further be obtained for a variety of cases, including single tails with different lengths and completely overlapping tails, using this technique.

Discriminating among the lengths of RNA tails is also possible with this technique using modified dwell times. Figure 7.12e presents an error bar plot of dwell times for different tails before any modification is introduced. A decision boundary is defined in the plot by bisecting the midpoints of the error bars for consecutive tails. However, it is observed that the dwell times of a particular tail can extend beyond its individual region, making it difficult to differentiate tail lengths using normal dwell times. This effect arises from a decrease in the biomolecule velocity when overlapping tails are present inside the pore, which increases the dwell time of the tails. To account for this change in velocity, the dwell time is modified using the following equation:

$$t_{norm} = \frac{v_{ol}}{v_{dna}} t_{ol} + \frac{v_{nol}}{v_{dna}} t_{ncl} \tag{7.25}$$

where t_{norm} is the modified dwell time, t_{nol} and t_{ol} are the durations in which a tail is present inside the pore in non-overlapping and overlapping stages, respectively. Similarly, v_{nol} and v_{ol} are the velocities during the translocation of the biomolecule with non-overlapping tails and overlapping tails inside the pore, respectively, and v_{dna} is the biomolecule velocity when only the DNA resides inside the pore. The error bars obtained using these modified dwell times are shown in Fig. 7.12f. We can clearly see from this figure that the modified dwell times are more centralized. The decision boundaries are now more distinct, allowing for the differentiation of RNA tail lengths. This technique can also be applied to longer dsDNA with more than two tails.

7.4 Multilayer Membrane Structures

One of the notable advantages of the solid-state membrane nanopore is its capability to form a multilayer heterogeneous nanopore by stacking different nanopores in a concentric manner. This multilayer design has various superiority such as controlling the biomolecule movement in the nanopore, modulating the electronic current along the solid-state membrane, reducing conformational noises, and achieving statistically significant sensing signals. In this section, we will focus on the discussion on multilayer nanopore biosensors.

7.4.1 DNA quenching in solid-state nanopore sensor

One of the major challenges remaining in solid-state nanopore sequencing is the low signal-to-noise ratio (SNR) of the detected electric signal because of noises from thermal fluctuations of DNA bases, ions, and water inside SSNs, limiting the reliability of the detection. Therefore, it is meaningful to seek methods that can introduce certain controls of the motion and translocation velocity of the DNA inside the SSNs. In this section, we describe a solution to use a multilayered membrane transistor that possesses a motion-control electrode layer to control molecule motions [28].
On top of the device structure described in Sect. 7.2, a motion-control layer biased at a voltage, V_C is above the top oxide layer and isolated from the sensing membrane (such as graphene). V_C, together with source (V_S) and drain (V_D) voltages across the sensing membrane use a common ground. With such a structure, the electrostatic potential and the electric field can first be simulated, which can then be incorporated into the MD simulations. For the same SSN with varying positive and negative V_C, the center of mass (CM) positions of the DNA for each frame in the last 50 ns of the 70 ns simulations are shown in Fig. 7.13a, b, respectively. As shown in Fig. 7.13a, with 0V of bias, the CM almost scattered uniformly inside the pore (pole diameter = 2.4 nm). The center point of the scattered results is slightly off by the pore center by 0.2 Å, which is believed to be an artifact of the limited amount of simulation times. On the other hand, with greater positive V_C, the CM is more and more concentrated to a specific location, indicating the motion of the DNA is more confined with a higher positive voltage. The location is also not exactly the pore center due to the slight random conformational change in response to applied voltages. With negative voltages, however, as shown in Fig. 7.13b, there is very little control over the motion of the DNA.

The reason behind the difference in the behaviors under positive and negative bias is the shaping of DNAs. For positive bias, the control of DNA can be perceived by the backbone spread in the overlapped DNA conformations (obtained from MD simulation), which is thinner than that in a zero-bias case. The stabilization comes from the attraction of the positively charged pore surface and negatively charged DNA backbone. For negative bias, a thicker backbone spread in the DNA conformations is observed. Specifically, the DNA is compressed laterally because of the repulsion between the negatively charged pore surface and the DNA backbone. Because of the natural asymmetry and material softness of the DNA, it does not achieve a perfect symmetric configuration around the pore center. Instead, the CM locations are scattered randomly near the pore center as shown in Fig. 7.13b.

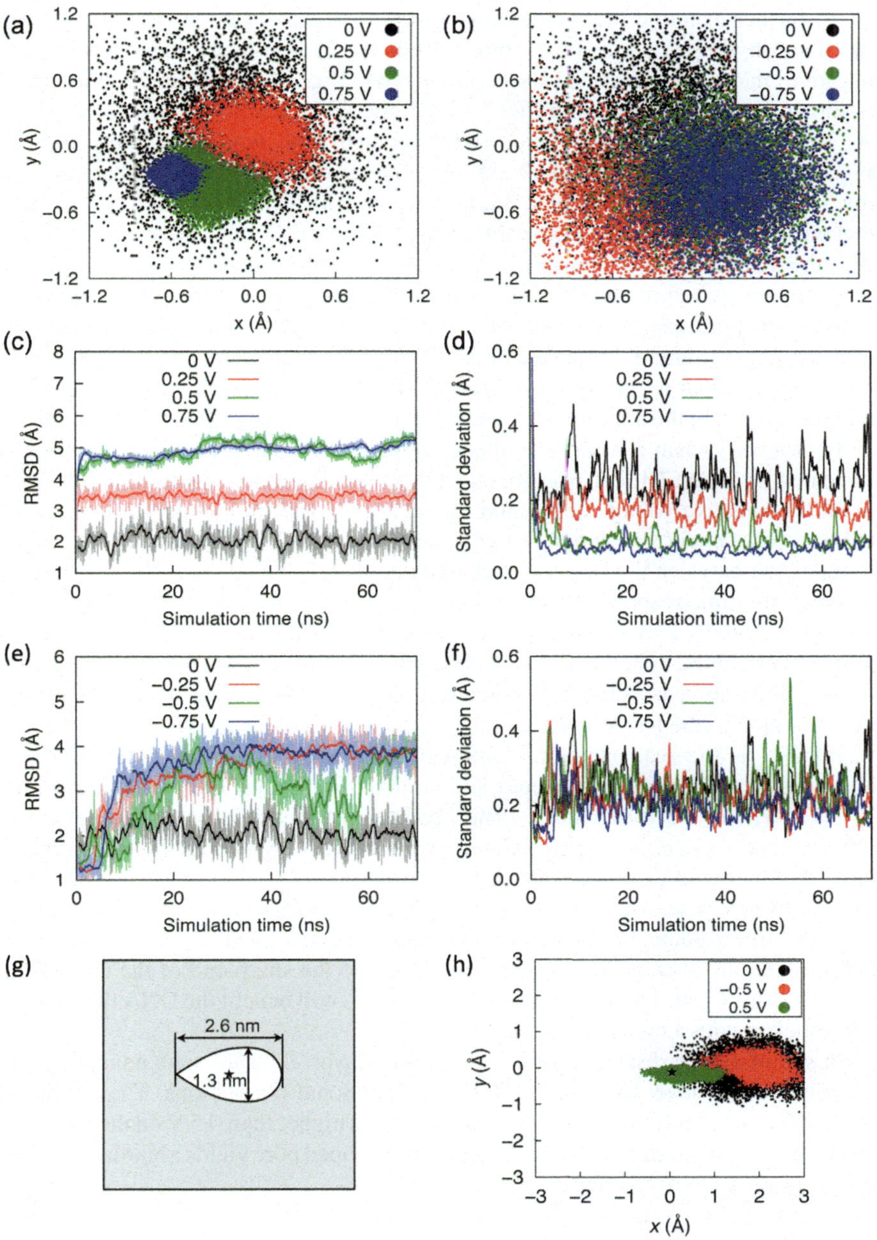

Fig. 7.13 **a**, **b** 50 ns of MD simulation of center of mass of the DNA with **a** different positive V_C, and **b** different negative V_C. **c**–**f** RMSD of all the non-hydrogen atoms of the 6th–17th of the 22 base pairs under **c** positive voltage and **e** negative voltage. The corresponding standard deviations are shown in **d** for positive voltage and **f** for negative voltage. **g** The geometry of a teardrop-shaped pore. **h** 50 ns of MD simulation of the center of mass of the DNA in the teardrop-shaped pore with 0, − 0.5V, and 0.5 V of voltage. Adapted with permission from [28]. Copyright 2016 American Chemical Society

To further examine the issue, the root-mean-square deviation (RMSD) profiles of all non-hydrogen atoms of the 6th–17th of the 22 base-pairs under positive and negative charges are simulated and summarized in Fig. 7.13c–f. With 0 V bias, which is shown as the black curve in Fig. 7.13c, d, the DNA RMSD fluctuates close to 2 Å. The moving standard deviation (STD), which is shown as the black curve in Fig. 7.13d, f, ranges from 0.2 to 0.5 Å. With positive bias, the RMSD increases and the STD decreases. In particular, the increase in RMSD saturates at 0.5 V with the fact the 0.75 V case has a similar value as the 0.5 V case. For the decrease in STD, on the other hand, the value continuously goes down with increasing positive bias. The reason, same as before, in the lateral expansion of the DNA induced by the attraction between the pore surface and the DNA backbone. The saturation of RMSD basically indicates that the lateral expansion of the DNA has saturates. However, the decreased STD suggests that higher positive bias still introduces stronger stabilization. Such effect is a result of the adhesion of the DNA to the pore surface (because of attraction).

In contract, when a negative voltage is applied, no stabilization is observed as indicated from Fig. 7.13e, f. Specifically, large fluctuations are observed in Fig. 7.13e in all the negative voltage cases, and the STD does not show great differences for the four cases under simulation in Fig. 7.13f. The reason, as explained before, is due to repulsion between the negatively charged pore surface and the DNA backbone.

Next, the time response of the system is examined with the voltage turning on and off repeatedly every 30 ns and the corresponding RMSD and STD is calculated. The results can be concluded that within 1 ns after turning on/off the voltage, the DNA stabilization is achieved, indicating the potential for real-time and precise manipulation to the DNA motion.

Finally, as the real nanopore cannot be fabricated to a perfect circle pore, a teardrop-shaped pore is considered and similar simulations are performed with 0 V, − 0.5 V, and 0.5 V of voltage. The geometry of the teardrop-pore is shown in Fig. 7.4a and the scattered CM is summarized in Fig. 7.13h. From the results, it can be shown that with negative voltage, the distribution of CM locations is almost the same as the 0 V case, which is consistent with the previous results with circular pores. With positive voltage, on the other hand, a strong DNA localization is observed as before, with the average location moving closer to the sharp end of the teardrop as shown in Fig. 7.13g. To sum up, the positive voltage will benefit the DNA fluctuations control even with a non-circular pore.

In summary, we discussed the usage of multilayer 2D solid-state nanopore with a biased control layer to quench DNA conformational fluctuations. The optimized applied bias to stabilized DNA is a positive voltage higher than 0.5 V on the nanopore electrodes. We also see that the use of a teardrop-shaped pore yields a teardrop-shaped distribution of DNA positions.

7.4.2 2D solid-state nanopore multi-probe sensor

In addition to using an electrical controlling gate to quench the conformational noise, we also analyze the noise reduction in stacked 2D solid-state nanopore [4]. It is shown that the signal-to-noise ratio is improved in such stacked nanopore sensor. The multilayer nanopore is also capable of providing multiple statistically significant sensing signals for a single translocating event to further achieve identification of homopolymers. In this section, we introduce the ability of stacked structures made of alternating 2D semiconducting and insulating monolayers to selectively sense biomolecules by transverse electronic current detection

The proposed stacked 2D solid-state nanopore consists of alternating semiconducting (MoS_2) layers and insulating (hBN) layers with concentric nanopores of diameter 1.6 nm as shown in Fig. 7.14a. The multilayer nanopore sensor is placed in the electrolyte the same way as the previously discussed single-layered nanopore sensor, where a transmembrane bias is applied to drive the DNA to translocate through the nanopore. Due to the existence of the insulating layers, the current across each semiconducting MoS_2 layer independently measures the DNA translocation event. In the simulation, we thread four homopolymers of the four different DNA bases to translocate though the nanopore, and for each translocation event, we simulated the in-plane electronic current signal responses in each MoS_2 layers.

We first observed that, rather than translocating linearly through the multiple layers, the biomolecule strand first forms a cluster on the top hBN layer. Hence, the translocation of the first 20 bases takes place within 2.5 ns, followed by a constant translocation rate for the rest 30 bases. This behavior is due to the high hydrophobicity of the hBN layer combined with the increased pore resistance due to the stacked layers; both dominate the access resistance by enabling the cluster formation inducing the non-linear ssDNA translocation. In the following analysis, we focus on the electronic signal of the last 30 bases that are translocated linearly through the multilayer nanopores.

Significant noise reduction (as much as 51%) were observed in the electronic current of in the third MoS_2 layer compared with the first MoS_2 layer. To identify the source of the noise reduction, we show in Fig. 7.14b a scattered plot and the corresponding distribution (in x- and y-axis) of each nucleotide center of mass (CoM) as it translocates through the multi-layer membrane. It is clear that the CoM distribution of the translocated homopolyer is increasingly confined as the nucleotide goes through each MoS_2 layer. The Van der Waals interactions between the DNA and each monolayer of the stacked membrane weaken the conformational noise, which results in a decreased standard deviation of the in-plane current signals obtained in the second and third MoS_2 layers.

Another key advantage of the proposed multilayer nanopore sensor is to achieve multiple independent electronic current sensing signals of a single translocation event. However, these signals are characterized by a time lag between neighboring layers. In order to identify these time lags between layers and to synchronize

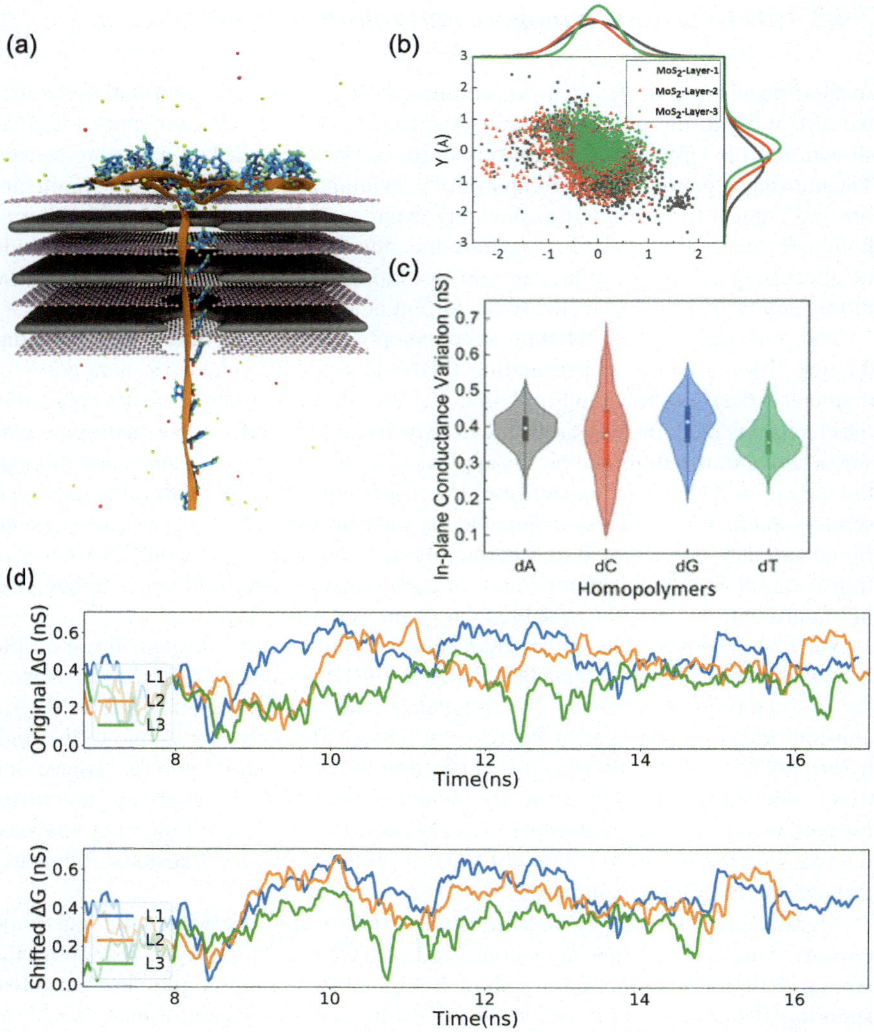

Fig. 7.14 **a** Schematic of the stacked nanopore bio-transistor with three MoS_2 layers (shown in black) encapsulated with hBN monolayers. **b** Scatter plot of the center of mass of each nucleotide in the x-y plane with the corresponding distribution along each axis as the nucleotide translocated through the successive MoS_2 layers. **c** Mean values of the conductance variation of each homopolymer, with error bars showing the standard error of the dataset. **d** The original (top) and time-lag-corrected (bottom) transverse conductance variations in all three layers during the translocation of poly-$(dC)_{25}$. Reproduced from [4] with permission. Copyright 2023 American Chemical Society

the successive transverse conductance signals, we performed a time-lagged cross-correlation (TLCC) between the in-plane currents originating from neighboring layers for the last 30 bp of each homopolymer translocating with constant velocity. The Pearson correlation coefficient (PCC) is calculated as a function of time offset between layers. The time offset that maximized the PCC is determined as the actual time offset between the signals in the neighboring layers and is used to synchronize the sensing signals. The synchronized poly-(dC) signals of the three layers are displayed in Fig. 7.14d bottom, which exhibits a clearer common behavior among the three signals than the raw data presented in Fig. 7.14d top.

After performing the synchronization of the in-plane current signal in different layers of all four homopolymers, we performed statistical visualization and analyses on synchronized signal data to idenfity the four homopolymers. Figure 7.14c shows the violin plot of the processed in-plane conductance variation of the four homopolymers with their mean values indicated as white circles and the 25/75 percentiles represented by the grey boxes. The distribution of poly-$(dC)_{50}$ has a broader spread compared to the other homopolymers. Furthermore, the in-plane conductance data of poly-$(dG)_{50}$ is left-skewed, whereas poly-$(dT)_{50}$ is right-skewed, and the other two homopolymers are relatively symmetrical.

We conducted the two-sample t-test between the signals of all four different homopolymers, where the null hypothesis is that they share the same mean. The resulted high t-scores (with absolute value greater than 5) and small p-values (smaller than 1×10^{-7}) reject the null hypothesis and confirms that the electronic current signals of all the four homopolymers have distinct statistical characteristics.

In summary, the electronic current variations in multi-probe nanopore transistor consisting of three semiconducting MoS_2 monolayers encapsulated by hBN materials boost the SNR by 51% due to the natural quenching of conformational fluctuations during biomolecule translocations. Note that this methodology is different from Sect. 7.4.1 where the conformational quenching was electrostatically induced to the translocating molecule. The synchronization of the electronic signals recorded at each MoS_2 probe is obtained by maximizing their corresponding correlation coefficient, combined with statistical testing, enables the distinguishability, and ultimately the identification of homopolymers. It is anticipated that the present approach will provide guidelines for designing a highly accurate 2D nanopore transistor biosensor capable of mapping the biomolecular structures while sequencing long strands of DNA in realtime.

7.5 Scalability and Parallel Detection

The potential for large-scaling multiplexing is one of the advantages of solid-state nanopores that lacks detailed investigation. In this section, we focus on the paralle DNA detection by 2D solid-state multipore system and demonstrate that the in-plane transverse current signals are immune to the crosstalk effects arising from the simultaneous translocations of target biomolecules [3].

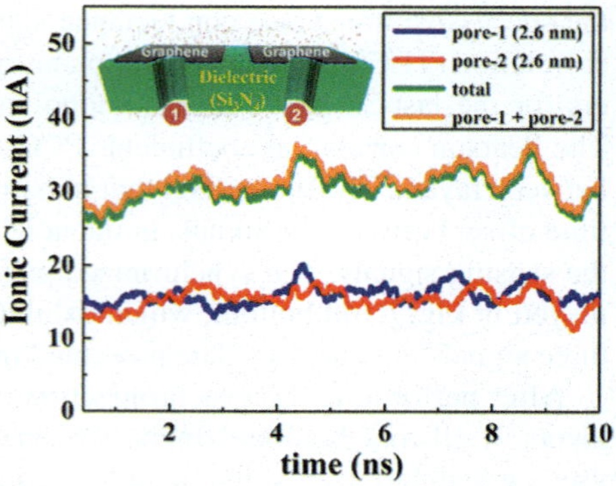

Fig. 7.15 Ionic current calculated for pore-1 (blue), pore-2 (red), the entire system, and sum of pore-1 and pore-2. Inset: schematic off the double-pore configuration. Reproduced from [3] with permissions

As a proof-of-concept, a two-pore setup is considered as shown in the inset of Fig. 7.15. Each membrane measures 9 nm by 9 nm with a 2.6 nm diameter circular pore in the center. The membranes are integrated on a Si_3N_4 substrate and submerged in a 1.0 M KCL solution. The ionic current for each pore is first calculated, which has a similar magnitude of 15 nA. The sum of the ionic currents for pore 1 and pore 2 is shown as the orange curve together with the total ionic current as the green curve, which consists of both graphene membranes. From the results, it is clearly shown that the ionic current in such a system can be characterized as the sum of that of each individual pore. However, a similar double-pore setup with one pore still at 2.6 nm diameter but another at 3.6 nm diameter shows different results. Specifically, the ionic current through the bigger pore is greater than that of the smaller pore only during the first 5 ns, after which converges to a similar amplitude. The reason is the current crosstalk between the two pores as they are merged into the same solution. This effect is naturally overlooked in previous simulations because the pore sizes are the same.

Furthermore, the electronic in-plane current signals of the graphene nanopores are calculated using the g-QPC model discussed in Sect. 7.3. The parameter and details of the calculation can be found in [3]. The ionic and transverse current for three cases are simulated for a double-pore setup with the same 2.6 nm pore: (i) simultaneous DNA translocation through both pores, (ii) partically time-overlap translocation of two DNAs through the two pores and (iii) totally separated translocation of two DNAs through the two pores. Illustration of the processes is provided in Fig. 7.16a. For events (i), (ii), and (iii), the corresponding ionic and transverse currents are shown in Fig. 7.16b, c, respectively. For case (i), the total ionic current [Fig. 7.16b(i)] cannot distinguish the two simultaneous events. However, the transverse current can clearly distinguish the two events as shown in the difference in current traces in Fig. 7.16c(i). Therefore, only the transverse current has the ability to detect individual translocation through each pore. To further examine this conclusion, corresponding results for case (ii) and (iii) are also provided in Fig. 7.16b, c. Clearly, for case (ii), the transverse current trace shows the difference in the start time of two translocations and several

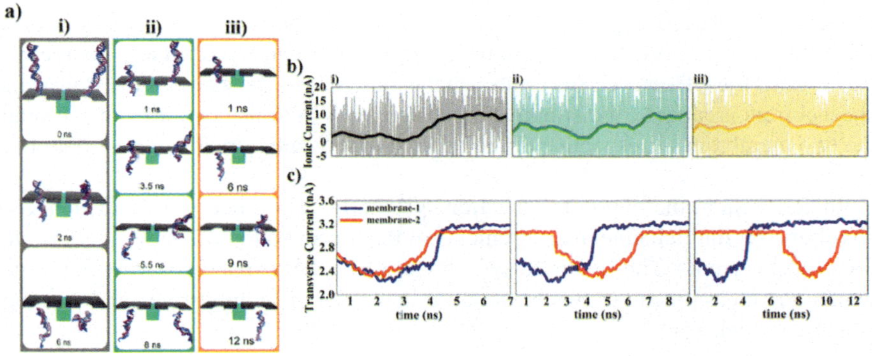

Fig. 7.16 **a** Illustration of three cases with a double-2.6 nm-pore configuration, (i) simultaneous translocation of two DNA in each pore (ii) partially time-overlap of translocation of two DNA in each pore and (iii) completely separate two events of translocation of two DNA in each pore, **b** the corresponding ionic current in three cases (light: total ionic current, dark: average ionic current). **c** Transverse currents for each membranes. Reproduced from [3] with permissions. Copyright 2018 American Chemical Society

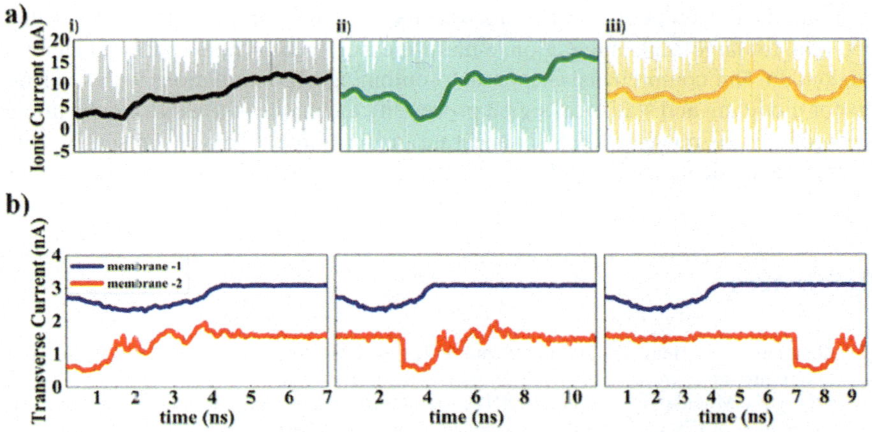

Fig. 7.17 For a double-pore configuration with one pore diameter of 2.6 nm and another pore diameter of 3.5 nm, **a** the corresponding ionic current in three cases (light: total ionic current, dark: average ionic current) and **b** the transverse currents for each membranes in the three cases. Reproduced from [3] with permissions. Copyright 2018 American Chemical Society

seconds of overlapping time. For case (iii), it clearly shows that the first pore finishes DNA translocation at $\sim$ 5 ns whereas the second pore starts the DNA translocation at $\sim$ 7 ns.

Finally, similar analysis was performed on another two-pore system, with one pore of 2.6 nm diameter and another of 3.6 nm diameter. The geometry of the QPC is changes accordingly to optimize the sensitivity (details can be found in the paper). For the same three cases (i), (ii), and (iii), the corresponding ionic current and transverse current are shown in Fig. 7.17a, b, respectively. The results are similar to that of the Fig. 7.16b, c, except that the transverse current response of DNA translocation for the 3.6 nm pore is different from the 2.6 nm pore because of different geometries and sensitivities. However, the main conclusion remains the same as the transverse sheet current is capable of distinguishing different DNA translocation events through the two membranes irrespective of the relative timing and pore geometry nonhomogeneity.

In conclusion, the two-pore detection system discussed here shows evidence of simultaneous multichannel measurements of the translocation of DNA via ionic and transverse currents. The two-pore system can be further expanded to a larger scale, which has great potential to achieve rapid detection of biomolecules and possibly genome sequencing.

7.6 Conclusion

In this chapter we emphasized the advantages of using the in-plane current over the traditional ionic current in providing more accurate and real-time bio-information, enabling the detection of detailed biomolecule structures of interest, as well as their potential for large-scale integration in multi-measurement in both spatial and temporal domain. Our comprehensive approach combining MD simulations with semiconductor modeling and statistical signal processing techniques therefore can be implemented into new powerful tools for accurate and real-time biomolecular sensing with promising applications of solid-state nanopore systems in personal nano-medicine.

References

1. Aksimentiev A, Heng JB, Timp G, Schulten K (2004) Microscopic kinetics of DNA translocation through synthetic nanopores. Biophys J 87(3):2086–2097
2. Athreya N, Milenkovic O, Leburton JP (2020) Interaction dynamics and site-specific electronic recognition of DNA-nicks with 2D solid-state nanopores. NPJ 2D Mater Appl 4(1):1–8
3. Athreya NBM, Sarathy A, Leburton JP (2018) Large scale parallel DNA detection by two-dimensional solid-state multipore systems. ACS Sensors 3(5):1032–1039
4. Athreya NBM, Xiong M, Chakraborty R, Crawford-Eng HT, Leburton JP (2023) Molecular dynamics and electrical modeling of multi-layered nanopores of MoS_2 and hexagonal boron nitride for biosensing of dna. ACS Appl Nano Mater 6(6):4406–4416
5. Bayley H (2015) Nanopore sequencing: from imagination to reality. Clin Chem 61(1):25–31
6. Chakraborty R, Xiong M, Athreya N, Tabatabaei SK, Milenkovic O, Leburton JP (2023) Solid-state MoS_2 nanopore membranes for discriminating among the lengths of RNA tails on a double-stranded DNA: a new simulation-based differentiating algorithm. ACS Appl Nano Mater 6(6):4651–4660
7. Chen P, Gu J, Brandin E, Kim YR, Wang Q, Branton D (2004) Probing single DNA molecule transport using fabricated nanopores. Nano Lett 4(11):2293–2298
8. Deamer D, Akeson M, Branton D (2016) Three decades of nanopore sequencing. Nat Biotechnol 34(5):518–524
9. Feng J, Liu K, Bulushev RD, Khlybov S, Dumcenco D, Kis A, Radenovic A (2015) Identification of single nucleotides in MoS_2 nanopores. Nat Nanotechnol 10(12):1070

10. Foloppe N, MacKerell AD Jr (2000) All-atom empirical force field for nucleic acids: I. parameter optimization based on small molecule and condensed phase macromolecular target data. J Comput Chem 21(2):86–104
11. Girdhar A, Sathe C, Schulten K, Leburton JP (2013) Graphene quantum point contact transistor for DNA sensing. Proc Nat Acad Sci 110(42):16748–16753
12. Girdhar A, Sathe C, Schulten K, Leburton JP (2015) Tunable graphene quantum point contact transistor for DNA detection and characterization. Nanotechnology 26(13):134005
13. Gracheva ME, Aksimentiev A, Leburton JP (2006) Electrical signatures of single-stranded DNA with single base mutations in a nanopore capacitor. Nanotechnology 17(13):3160
14. Gracheva ME, Xiong A, Aksimentiev A, Schulten K, Timp G, Leburton JP (2006) Simulation of the electric response of DNA translocation through a semiconductor nanopore-capacitor. Nanotechnology 17(3):622
15. Graf M, Lihter M, Altus D, Marion S, Radenovic A (2019) Transverse detection of DNA using a MoS_2 nanopore. Nano Lett 19(12):9075–9083
16. Hall AR, Scott A, Rotem D, Mehta KK, Bayley H, Dekker C (2010) Hybrid pore formation by directed insertion of α-Haemolysin into solid-state nanopores. Nat Nanotechnol 5(12):874–877
17. Han SJ, Jenkins KA, Valdes Garcia A, Franklin AD, Bol AA, Haensch W (2011) High-frequency graphene voltage amplifier. Nano Lett 11(9):3690–3693
18. Heerema SJ, Vicarelli L, Pud S, Schouten RN, Zandbergen HW, Dekker C (2018) Probing DNA translocations with inplane current signals in a graphene nanoribbon with a nanopore. ACS Nano 12(3):2623–2633
19. Humphrey W, Dalke A, Schulten K (1996) VMD: visual molecular dynamics. J Mole Graph 14(1):33–38
20. Jones JE (1924) On the determination of molecular fields—i. from the variation of the viscosity of a gas with temperature. Proc R Soc Lond Ser A Containing Pap Math Phys Char 106(738):441–462
21. Kasianowicz JJ, Brandin E, Branton D, Deamer DW (1996) Characterization of individual polynucleotide molecules using a membrane channel. Proc Nat Acad Sci 93(24):13770–13773
22. Krasnozhon D, Lembke D, Nyffeler C, Leblebici Y, Kis A (2014) MoS_2 transistors operating at gigahertz frequencies. Nano Lett 14(10):5905–5911
23. Li J, Gershow M, Stein D, Brandin E, Golovchenko JA (2003) DNA molecules and configurations in a solid-state nanopore microscope. Nat Mater 2(9):611–615
24. Liu K, Lihter M, Sarathy A, Caneva S, Qiu H, Deiana D, Tileli V, Alexander DT, Hofmann S, Dumcenco D et al (2017) Geometrical effect in 2D nanopores. Nano Lett 17(7):4223–4230
25. Mandelkern M, Elias JG, Eden D, Crothers DM (1981) The dimensions of DNA in solution. J Mole Biol 152(1):153–161
26. Phillips JC, Braun R, Wang W, Gumbart J, Tajkhorshid E, Villa E, Chipot C, Skeel RD, Kale L, Schulten K (2005) Scalable molecular dynamics with name. J Comput Chem 26(16):1781–1802
27. Qiu H, Sarathy A, Leburton JP, Schulten K (2015) Intrinsic stepwise translocation of stretched ssDNA in graphene nanopores. Nano Lett 15(12):8322–8330
28. Qiu H, Girdhar A, Schulten K, Leburton JP (2016) Electrically tunable quenching of DNA fluctuations in biased solid-state nanopores. ACS Nano 10(4):4482–4488
29. Qiu H, Sarathy A, Schulten K, Leburton JP (2017) Detection and mapping of DNA methylation with 2D material nanopores. NPJ 2D Mater Appl 1(1):1–8
30. Sarathy A, Qiu H, Leburton JP (2017) Graphene nanopores for electronic recognition of DNA methylation. J Phys Chem B 121(15):3757–3763
31. Sarathy A, Athreya NB, Varshney LR, Leburton JP (2018) Classification of epigenetic biomarkers with atomically thin nanopores. J Phys Chem Lett 9(19):5718–5725
32. Sathe C, Zou X, Leburton JP, Schulten K (2011) Computational investigation of DNA detection using graphene nanopores. ACS Nano 5(11):8842–8851
33. Schneider GF, Kowalczyk SW, Calado VE, Pandraud G, Zandbergen HW, Vandersypen LM, Dekker C (2010) DNA translocation through graphene nanopores. Nano Lett 10(8):3163–3167

34. Shaw DE, Deneroff MM, Dror RO, Kuskin JS, Larson RH, Salmon JK, Young C, Batson B, Bowers KJ, Chao JC et al (2008) Anton, a special-purpose machine for molecular dynamics simulation. Commun ACM 51(7):91–97
35. Storm AJ, Storm C, Chen J, Zandbergen H, Joanny JF, Dekker C (2005) Fast DNA translocation through a solid-state nanopore. Nano Lett 5(7):1193–1197
36. Tabatabaei SK, Wang B, Athreya NBM, Enghiad B, Hernandez AG, Fields CJ, Leburton JP, Soloveichik D, Zhao H, Milenkovic O (2020) DNA punch cards for storing data on native DNA sequences via enzymatic nicking. Nat Commun 11(1):1–10
37. Wanunu M (2012) Nanopores: a journey towards DNA sequencing. Phys Life Rev 9(2):125–158
38. Xiong M, Graf M, Athreya N, Radenovic A, Leburton JP (2020) Microscopic detection analysis of single molecules in MoS_2 membrane nanopores. ACS Nano 14(11):16131–16139

Index

J.-P. Leburton (ed.), *Solid State Nanopores*, Nanostructure Science and Technology, https://doi.org/10.1007/978-3-031-42336-9

The manufacturer's authorised representative in the EU is Springer Nature Customer Service Centre GmbH, Europaplatz 3, 69115 Heidelberg, Germany. If you have any concerns regarding our products, please contact ProductSafety@springernature.com

Printed and bound by CPI Group (UK) Ltd, Croydon, CR0 4YY
01/10/2026
02225902-0003